MITA

Mobile Internet Technical Architecture

Technologies and Standardization

IT Press

Published by
Edita Publishing Inc.
IT Press
P.O.BOX 760
FIN-00043 EDITA
FINLAND

Distributor information:

For information about distribution, please visit our Web site at **http://www.itpress.biz/distributors**

Copyright © 2002 Edita Publishing Inc. All rights reserved.

Access the latest information about IT Press books from our World Wide Web site at **www.itpress.biz**.

All rights reserved. No part of this book may be reproduced or transmitted in any form or by any means, without the written permission from the Publisher.

ISBN 951-826-668-9

Printed by Gummerus Inc., Finland.

This document is an approved part of Nokia Mobile Internet Technical Architecture release.

Nokia Mobile Internet Technical Architecture, aims to provide seamless interoperability between all interaction modes, any network environment and, with any type of access. The ultimate objective of the initiative is to create a user-friendly Mobile Internet experience for everyone.

In developing a clear technical architecture for the Mobile Internet, Nokia aims to limit the complexity of the inherently technical environment; consumers do not want to worry about the underlying technologies.

An open solution benefits all; profitable business scenarios call for interoperability, short development cycles, large volumes and, most of all, global reach. Unless there is a commonly accepted architectural solution, markets will be fragmented as well as requiring separate parameters, and the total volume will be much smaller than in a single global market.

Nokia Mobile Internet Technical Architecture (MITA) is defined by Nokia Mobile Software unit. Any comments concerning this document can be sent to mita.feedback@nokia.com. Please visit the official MITA Web site at: www.nokia.com/mita

The contents of this document are copyright ©2002 Nokia. All rights reserved.

A license is hereby granted to a copy of this document for personal use only. No other license to any other intellectual property rights is granted herein. Unless expressly permitted herein, reproduction, transfer, distribution or storage of part or all of the contents in any form without the prior written permission of Nokia is prohibited.

The content of this document is provided "as is ", without warranties of any kind with regards its accuracy or reliability, and specifically excluding all implied warranties, for example of merchantability, fitness for purpose, title and non-infringement.

In no event shall Nokia be liable for any special, indirect or consequential damages, or any damages whatsoever resulting from loss of use, data or profits, arising out of or in connection with the use of the document.

Nokia reserves the right to revise the document or withdraw it at any time without prior notice.

Nokia and Nokia Connecting People are registered trademarks of Nokia Corporation. Nokia product names are either trademarks or registered trademarks of Nokia. Other product and company names mentioned herein may be trademarks or trade names of their respective owners.

Table of Contents

I PROLOGUE 1

Prologue 3
Mikko Terho

II INTRODUCTION 5

The Mobile World 7
Marko Suoknuuti

Introduction 8
- Mobile World Challenge 8
- Interaction Modes 8
  - Messaging 9
  - Browsing 9
  - Rich Call 9
- Network Environments 9
- Identity 9

Mobile Internet Technical Architecture 10
- MITA Objectives 11
- MITA Work Phases 11
- MITA Methodology 13
  - Architecture Modeling Principles 13
  - Architecture Concept Models 14
  - Architecture Implementation Models 14
  - Architecture Specifications 15
  - Reference Implementations 15

Compliance with MITA Principles 16
- Providing Multiple Access Capable Solutions 16
- Securing the Highest Quality Interworking Between Networks 17
- Securing the Best Interworking Between Multiple Addresses and Identities ... 17

Applications and the Technical Architecture 17

Mobile Internet Business Architecture 18
- Middleware Benefits 19

MITA Deliverables – End-to-End Solutions 20

Conclusions 21

Introduction to MITA Specifications ... 23

Marko Suoknuuti, Juha Lampela

Architecture Specifications ... 23
UI and Application Framework in MITA ... 24
Rich Call in MITA ... 24
Browsing in MITA ... 24
Messaging in MITA ... 25
Internet Protocols in MITA ... 25
Operating Systems and Platform Support in MITA ... 26
Multiple Access in MITA ... 26

Key System Specifications ... 27
Naming, Numbering and Addressing in MITA ... 27
Presence in MITA ... 27
Reachability in MITA ... 27
Access Independent Connectivity in MITA ... 28
Service Discovery in MITA ... 28
Location in MITA ... 28
Device Management and Data Synchronization in MITA ... 29
Content Formats in MITA ... 29
Content Adaptation in MITA ... 29
Operation Support Systems in MITA ... 30
Privacy in MITA ... 30

Architecture Frameworks ... 30
Directories in MITA ... 31
Security in MITA ... 31
Mobile Internet Interfaces in MITA ... 32
Quality of Service in MITA ... 33

Introduction to the Systems Software Architecture ... 35

Edited by Marko Suoknuuti

Design Principles of the Systems Software Architecture ... 35
Systems Software Architecture ... 36
Terminal ... 37
Service Provisioning Infrastructure ... 38
Application Programming Interfaces ... 38
Applications and Content ... 38
Mobile Web Service Interfaces ... 39
Mobile Service Brokers ... 40
Network Servers ... 41

Introduction to Service Enablers 43
Edited by Marko Suoknuuti

    Multimedia Messaging Service 44
    Java™ Technology 44
    Browsing 45
    Device Management 45
    Device Profile 46
    Delivery 47
    Identification/Authentication 47
    Mobile Payment 48
    Mobile Digital Rights Management 49
    Presence 49
    Instant Messaging 49
    User Profile Management 50
    Data Synchronization 50
    Streaming 50
    Location 51
    Group Management 51
    Conclusions 51

III WEB DOMAIN TECHNOLOGIES 53

Part 3.1 Platform Layer 55

Linux 57
Mika Grundström, Tapio Tallgren

    Origins of Linux 58
    Linux Architecture 59
        Features 59
        Monolithic 59
        Modules 60
            Kernel HTTPD Accelerator 61
        Distribution Packages 61
            Installation Help 62
            Upgrades 62
            Automatic Settings 62
        Real-Time Support 62
        Security 63
        Networking 63
            Networking Functions 64
            Routing 64
            Firewall 64

- Development Cycle 64
- Conclusions 65
- References 65

Part 3.2 Mobile Internet Layer 67

Basic Internet Protocols 69
John Loughney

- Transport Protocols 69
  - Transmission Control Protocol 69
  - User Datagram Protocol 70
  - Stream Control Transmission Protocol 70
- Related Protocols 71
  - Domain Name System 71
  - Dynamic Host Configuration Protocol 71
- Common Internet Application Protocols 71
  - HyperText Transfer Protocol 72
- File Transfer Protocol 73
  - Simple Mail Transfer Protocol 73
  - Multipurpose Internet Mail Extensions 73
  - Network News Transfer Protocol 73
  - Terminal Emulation Protocol 74
- References 74

Internet Protocol Version 6 75
John Loughney

- IP Headers 75
- IPv6 Header Handling 76
- Other Enhancements 77
- IPv6 Address Types 77
- Conclusions 79
- References 79

Multimedia Sessions in the Web Domain 81
Markus Isomäki

- Internet Multimedia Architecture 81
- Session Initiation Protocol 84
- Conclusions 87
- References 87

Internet Multicast and Services 89
Mika Grundström

IP Multicast 89
Managing a Group 90
Layer 2 Broadcast and Multicast 91
Service Discovery 92
Services 92

Quality of Service in the Web Domain 93
Kalevi Kilkki, Jussi Ruutu

Technologies 93
Best Effort 94
Integrated Services 94
Differentiated Services 95
Multiprotocol Label Switching 96

Java™ in the Web Domain 97
James Reilly, Kari Systä

Benefits of Java Technology 97
A Brief History 99
Core Java Platforms 99
Java 2 Platform, Standard Edition 100
Java 2 Platform, Enterprise Edition 101
Java Platform Support for Web Services 104
Conclusions 105
References 105

Part 3.3 Application Layer 107

Web Services in the Web Domain 109
Suresh Chande, Markku Laitkorpi, Murali Punaganti

Overview 109
Key Participants 110
Basic Web Service Process 111
Web Service Technologies 112
Data Encapsulation 112
Communication Framework 112
Simple Object Access Protocol 113
XML Protocol 114
Service Description 115

Service Brokers	117
The Universal Description, Discovery and Integration	117
UDDI Service APIs	118
Publishing Web Services Locally	119
Electronic Document Interchange Based Web Services	119
Advanced Web Services	120
Transactions	120
Workflow and Conversations	121
Security Issues	122
Conclusions	123
References	124

Semantic Web 127

Ora Lassila

Role of Ontologies	127
Languages of the Semantic Web	128
Resource Description Framework	128
DARPA Agent Markup Language	130
W3C Web Ontology	130
Semantic Web as a Web of Services	130
Web Services Using DAML	130
Ultimate Form of Interoperability	132
Conclusions	133
References	133

IV MOBILE DOMAIN TECHNOLOGIES 137

Part 4.1 Platform Layer 139

The Symbian Platform 141

Michael Przybilski

Motivation	141
Overview of the Symbian Platform	142
Symbian Operating System	143
Generic Technologies	143
Device Family Reference Designs	145
Application Suite	147
Application Development	147
Conclusions	148
References	148

Carrier-Grade Linux for the Mobile Domain 149
Tapio Tallgren

Support for Replication 150
Real-Time 151
Embedded 152
Monitored 152
Fault Resistant 152
Conclusions 153
References 153

Wideband Code Division Multiple Access Technology 155
Harri Holma, Antti Toskala

Air Interfaces for IMT-2000 155
Spectrum for Universal Mobile Telecommunications System 156
Radio Access Network Architecture 158
WCDMA Basics 159
Physical Layer and Mobile Device Capabilities 161
High Speed Downlink Packet Access 164
Air Interface Performance 165
Conclusions 166
References 167

GSM/EDGE Radio Access Network 169
Jose Gimenez, Eero Nikula, Javier Romero

Architecture 170
- Legacy Interfaces 170
- New Interfaces 171
Quality of Service 173
Radio Protocols 174
- Packet Data Control Protocol 174
- Radio Resource Control 175
- Radio Link Control 176
- Medium Access Control 177
Physical Layer 177
GERAN Performance 179
- Incremental Redundancy 179
- EGPRS Link Adaptation 180
Conclusions 182
References 182

Wireless Local Area Networks ... 185
Jouni Mikkonen, Markku Niemi

Frequencies ... 185
2.4 GHz Industrial, Scientific and Medical Frequency Band ... 186
5 GHz Frequency Band ... 186
IEEE 802.11 ... 187
Reference Model ... 187
Architecture ... 188
Services ... 189
Medium Access Control Layer ... 191
Basic Channel Access ... 191
Physical Layer ... 192
Frequency Hopping Spread Spectrum ... 193
Direct Sequence Spread Spectrum ... 193
Infrared ... 193
Conclusions ... 193

Bluetooth ... 195
Holger Hussmann

Connectivity Scenarios ... 195
Technical Overview ... 196
Lower Layers ... 196
Middleware Protocols ... 197
Profiles ... 199
Future Developments ... 199
Conclusions ... 200

Digital Video Broadcasting – Terrestrial Network ... 201
Janne Aaltonen, Mika Grundström, Harri Hakulinen, Holger Hussmann, Pekka Talmola

Technical Overview ... 201
Radio Characteristics ... 201
Mobility ... 202
Radio Network Design Issues ... 202
Datacasting Profiles ... 203
IP Multiprotocol Encapsulation ... 204
IP Datacasting ... 205
Enabling multicast ... 205
Hybrid solution ... 205
Conclusions ... 206
References ... 206

Part 4.2 Mobile Internet Layer 207

Mobility Support for IPv6 209
Charles Perkins

Mobile IP 210
IPv6 Design Points 212
Router Advertisement 212
Address Autoconfiguration 212
Security 213
Destination Options 214
ICMPv6 Improvements to Encapsulation 215
Mobile IPv6 Protocol Overview 216
Detection 216
Address Configuration 216
Binding Update 217
Tunneling 217
Route Optimization 218
Ingress Filtering 219
Home Agent Discovery 220
Renumbering 221
Recent Directions 222
Seamless Mobility 223
Context Features for Transfer 224
Context Transfer Framework 225
Localized Mobility Management 226
Binding Security Association Establishment 226
Mobile IPv6 Status 228
Conclusions 228

Multimedia Sessions in the Mobile Domain 231
Markus Isomäki

Characteristics of the Mobile Domain 231
3GPP IP Multimedia Subsystem 231
Conclusions 234
References 234

Java™ in the Mobile Domain 235
James Reilly, Kari Systä

Mobile Devices and Resource Constraints 235
Java 2 Platform, Micro Edition 236

Mobile Information Device Profile Version 1.0	237
Mobile Information Device Profile Version 2.0	238
Midlet Provisioning	239
Midlet Networking	240
Mobile Media	241
Conclusions	242
References	242

Data Synchronization in the Mobile Domain 243
Teemu Toroi

SyncML Technology	244
Security	245
Interoperability	245
Conclusions	245

Middleware in the Mobile Domain 247
Kimmo Raatikainen

Application Requirements	247
Internet Protocols and Middleware	248
Programming Models	249
Next Steps	250
Conclusions	251
References	251

Authentication Methods and Technologies 253
John Loughney

Password Authentication	254
Onetime Passwords	254
Hardware Token Based Authentication	255
RSA SecurID™	255
Radio Frequency ID	255
Symmetric Key Based Solutions	255
Smart Card Authentication	255
Subscriber Identity Module Authentication	256
Kerberos Authentication	256
Public Key Cryptography	258
Biometric Authentication	259
Remote Authentication Dial-In User Service	259
Diameter	260
Single Sign-On	260

Conclusions	262
References	262

Part 4.3 Application Layer ... 263

Browsing in the Mobile Domain ... 265
Franklin Davis, Asko Komsi

Overview	265
XHTML Basic	266
XHTML Mobile Profile	266
Key Features and Capabilities	267
Well-Formed Content	267
Valid Content and the Document Type Definition	268
Cascading Style Sheets	269
Browser Examples	270
Transformations	271
XHTML is the Future	271

Messaging in the Mobile Domain ... 273
Chris Bouret, Stephane Coulombe

Multimedia Messaging ... 275
Stephane Coulombe, Guido Grassel, Peter Hjort

Short Message Service	275
Multimedia Messaging Service	276
Standards	278
Network Architecture	278
Inter-Network Routing	280
Message Encapsulation	281
Applications	282
Message Adaptation	282
Interoperability With Internet Applications	282
Interoperability Between Mobile Devices	283
Pervasive Mobile Device Environment	283
Automatic Adaptation	283
Message Adaptation Use Cases	284
Mobile Device Capabilities Negotiation	284
Conclusions	285
References	285

Instant Messaging and Presence 287
Chris Bouret, Jari Kinnunen, Krisztian Kiss, Pekka Kuismanen, Mikko Lönnfors

Session Initiation Protocol 287
Services 288
Flow Examples 289
Service Creation Model 290
Current Status and Future of the Technology 291
SIP IM/P in IETF 291
SIP IM/P in 3GPP 293
Conclusions 293
References 293

Wireless Village Interoperability Framework 295
Janne Kilpeläinen

System Architecture 295
Protocol Suite 296
Reference 297

Public Key Infrastructure 299
Markku Kontio

Motivation 299
Security Services 299
Cryptography 300
Encryption 301
Symmetric Encryption 301
Asymmetric Encryption 301
Cryptographical Hash Algorithms 302
Message Authentication Codes 303
Digital Signatures 304
Digital Signatures With Hash Algorithms 304
A Few Hard Assumptions 305
PKI Technology 306
The History of PKI 306
Digital Certificates 306
Issuing a Certificate 307
Using a Certificate 307
Certificate Details 308
Certificate Revocation 308
Transitive Trust 309
Attribute Certificates 310

Applications Using PKI 310
Secure Web Applications 310
Secure E-Mail 310
Virtual Private Networks 310
Wireless Public Key Infrastructure 311
Conclusions 311
References 312

Digital Rights Management 313
Kimmo Djupsjöbacka

System Models 314
Basic System Model 315
System Model Variations 316
    Content Delivery and Encryption 317
    Voucher Creation and Rights Assignment 317
    Variations in Usage Right Definitions 318
Technologies Related to DRM 318
Security Technologies 319
    Access Control 319
    Session or Connection Protection 319
    Symmetric Encryption 319
    Asymmetric Encryption 319
    Tamper Resistance 320
    Watermarking 321
    Digital Fingerprints 321
Content Distribution Methods 321
    Physical Media 321
    Legacy Internet 322
    Over the Air 322
    Peer-to-Peer 322
    Streaming 322
    Broadcast 323
Standardization 323
References 324

Mobile Payment 325
Sanna Raitanen, Tuomo Virkkunen

Mobile Commerce Services 325
Transaction Environments 326
Remote Environment 326
Local Environment 327
Personal Environment 327
Payment Concepts 327

Security in Mobile Transactions 328
- Security in Macro Payment Transactions 329
- Wireless Identity Module 329

Location Technologies in the Mobile Domain 331
Ville Ruutu

Mobile Location Methods 332
- Cell Coverage 332
- Received Signal Levels 333
- Angle of Arrival 334
- Timing Advance 334
- Round Trip Time 335
- Global Positioning System 336
- Time of Arrival 337
- Enhanced Observed Time Difference 337
- Observed Time Difference of Arrival – Idle Period Down Link 339

V MOBILE INTERNET STANDARDIZATION 341

Standardization in the Mobile World 343
Jussi Ruutu, Marko Suoknuuti

Standardization Challenges 344
- How to Extend Vertical Network Standardization with End-to-End Aspects 344
- How to Focus Standardization Work Items and Manage Growing Complexity ... 346
- How to Reduce Standardization Options 347
- How to Enable Different Technology Evolution Speeds at Different Layers 347
- How to Enable More Added Value to Consumers 348
- How to Support Application Development in Standardization 349

Principles for Standardization in the Mobile World Era 349

Open Mobile Alliance™ 351
Johanna Rautasalo, Seppo Aaltonen, Kati Riikonen, Jouni Toijala

Scope of the Alliance 351
Commitment to Interoperability 351
Benefits of the Open Mobile Alliance 352
Reference 353

Wireless Application Protocol Forum 355
Asko Komsi

Mission 355
Organization 356
- WAP Specification 356

Reference 358

Third Generation Partnership Project ... 359
Atte Länsisalmi

- Global Initiative ... 359
- Organization ... 359
- Organizational Partners ... 360
- Market Representatives ... 361
- Observers and Guests ... 362
- Individual Members ... 362
- Support Functions ... 362
- Project Coordination Group ... 363
- Technical Specification Groups ... 363
  - Structure of the Technical Work ... 363
  - TSG Services and System Aspects ... 363
  - TSG Terminals ... 364
  - TSG Core Network ... 364
  - TSG Radio Access Network ... 365
  - TSG GSM/EDGE Radio Access Network ... 366
- Working Procedures and Methods ... 367
- References ... 368

The Internet Engineering Task Force ... 369
John Loughney

- Organization of the IETF ... 369
- Working Process ... 370
- Related Organizations ... 371
- Mobile Internet in the IETF ... 371
- References ... 373

International Telecommunications Union ... 375
Matti Alkula

- History ... 375
- Structure of the ITU ... 376
- The Sectors of the ITU ... 378
  - Radiocommunication Sector ... 378
  - Telecommunication Standardization Sector ... 380
  - Telecommunication Development Sector ... 385
- ITU Reform ... 386
- References ... 389

European Telecommunications Standards Institute 391
Kari Lång

ETSI Structure 392
Technical Organization 392
Highlights of ETSI Technical Activities 394
Methods for Testing and Specification 394
Broadband Radio Access Networks 394
Telecommunications and Internet Protocol Harmonization Over Networks 395
Third Generation Partnership Project 396
Other ETSI Activities 396
Global Cooperation from the ETSI Perspective 397
Support Functions 397
Competence and Service Centers in ETSI 398
Information Technology in Specifications Creation 399
References 399

The Institute of Electrical and Electronics Engineers 401
Mika Kasslin

Working Groups 401
IEEE 802 402
Organization 402
From an Idea to a Standard 403
Wireless 802 Working Groups 404
802.11 Wireless Local Area Network 404
Enhancements to the Basic 802.11 405
802.16 Broadband Wireless Access 406
802.15 Wireless Personal Area Network 406
Reference 406

Wireless Ethernet Compatibility Alliance 407
Markku Niemi

Testing Process 407
Organization 408
Reference 408

Bluetooth SIG 409
Holger Hussmann

Organization 409
Bluetooth Architecture and Review Board 410
Bluetooth Qualification Review Board 410
Bluetooth Test and Interoperability Committee 410

Marketing Committee 411
　　　Regulatory Committee 411
　Bluetooth Qualification Program 411
　Reference 412

Open Source Development Lab and Carrier Grade Linux Work Group 413
Mika Kukkonen, Ville Lavonius

　Carrier Grade Linux Work Group 415
　Scope of Carrier Grade Linux Work Group 417
　Conclusions 418
　References 419

Service Availability Forum 421
Timo Jokiaho

　Open Standards for Service Availability 421

Java™ Community Process 423
Pentti Savolainen

　JCP Agreement and Process 423
　　　Document 423
　　　Membership 423
　　　Process 423
　　　Guidance Body 424
　Life Cycle of a Specification 424
　Specification Lead Position 426
　Intellectual Property 426

World Wide Web Consortium 429
Ora Lassila

　Operation and Organization 429
　Current Activities of Interest 430
　　　Hypertext Markup Language 430
　　　Cascading Style Sheets 430
　　　Web Services 431
　　　Semantic Web and Resource Description Framework 431
　　　Graphics and Multimedia 431
　　　Device Independence 431
　Conclusions 432
　Reference 432

Web Services Fora 433
Michael Mahan

OASIS & UN/CEFACT 433
UDDI.org 434
Web Services Interoperability Organization 434
DAML-S 435
RosettaNet 435

Object Management Group 437
Kimmo Raatikainen

OMG Modeling Specifications 437
Recent and Forthcoming Enhancements 438
Reference 438

Liberty Alliance 439
Senthil Sengodan

Membership and Organization 439
Phased Approach 440
Liberty Applicability and Architecture 440
Conclusions 441
Reference 442

Location Inter-Operability Forum 443
Juhani Murto

Reference 443

Mobile Commerce Fora 445
Piotr Cofta

Mobile Electronic Transaction Initiative 445
Mobey Forum 446
References 447

Wireless Village 449
Janne Kilpeläinen

IMPS Solution 449
Wireless Village Organization 450
    Technical Specification Groups 450
    Working Procedures and Methods 451
    Decision Making in TechCom 451
    Issues in TechCom 451

VI EPILOGUE 453

Epilogue 455
Marko Suoknuuti

VII APPENDIX 457

Appendix A: Glossary 459

Index 485

Acknowledgements

The past two years have been an exciting time with the challenges of the Mobile Internet and its defining Technical Architecture. We have seen the summit of the Internet boom and lately faced a more demanding period in the mobile industry. In the interim, the definition of the Nokia Mobile Internet Technical Architecture has matured to a complete release in the form of this book series.

I would like to thank the people involved in this effort:

- All the contributors and teams for their remarkable efforts in creating the content
- The Editors' Board: Jyrki Kivimäki, Juha Lampela, Hanna Passoja-Martikainen, Sari Päivärinta, Krister Rask and Jussi Ruutu for their excellent work in keeping the book process on schedule and in shape
- Sami Inkinen and Mikko Terho for their support and guidance en route to the complete release
- Jani Ilkka, Juha Kaski and the rest of the IT Press team for doing an outstanding job with the publishing process
- Nely Keinänen, Anna Shefl and Michael Jääskeläinen for doing a great job with the language revision and proofreading large amounts of technical material on a very strict schedule
- Eija Kauppinen and her team at Indivisual for numerous high-quality illustrations

And finally:

- My wife, Helena, for her understanding and support, especially in the last few months when the book process has occupied my time

Marko Suoknuuti
Chair of the Editors' Board

The Authors/Contributors:

Antti Toskala
Asko Komsi
Atte Länsisalmi
Charles Perkins
Chris Bouret
Eero Nikula
Franklin Davis
Guido Grassel
Harri Hakulinen
Harri Holma
Holger Hussmann
James Reilly
Janne Aaltonen
Janne Kilpeläinen
Jari Kinnunen
Javier Romero
Johanna Rautasalo
John Loughney
Jose Gimenez
Jouni Mikkonen
Jouni Toijala
Juha Lampela

Juhani Murto
Jussi Ruutu
Kalevi Kilkki
Kari Lång
Kari Systä
Kati Riikonen
Kimmo Djupsjöbacka
Kimmo Raatikainen
Krisztian Kiss
Markku Kontio
Markku Laitkorpi
Markku Niemi
Marko Suoknuuti
Markus Isomäki
Matti Alkula
Michael Mahan
Michael Przybilski
Mika Grundström
Mika Kasslin
Mika Kukkonen
Mikko Lönnfors
Mikko Terho

Murali Punaganti
Nicole Cham
Ora Lassila
Pekka Kuismanen
Pekka Talmola
Pentti Savolainen
Peter Hjort
Piotr Cofta
Sanna Raitanen
Senthil Sengodan
Seppo Aaltonen
Stephane Coulombe
Suresh Chande
Tapio Tallgren
Teemu Toroi
Timo Jokiaho
Tommi Raivisto
Tuomo Virkkunen
Ville Lavonius
Ville Ruutu
Vipul Mehrotra

I PROLOGUE

- 0 Prologue

Prologue

The next few years will see the convergence of mobile communications and the Internet resulting in new services, new business models and new business opportunities.

The Internet has become an everyday source of information, entertainment and other services for millions of people around the world. In our vision, the Internet will become mobile and content will go wireless, as voice communication has done. The future is no longer about the Internet but about services on the Mobile Internet. This is a new environment, called the Mobile World, where the evolution paths of mobile communications and the Internet have converged. The Mobile World provides innovative services for consumers, and the development of these services is driven by consumer behavior.

The Mobile World can be perceived as a holistic, evolving environment of the future, but also as an environment on a personal level that enables people to shape their Mobile World through personalized communication services. In the Mobile World, delivering targeted, timely information and services is essential.

The key to commercial success in the Mobile World depends on three main factors:

- o   Understanding consumer needs, lifestyles and attitudes, and consequently being able to provide a matching combination of service portfolios and products,
- o   Readiness of the business system to enable service consumption, and
- o   Being able to match technical architectures to evolving consumer needs and to the evolution of business systems.

Those who deliver product categories and platforms with the right services and technologies will be the victors in the Mobile World.

The winning technical architecture in the Mobile World will enable seamless interoperability between key applications, network environments and the user identity/addressing system, limit the complexity of the technical environment supporting services so that consumers do not need to be concerned with the underlying technologies, support open technologies, standards and relevant initiatives, which support and facilitate the deployment of global technologies and services, and stimulate market growth.

The Nokia solution addressing these demands is the Mobile Internet Technical Architecture (MITA). For the players in the field, MITA is the essential framework for creating user-friendly Mobile World experiences. It is an architecture, which comprehensively enables networks to be driven by services. MITA addresses relevant technologies and standardization forums for the Mobile Internet, the current solutions and developer environments provided by Nokia, specifies design principles and technology visions for the Mobile Internet, and describes reference implementations.

Prologue

Long term MITA technical visions and technology evolution scenarios are complemented with a Systems Software Architecture (SWA) that specifies a near term application and service architecture for the Mobile Internet. It has been a contribution of Nokia to open global standardization of mobile services (e.g., Open Mobile Alliance).

The Mobile Internet Technical Architecture is built on existing technologies, which will form a solid basis for future innovations, development work and products. A technologies section will look at the selected set of the technologies that already exist either in the research environment, in the product development phase or in actual products. The technologies can be roughly divided into those for the Web Domain and those for the Mobile Domain, based on their background and origin. Technologies in the Mobile Domain have been mainly created for the use in mobile networks, while technologies for the Web Domain have emerged for Internet services.

One of the characteristic features of the communications industry is a strong need for standardization work to ensure interoperability between products from various vendors. This is true for both the Web and the Mobile domains. There is no reason to assume that the future Mobile World will depend less on standardization. This section of the book begins by discussing some of the future challenges for standardization, while the rest of the section introduces various standardization forums.

The Mobile World is evolving as we speak. New consumer needs are arising, and innovation on the Internet is taking place at an accelerating pace. Thus, it is clear that technical architectures also need to be continuously developed. Nokia invites all parties of the Mobile Internet to join and contribute to the development of the Mobile World. For this purpose, we would greatly appreciate input from our readers.

Mikko Terho
Senior Vice President

II INTRODUCTION

o The Mobile World

o Introduction to MITA Specifications

o Introduction to the System Software Architecture

o Introduction to Service Enablers

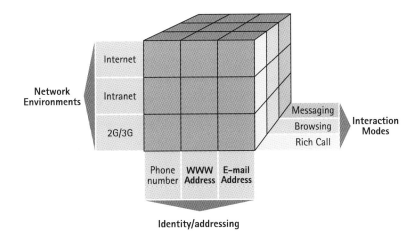

The Mobile World

The next few years will see the convergence of mobile communications and the Internet resulting in various new technologies, new business models and business opportunities.

The Mobile Internet is not simply the Internet of today accessed from a mobile device. We will not spend our time browsing Internet pages for content as we do today, although this will still be possible. Instead, we will use applications to access content, make transactions, do business, link up with friends and family, play games, watch videos, and listen to and download music. More importantly, we will use the Mobile Internet to help control our lives and give us more time to do the things we enjoy.

We are moving towards a Web-based business model where mobility and the Internet are unified. This will not happen overnight and will require new competencies from all parties involved in the industry. Understanding mobility and the unique characteristics of mobile business will be vital in building the networks and services of the future. Success in the Mobile World will be about speed: speed of application and mobile device development for refined consumer segments, speed of new service creation, and speed of cost-optimized network development and roll-out. It will be about the control of services and content residing in your own hands.

From a consumer's perspective, the most tangible element is the mobile device, which provides access to all of these services. To ensure the success of services in the Mobile World, they must be highly user-friendly. Actually, the user should be able to ignore the underlying technologies and purely enjoy the richness of the supplied services, regardless of the access method.

Along with the continually growing importance of mobile devices as life management tools, the Mobile World will become much more than a source for a quick weather information update. The Mobile Internet is not a fixed entity, yet it is starting to formulate its existence with the services and applications to be linked to it. Hence, as the business and technical environments develop, the available architectures will need to match them accordingly.

In the Mobile World, delivering targeted, timely information and services is essential. Consumer behavior drives the development of these applications and services. The key to commercial success lies in understanding consumers, their lifestyles and attitudes, and in creating the product-service combinations which match their wants and needs. Those who deliver winning product categories and platforms with the right technologies will succeed in the Mobile World.

The Nokia solution to cover these demands is the Mobile Internet Technical Architecture (MITA). For the players in the field, MITA is the essential framework for creating user-friendly Mobile World experiences. MITA supports network evolution to the Mobile Internet for both voice and data and is an architecture which comprehensively enables networks to be driven by services. The main target of MITA is to provide seamless interoperation in the consumption of content in all networks and create open and non-fragmented markets with maximum access to all.

Introduction

Nokia is proceeding towards a vision of the Mobile World. In this vision, the majority of all personal communication will be wireless - be it in the form of phone calls, messaging, browsing or images.

The perception of the Internet has been technology focused, i.e., it has been assumed that people wish to access the Internet and huge amounts of data while on the move. Yet, what people will do is something very different. To begin with, in the Mobile World, the Internet will become practically invisible as people are not concerned about "accessing the Internet" but about using services and their favorite applications. A special focus will be on consuming information and producing information to share with others. The consumption will apply to mobile brands, products, information, and advertising.

Consumers want value for money, ease of use, and a diverse choice of personally relevant services. Without an open solution, they are forced into a confusing and fragmented world of proprietary services and terminals. Therefore, the aim will be to limit the complexity of the technical environment so that consumers can enjoy services without worrying about the underlying technologies. As a result, a unified architecture will benefit all users in the Mobile World.

Mobile World Challenge

The Mobile World vision assumes that consumers will use different types of devices for connecting to multiple sets of services via various access networks. These services should be reachable in a unified way regardless of the access technology or mobile device. Naturally, there will always be slight differences between user experiences. However, it is expected that these will be made as transparent as possible.

The above requirements describe technical challenges, which need to be solved before the objectives of the Mobile World can be completely achieved. Interaction Modes, Identity and Network environments are the main elements that the industry needs for constructing an environment for tackling the Mobile World vision. This environment can be described as the Mobile World challenge.

Also, from the MITA perspective, this setting presents those areas that require close consideration in order to achieve a fully functional technical architecture for the Mobile World.

Interaction Modes

Today, consumers have the possibility of making phone calls, exchanging short messages, e-mail or images, and browsing data on the Internet. It is assumed that these communication modes are a baseline for services in the Mobile World. However, the convergence of communication technologies into IP-based solutions enables new services and extensions into baseline communication services.

The Mobile World Challenge addresses this evolution by classifying consumers' communication needs into three types of interaction modes to simplify the definition of basic requirements for applications, terminals, and service provisioning technology.

Messaging

Messaging is non real-time, client-server based communication in which an intermediary server is involved in the communication sequence. The intermediary server stores and/or processes messages before they are delivered to the destination. The intermediary server can be based on store and forward (e.g., Short Message Service and Multimedia Messaging Service), store and retrieve (e.g., e-mail) or even store and push functionality. The intermediary server may also provide queuing services.

Browsing

Browsing is nearly real-time one-way or two-way communication between a source and the destination. It includes one-way audio and video streaming. Differing from Rich Call, Browsing has no delay limitations for communication. In third generation networks, Browsing will be based on an evolution from today's Wireless Application Protocol (WAP) based browsing to graphics and multimedia-enriched browsing.

Rich Call

The Rich Call interaction mode maps to communication using a two-way real-time component. The real-time requirements mean that there are limitations for end-to-end and round-trip delays, and for jitter. Rich Call refers to voice and video calls in their simplest form, and in more complicated cases, concurrent communication (e.g., file transfer is combined with a real-time communication element).

Network Environments

The Internet boom started around the mid 1990s. Since then, an increasing number of corporations have connected their Intranets to the Internet, and a growing number of consumers use Internet services. Similarly, consumers have found attractive mobile services for use in their daily lives.

These three Network environments: mobile networks, the Internet and Intranets are the environments where most of the user interaction will take place in the Mobile World. Moreover, these different environments need to work seamlessly together and enable evolution towards unified service networks.

Identity

Today, each communication mode and its applications have specific addressing mechanisms (e.g., phone calls require numbers, messaging uses e-mail addresses). Luckily, applications have evolved and enable address books for mapping technical addresses to formats which are easier for consumers to manage (e.g., names).

An effective addressing system is vital for handling and simplifying the multitude of e-mail addresses, phone numbers and Web addresses of individuals and organizations. This will be even more true in the future, as the ways of communicating will expand and individuals will be using several different types of devices. The consumer should be faced with as few differences as possible between the addressing methods.

These Mobile World challenges are illustrated in the figure below, which models the interworking scenarios of networks, identities and interaction modes. It is a simple diagnostic framework for highlighting the issues involved in interworking between the layers on the Mobile Internet.

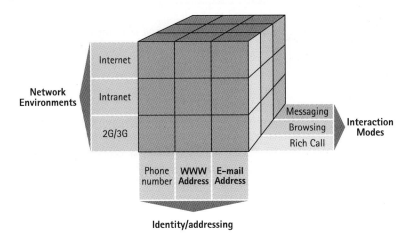

Mobile World Challenge

Mobile Internet Technical Architecture

To make the Mobile World a reality, correct technology choices are required if a technical architecture capable of supporting the myriad of services is to be built. For this reason, Nokia created MITA as a methodology to model the new environment and to understand the inherent technical issues in building the optimal technical architecture. Nokia is using this approach to assist players in the field in making the right technology choices to deliver the optimal consumer experience.

Openness and interoperability are central elements in the MITA approach. Situational analysis is not biased toward any network environment - fixed or mobile - and this maximizes inclusivity; Nokia recognizes the importance of collaboration in this new business environment and MITA gives all the players equally open access to the emerging architecture so that the ultimate goal - a world in which subscribers can utilize the seamless functionality of all Mobile Internet related applications and services - is achieved with the active participation of the whole industry. To achieve this goal, a mobile service market must be created, which offers more business for all companies involved, and provides better and a larger number of services for consumers.

Because all the elements have been taken into consideration, the outcome is highly beneficial to all parties including consumers, carriers, application developers, service providers, infrastructure providers, etc. Only through an open approach from all key players can the industry ensure a seamless user experience and rich service offering on a global scale. At the same time, this approach ensures open competition and rapid growth of the Mobile Internet market.

MITA Objectives

The primary objective is to provide a user-friendly Mobile Internet experience, with browsing, messaging and rich calls working seamlessly in any network environment and with any type of access. Hiding different addressing mechanisms and minimizing the impact of different access technologies from the consumer improves usability. These essential aspects of a user-friendly mobile entity are collected into the following figure presenting the scope of MITA.

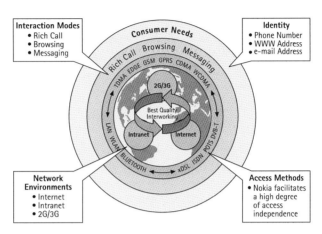

Scope of the Mobile Internet Technical Architecture

In order to lay a solid foundation for the evolution path towards future MITA compliant solutions, the development of current solutions must consider future needs. This backward compatibility will provide valuable building blocks for the Mobile Internet, which will evolve into something much more versatile than it is today. Rapid development is already seen and we can only estimate the future impact of a comprehensive technical architecture for the Mobile Internet.

To ensure far-reaching accessibility, Nokia supports open interfaces bringing new options for the Mobile Internet. Therefore, the Internet itself will not be changed but utilized in more versatile ways, while mobility has a principal role. The value of the Mobile Internet will increase if users are able to access services independent of location and network environment. The Mobile Internet itself will not be a tightly set entity and, hence, the preferred business and technical architectures will need continual revision as the industry and technologies advance.

MITA Work Phases

When the first analysis of the technical architecture was completed, we concluded that following the typical design path (i.e., begin with existing technologies and R&D activities and search for the optimal roadmap towards the Mobile World environment in gradual steps) is not the optimal route to the best technical choices. Because of this, we studied other alternatives, concluding that by following the development cycle in reverse, we could identify new and potentially revolutionary issues on time and take the actions needed to tackle them as a part of normal evolutionary roadmaps.

Introduction

Based on the above conclusion, MITA work follows three primary phases:

1. Set targets
2. Align with environment
3. Influence roadmaps

The first phase studies current products and technologies, reviews Nokia roadmaps and the products and technologies that are emerging within a 3 year time frame and compares them to problems in the Mobile World and MITA targets. An outcome of this phase is a technical vision which provides direction for the technical architecture and gives it its first model.

In the second phase, the technical architecture model is aligned with the existing R&D roadmaps at Nokia and the activities in standardization bodies. The result is a more detailed technical architecture and an initial implementation view.

In the third phase, the implementation view is finalized and the required R&D activities are addressed for product development.

Every phase has a different relationship with the time frame, as illustrated in the next figure. The first phase addresses MITA on long-term issues, the second phase shifts the focus to mid-term aspects and the third phase completes MITA work with short-term roadmaps, product development plans and reference implementations.

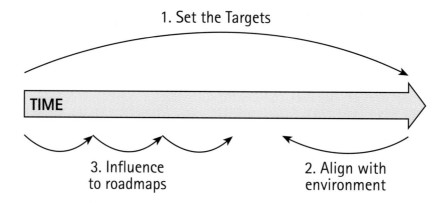

MITA work phases

MITA has already completed the first phase and the key results have been collected in these MITA books. Second phase work has begun with updated work items. The following figure updates the MITA work process for the second phase.

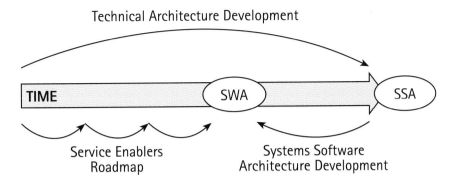

MITA work items in the second phase

The main task in the second phase is to develop a Systems Software Architecture (SWA) for Service Enabler implementations. Another task is to develop a technical architecture into the Subsystem Architecture (SSA), which describes MITA Subsystems and Mobile Internet Interfaces in more detail. The third task is to update technical visions with the latest research results.

MITA Methodology

At the core of the MITA approach is a neutral methodology to help the constituents of the Mobile World understand the critical technical issues in the Mobile Internet. This is achieved through abstract models which describe a conceptual framework for the Mobile Internet environment in progressively more detail and highlight the challenges of interworking between them. MITA tools consist of:

o   Architecture modeling principles
o   Architecture concept models
o   Architecture implementation models
o   Architecture specifications
o   Reference implementations

Collectively, the MITA tools are used to provide a consistent, unbiased situational analysis and a direction for making technology decisions. And, as the approach is technology neutral, it is flexible and ensures that a consistent approach is applied as new technologies evolve.

Architecture Modeling Principles

The technical architecture modeling is based on principles which should enable the creation of a future-proof end-to-end architecture with high possibilities for the modularity of components in the technical architecture.

The principles for modeling can be listed as follows:

- The architecture divides into independent subsystems while fulfilling business needs
- The architecture is open, modular, and hierarchical, utilizing a layered approach
- The architecture can tolerate varying rates of technological change in individual components. Change in one component should not render the whole architecture useless.

The above principles dictate that layered models are applied to element design, network modeling, and identity structuring. A generic content delivery model is defined for a specification of Interaction Modes and related subsystems. In a similar way, access independent requirements are collected into and specified in a model of an Access Independent Interface.

Architecture Concept Models

For the architecture, there are two approaches worth closer examination. From a business architecture viewpoint, there is the Mobile Internet Business Architecture (MIBA) and MITA can be used as the concept model for a more technical approach.

As a concept model, MIBA defines the interaction between different architecture elements. In the MIBA model, the terminal segment provides the consumer with access to services, the network segment provides terminals with connectivity to service networks, and the server segment provides content and services for consumers.

Within each segment (i.e., content, connection and consumption) the MITA approach models each interacting entity as an element that can be described as consisting of three abstract layers: Application, Mobile Internet, and Platform.

Describing MITA elements in terms of constituent layers enables us to identify the interfaces required for interworking between entities in each segment. Interworking is based on protocols implemented on the Mobile Internet layer and protocol-enabled content exchange between applications on the Application layer.

By mapping the MITA element to the MIBA concept model, a conceptual end-to-end view of MITA is achieved. The MITA end-to-end view is a high-level reference model for more detailed technical architecture design. All architecture elements in MITA can be derived from the same baseline MITA element.

Architecture Implementation Models

The MITA Subsystem Architecture introduces an implementation view of the technical architecture. It is another viewpoint into the technical architecture, encapsulating a group of functionalities into independent, well-isolated subsystems.

In general, a primary requirement is that an implementation of any subsystem can be replaced with another implementation, requiring no or only a minimum amount of modifications to other subsystems or applications utilizing the services of the replaced subsystem. In a similar way, the implementation technology of a subsystem may be completely changed without influencing other related subsystems. The modular MITA implementation model enables a flexible evolution of product implementations according to MITA specifications.

Architecture Specifications

Technical architecture work is divided into MITA specifications, to create technical work items of manageable scope and to enable comparison with other models on the specification level. The MITA specifications are divided into three groups:

- Architecture specifications
- Key system specifications
- Architecture frameworks

The architecture specifications are specifications that have a more permanent nature, whereas the key system specifications represent issues where understanding end-to-end system aspects prior to technology selections is important. The third set of specifications includes frameworks for complicated issues (e.g., security or Mobile Internet Interfaces).

Reference Implementations

Technical architecture development has multiple parallel tracks: reviews of existing products and solutions, analyses of the current actions in standardization, reviews and studies of the latest research results, conceptual modeling actions, and the production of technical visions and specifications. However, these tracks as such do not make any practical implementations or references.

A combination of the Mobile Domain and the Web Domain is by no means without challenges. The resulting environment will contain many relatively complicated components. Without thorough prototyping, an adequate result could be hard to obtain. The experiences and competences obtained from prototyping can be fed quickly back to other parts of architecture development, for example, to protocol developers, which saves development time and requires less work to reach the targets.

However, it is not simply a question of prototyping individual components, e.g., protocols. An equally essential part of prototyping is the integration of various parts to the overall system. While the careful definition of interfaces between various layers can make integration easier, real, hands-on experience is one of the keys to success. This is because reference implementations provide a bridge between the technology visions and the real world, a world that is sometimes full of technical limitations and complicated dependencies. What works in theory may not work well in practice, as additional components affect the overall performance of the system. For example, two technologies can, in a tightly integrated environment, interact in a manner that was not foreseen by the developers of the individual technologies. Essentially, these side effects are the price of increased complexity in future communication systems.

Thus, any far-reaching process, such as MITA work, must be able to face the facts and be able to make the correct conclusions at the right time. Implementations are the right tool for this. They serve as a reality check and as insurance against unexpected problems.

For these reasons, MITA work includes a strong effort to produce so-called MITA Reference Implementations. These do not form a complete MITA system, but are a summary of key technologies needed to build the Mobile World. Note that MITA reference implementations are a continuous development process that adopts new technologies when they are ready to be prototyped.

Compliance with MITA Principles

As described in the previous sections, the evolution towards the Mobile World brings new requirements and various challenges. For product and solution developers, MITA is a framework for building product and solution architectures, as illustrated in the following figure.

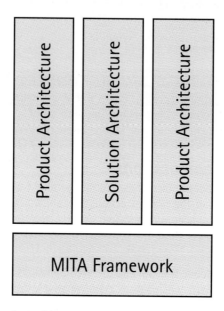

MITA Framework and product architectures

When applying the MITA Framework and methodology to product and solution implementations, three key criteria must be adhered to for compliance with MITA:

- o Providing multiple access capable solutions
- o Securing the highest quality interworking between networks
- o Securing the best interworking between multiple addresses and identities

The above criteria are a very high-level set of principles guiding developers up an evolution path to the Mobile World.

Providing Multiple Access Capable Solutions

Services should not be tied to access methods. If this were the case, new access methods would render the architecture useless. Therefore, a high degree of access independence is a key driver for the technical architecture. Multiple Access Capable solutions are required in order to achieve the global reachability of individuals as well as global access to content.

Securing the Highest Quality Interworking Between Networks

There will also be separate Network environments of mobile networks, the Internet and Intranets. The interworking of these is yet another factor to ensure seamless end-to-end services. These Network environments will be where most user interaction will take place in the Mobile World. To ensure smooth communication between these environments for the consumers, fluent interworking on the technical level is required.

Securing the Best Interworking Between Multiple Addresses and Identities

The reality of multiple access services will require handling multiple identities and addresses that have, historically, been closely tied to networks or applications. A MITA compliant solution recognizes this issue and proposes the best possible way to overcome the challenge.

The dissection of applications should be done in such a way that network requirements are identified, and decoupled from application specific requirements. Requirements for similar applications are grouped under the three Interaction Modes (i.e., Rich Call, Browsing and Messaging). It is important to consider the differences between Network environments, and how to enable applications to seamlessly work between any of the environments, i.e., to ensure service interoperability.

Applications and the Technical Architecture

Nokia classifies applications into four main categories: content, communication, productivity, and business solutions. Example services for each classification category are shown in the following table.

Category	Services
Content	News, Banking, Finance, Local Services, Buy and Sell, Travel, Music, TV, Lifestyle, Fun, Games, Astrology, Dating
Communication	Messaging, e-mail, Fax, Rich call
Productivity	Organizers, Personal Assistants, Tools
Business	Intranet and Extranet access, Information management, Enterprise Communication, Virtual private networks, Telematics

For instance, a game can be downloaded to a terminal from the net and played independently. Here the Interaction Mode is Browsing and the Network environment is the Internet. If the game is played interactively between several people who send messages while playing and, for example, solving game related tasks, then the Interaction Mode is Messaging and the Network environment is the Mobile network, possibly also interaction with the Internet. Finally, if the game is played in an on-line interactive environment, the situation is a Rich Call Interaction Mode in the Mobile network. These game situations highlight the fact that applications must be able to interact freely in all different Interaction Modes and over all Network environments.

In applying the MITA approach to application development, it becomes clear that applications which have been implemented directly on the Platform layer functions while having integrated Mobile Internet layer functions are unacceptable. MITA compatible applications follow the modular implementation model keeping functions between the MITA layers separated.

Mobile Internet Business Architecture

To be able to create a winning solution for the Mobile Internet Business Architecture, Nokia has identified the drivers behind it as well as both the short and long term ramifications which we need to focus on.

Clearly, we cannot overlook the maintenance of profitability, so we prefer solutions which facilitate increases in business revenue. In order to do this, we need to identify the key constituents in the value chain, along with their potential. Hence, consumers, developers and content providers all have to be linked to service providers who in turn link them into the synergy of Consumption, Connection and Content, as illustrated in the following figure.

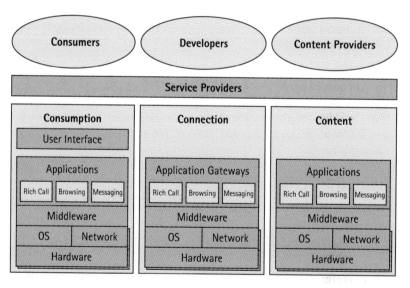

The Mobile Internet Business Architecture

The model contains the key constituencies on the top row, while a simplified architecture with three segments, Consumption, Connection and Content, is presented on the bottom. The key constituencies in the MIBA are consumers, developers, content providers and service providers. The consumer constituency covers both consumers and corporate users, while the service provider constituency covers all types of service providers, i.e., mobile service providers, Internet service providers, application service providers, or in some cases, companies acting as service providers for their own corporate users.

The consumption segment refers to consumers who will be accessing the content provided by a variety of content providers on the content segment. The connection segment (e.g., mobile network) provides a connection between these two segments, and the service providers will manage the connectivity network.

Once all these segments have reached seamless interoperability, all the constituencies will also reach their objectives. Consumers are looking for cost effective and sophisticated services within trusted and reliable coverage areas. Developers will be seeking big volumes and continuity. For content providers, the key issues will be cost efficiency and reachability on a massive scale. Furthermore, service providers are in the business to maintain and maximize their user base and to collect subscription and transaction fees to generate profits.

Middleware Benefits

Middleware solutions provide services for consumers by bringing together software solutions from application vendors and carriers. When the middleware architecture is based on MITA methodology, the following benefits are gained:

1. Support for different network access types from wide area mobile networks to alternate methods. This allows carriers and service providers to build intelligence into their services today and enable their evolution to future technologies.

2. The interfaces are based on open industry standards. Carriers can easily integrate the middleware to their current systems regardless of their network configuration. They can also easily manage their application portfolio. Service providers can provide content using the same standard interface and ensure that applications can utilize their content.

3. Application developers can develop their application to utilize the intelligence of middleware enhancing the features with interoperation capability. This will enhance the consumers experience and make users more willing to use services while on the move, through the Mobile Internet.

Other benefits of middleware for consumers include:

o **Single login** - user logs in once and all services become available

o **Terminal identification through the Mobile Internet layer** - Users will always receive content optimized for their terminal.

o **Preferences** - the user can choose areas of interest and only relevant options will shown (e.g., favorite restaurants, music).

o **Navigation support** - the user can use easy navigation

MITA Deliverables – End-to-End Solutions

It would be impossible to create winning solutions unless extensive end-to-end views guide the design of the technical architecture. These end-to-end views aim to cover all aspects of "Voice goes mobile," "Content goes mobile" and even "Servers become smart" so that overall functionality will be guaranteed.

The architecture elements requiring an end-to-end view are the interaction modes, network interconnectivity, key system elements and access methods. As we are in the middle of the development path towards the Mobile World, coexistent multiple views are already a reality and some new views will be added for completeness.

As the next wave of growth in the Mobile World is expected to come from mobile services, Nokia is actively working with open global standardization of mobile services. The mobile service industry has established a new forum to drive these mobile service standards, it is called Open Mobile Alliance (OMA).

Working within various standardization bodies, the mobile industry will cooperate to create non-fragmented, interoperable mobile services across markets, carries and mobile devices by identifying, endorsing and implementing open standards.

In order for Nokia to ensure the future compliance of these end-to-end views, we already have numerous solutions that fill the MITA model comprehensively and form the foundation for future applications and services. The following table lists many of these solutions and quite clearly illustrates the overall view that has carefully been taken into consideration when aiming to provide optimized creation and delivery of services.

Segment	Services
Consumption	Entertainment phones, Imaging phones, Voice/Messaging phones, Mediaphones, Communicators, Home Multimedia Terminals
Connect	Virtual Private Network solutions, Security appliances, Broadband gateways, Wireless broadband systems, Fixed broadband, xDSL, Wireless LAN, Mobile connection and control servers, Multi-technology radio access networks, Mobility gateways
Content	Mobile commerce, Content applications, Messaging and Community applications, Application developer community, Mobile entertainment services, Mobile Internet services middleware, Connectivity servers, Charging and billing solutions, Advanced call related services, Location services middleware

Conclusions

Nokia is developing a comprehensive technical architecture for the Mobile Internet. The Mobile Internet Technical Architecture aims to provide seamless interoperability between all interaction modes, any network environment and with any type of access. The ultimate objective of MITA is to create a user-friendly Mobile Internet experience for everyone. It will be done:

- by identifying the relevant interaction modes,
- by defining the key technologies required to support them, and
- by driving industry participation to develop a common Mobile Internet platform.

In developing a comprehensive technical architecture for the Mobile Internet, Nokia aims to limit the complexity of the inherently technical environment, as users do not want to worry about the underlying technologies. Nokia sees three key elements as fundamental to the Mobile Internet Technical Architecture: Identity, Interaction Modes and Network environments. Bringing these together and managing the challenges they pose is at the core of MITA and will ensure high-quality and seamless interoperability in end-to-end services.

An open solution benefits all; profitable business scenarios call for interoperability, short development cycles, large volumes and, most of all, global reach. Unless there is a commonly accepted architectural solution, markets will become fragmented as well as require separate parameters, and the total volume will be much smaller than in a single global market.

Introduction to MITA Specifications

Technical architecture work is divided into MITA specifications, to achieve technical teams of manageable scope and enable comparison to other models on a specification level. MITA specifications are divided into three groups:

o   Architecture specifications
o   Key system specifications
o   Architecture frameworks

The Architecture specifications are more permanent in nature, whereas the Key system specifications represent issues where understanding end-to-end system aspects prior to technology selections is important. The third set of specifications includes Architecture frameworks for broader system level issues.

Architecture Specifications

The architecture specifications are more permanent baseline specifications of MITA, covering the structure of the MITA element as illustrated in the next figure.

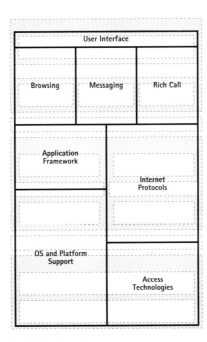

Architecture specifications and the MITA element

UI and Application Framework in MITA

The User Interface (UI) and Application Framework deal with issues related to the user interface and applications. A UI that is simple and easy to understand and use is seen as critical for the successful delivery of Mobile Internet services to the consumer. The importance of applications and their smooth and consistent interaction with the user interface is increasing, as application functionality increases. This becomes crucial when the size of mobile devices limits the display size and the number of user control mechanisms.

In the MITA era, various applications can be active simultaneously, thus providing the consumer with the possibilities to use multiple services at any time. This makes the use of services more convenient and flexible, but also imposes new challenges to introduce these solutions in a user-friendly way. Convenience and flexibility can be further improved by supporting application functionality, which allows seamless communication between applications to exchange any information needed to benefit the consumer.

As applications become downloadable and installable, security becomes increasingly important. Mobile Internet era devices will contain more and more personal and trusted information and functionality and the consumer must have confidence in the behavior of new applications to be downloaded; otherwise, consumer acceptance of new services may be reduced.

It is expected that a combination of revolutionary and evolutionary thinking is needed in order to create UI guidelines for the MITA era. We believe that these targets are best met by focusing the effort on a carefully selected (limited) set of UI styles and mapping the application functionality to the physical UI in an understandable and consistent way, thus minimizing the consumer learning curve and expediting the adoption of new applications and services.

Rich Call in MITA

Rich call is one of the three Interaction modes in MITA. Rich call covers applications which have as their cornerstone a real-time two-way communication component enriched with presence information, text, graphics, images and sounds before, during and after a call. Applications in this category are characterized by having demanding requirements for the underlying transport mechanisms and their integrity due to the real-time nature of the service. Typical applications in the Rich Call category are voice call and conference, video call and conference and real-time interactive games.

Rich call requires consistent behavior over various access networks and Network environments. The access network may impose some constraints over the Interaction mode, but it is desirable that the specific applications can scale their behavior to take this into account. The Network environments and their interworking are also required to be able to interconnect Rich call users residing in various types of networks.

Browsing in MITA

The Browsing Interaction mode covers applications which are interactive by nature, characterized by an interaction between a consumer and services in fixed or mobile servers and, in many cases, nearly real-time:

- o Online browsing (e.g., Markup style browsing interaction, interaction through executable programs and peer-to-peer browsing) o   Off-line browsing (e.g., Push of content to device, using the same markup languages and executables, but the interaction is limited to local servers and services)

Processed content may be displayable or metadata:

o   Displayable content includes various formats (e.g., text, graphics, animation and sound).

o   Metadata is not directly displayed, but can be used to expand the user experience by giving it a larger scope.

Browsing represents an interactive content consumption and manipulation experience. Initially, the consumer consumes content in an interactive manner, but the experience will soon expand into manipulation of the content (true interactive applications).

The browsing environment, and the browser in particular, will play a significant role as a platform, as the host of a browsing paradigm. The browser is enabled with Extensible Markup Language (XML), and can thus offer XML services to other applications. The concepts of embedded links and metadata in markup documents translate very well into any XML content or content that can be translated into an XML representation, be it contacts, calendar or generic data.

One of the key roles of the browser, as the most important component of browsing, is to serve as a platform for enhanced functionality to be used by other applications. This functionality may include multiple technologies, such as animation services, audio services, and video services. In order to flexibly offer this kind of functionality, dynamic installations and the configuration of software (for example plug-ins) become important.

Messaging in MITA

The Messaging Interaction mode covers applications which provide two-way, asynchronous delivery of content. For example, Short Message Service (SMS), Multimedia Messaging Service (MMS) and e-mail involve a separation between the submittal of content and the delivery to the recipient.

Emerging technologies (e.g., Session Initiation Protocol (SIP) and XML) and applications (e.g., news, chat and Internet Relay Chat (IRC)) address challenges to traditional messaging technologies (e.g., SMS and e-mail) causing new interworking requirements. In a similar way, interworking between MITA applications (e.g., click-to-call and click-to-chat) and the management and processing of the static and dynamic message content and attachments in various conditions address requirements for messaging subsystems.

Internet Protocols in MITA

The underlying fundamentals for the success of IP networking include architectural and technical simplicity, openness, and scalability. The open end-to-end architecture has enabled the efficient development of new services and applications, and fast deployment of them on the Internet. The IP protocols are being extended to globally support user, terminal, and service mobility, which will enable new useful services and applications for all Internet users. As an outcome of the extension, some existing Internet protocols may need modification to meet the new requirements.

The Internet protocols specification introduces the MITA Internet protocols architecture; it explains the functions and roles of the protocol stack layers and places the protocol layers in the layered MITA network model. This section also introduces the major Internet protocols implementing the protocol functionalities.

Operating Systems and Platform Support in MITA

As an entity, MITA is not about Operating Systems (OS); it is, however, about how operating systems and applications can co-operate with each other to provide a rich consumer experience in a consistent manner. MITA sets certain requirements on the underlying operating systems and support functionality provided by the operating system to the upper layers, such as, SDK libraries, application protocols, and network protocols.

In many cases, the underlying operating system libraries set the base for architectural paradigms. For example, the Linux operating system and Linux libraries and APIs are good examples of the procedural programming style. On the other hand, the Symbian operating system and Symbian OS libraries provide pure object-oriented interfaces. It should be noted that programming languages (e.g., C, C++ and Java) and programming paradigms are separate issues. However, in most cases, procedural programming environments have been extended with high functionality object-oriented designs.

MITA does not define a preferred operating system. Operating system selection is a business decision, not an architectural issue. However, the operating system should always be compatible with MITA requirements. These requirements are set by the upper layers and are collected into this specification for easy access. On one hand, this also means that MITA is independent of operating systems and programming paradigms. On the other hand, it means that the implementation of MITA in different platforms may vary according to platform capabilities.

Another part of the OS and platform support specification are issues related to baseline development environments, and the principles presented for operating systems also apply to them. This specification collects the generic requirements.

Multiple Access in MITA

A typical mobile device consumer is primarily interested in applications and services, while the actual access technology should be as transparent as possible and should not require any technical understanding, when using an application or subscribing to a service. Still, this transparency should not limit potential applications, as all of them might not work in an acceptable way over all access technologies. These technologies have different technical characteristics (e.g., available bandwidth, connection setup time, network delay and symmetric/asymmetric performance), which impact the consumer experience, i.e., the technologies differ in multiple attributes (e.g., data rates, QoS, handover capabilities, latencies).

The Access Technologies Specification defines the elements needed to allow access from a mobile device to an IP network. Due to the differing functionality of the various access technologies, the platform layer varies between different forms of access. The Access Independent

Interface is supposed to shield the Mobile Internet Layer from the underlying technology, but access specific functionality is required within the network protocols in order to facilitate some access-specific functionality.

Key System Specifications

The Key systems are end-to-end and cross-organizational building blocks of the Mobile Internet Technical Architecture.

Naming, Numbering and Addressing in MITA

Naming, Numbering and Addressing (NNA) concentrates on the names and addresses of individual people or devices and how names and addresses appear on the user interface. The consumer favored nickname information is translated into globally unique names and finally to routable addresses. The translations between globally unique names and the related directory infrastructure belong to the Naming, Numbering, and Addressing specification. The implications and requirements of the NNA environment (e.g., anonymous addressing and regulator aspects) are also studied in the specification.

Naming, Numbering, and Addressing provides one unified translation infrastructure for Rich call, Messaging and Browsing services. Interaction with presence services enables networked and dynamic resolution of naming layer addresses. On the mobile device, Naming, Numbering, and Addressing functionalities are located in the address book subsystem. However, it should be noted that the NNA services are not limited to the address resolution requirements of the rich call service, for any application may utilize these services.

Presence in MITA

Presence deals with the issues of providing dynamic information about the status and availability of consumers and mobile devices. Presence is a key system for a number of specifications. The messaging specification utilizes presence services to enable instant messaging between consumers.

It should be noted that one consumer could have multiple application level identities. How to manage these and how to dynamically associate those to the current Connectivity layer identities are issues which link the presence specification to the reachability specification.

Reachability in MITA

The reachability of devices and people deals with the issues related to the question of how a device or a person is reached through the Web and Mobile domains. A communication network is not a single network with seamless IP connectivity. A communication network consists of networks, such as IPv4 Internet, IPv4 intranets, IPv6 Internet, carrier IPv6 based mobile networks and Public Switched Telephone Network (PSTN). There are also different types of gateways (e.g., Network Address Translation (NAT), firewall or Session Initiation Protocol (SIP) entities) between networks. Seamless person-to-person and device-to-device connectivity over these fragmented networks is the scope of the Reachability Specification.

Access Independent Connectivity in MITA

The Access Independent Connectivity (AIC) specification deals with seamless access to personal services in all Network environments independently of their location and connectivity/access providers. The primary focus of the AIC specification is to provide continuous connectivity for interconnected MITA end-points. The Access Independent Connectivity specification provides access independent mobility management for the Connectivity layer.

Typically, mobility management is involved in connections between a device and a network. This statement is also valid for the AIC Specification. In MITA, mobility management has been divided between the AIC specification and the Access Technologies specification. The Reachability, Service Discovery, and Presence Specifications also deal with mobility issues.

Service Discovery in MITA

Service discovery consists of three types of discovery methods:

o Service discovery on ad-hoc networks
o Service discovery on the Internet
o A special type of service discovery is provided by the presence and virtual community navigation layer

The Service Discovery specification covers the Connectivity and Application layer service discovery technologies. The access layer-related service discovery technologies are covered in the Access Technologies specification.

The Service Discovery specification addresses requirements for the Access Independent Interface. When the AIC Subsystem identifies a need to change into a new access network domain, after proper authentication, the service discovery service for the Connectivity layer should be activated. In a similar way, the service discovery service provides control and event triggering interfaces for the Application layer subsystems and applications.

Location in MITA

The Location specification defines software and protocol interfaces for location services. In practice, this refers to interfaces receiving location information for applications and to applications that utilize location-based information. A goal in MITA is a mobile location services architecture and interfaces, which are able to deliver location information to applications regardless of the positioning technology and of the division of functionality between devices, networks and servers. The same architecture must be flexible for several business models. In other words, a simple business model should not dictate it.

Location information is also an important element in many other MITA specifications. For example, location can be used to automatically update a consumer's presence and reachability information, and it can also be utilized in service discovery for local services.

The mobile location services architecture in MITA actually consists of several different architectures, where the positioning infrastructures are different but location interfaces to applications and middleware, in between, should be the same.

Device Management and Data Synchronization in MITA

Device management, including both bootstrap and continuous device management, is very important. A good and correct configuration is a prerequisite for any application to actually be functional. Furthermore, it is to be expected that the complexity of mobile devices will increase with the addition of new and more powerful applications. At the same time, these increasingly complex devices will move into mass-market environments and it is not reasonable to assume that all the consumers will be willing to configure their applications manually. Thus, the need for device management will only increase.

It is highly likely that there will be multiple device management providers. Typically, each provider has access to a subset of the parameter space. For example, there might be different management entities for the device's look-and-feel, for the browsing environment and for the messaging environment. The managing entities, for the above three, could be the device manufacturer, service provider, and the company of the consumer, respectively.

Device management offers services to applications (e.g., provides them with configuration data). Similarly, device management also has its own user interface, for example, to allow the consumer to control and interact with the management process.

In MITA, data synchronization will cover at least the following issues:

- o Distributed applications and databases
- o Linkage between presence and contacts/calendar
- o Personal Information Management (PIM) related issues (e.g., contacts, calendar, to do lists)
- o Controlled content download and management, using smart messages or a device management mechanism

Content Formats in MITA

A content format refers to a convention of packaging content. Agreements must be made on content formats in order to build the necessary interoperability between various machines, devices and applications. Given the limitations of the processing environment in mobile devices, we have to select a certain reasonably small set of content formats to be promoted and supported in Mobile Internet offerings. Content formats are to be agreed on in the areas of audio, still images and vector graphics, video and general-purpose documents.

Content Adaptation in MITA

Content adaptation refers to the manipulation of content to make it suitable for specific machines, devices and applications. Although agreement on common content formats is the desirable solution, market segmentation, devices of different generations or categories with varying capabilities (e.g., processing power, display resolution, memory and bandwidth), and the unavoidable introduction of new formats are all true obstacles for interoperability. Content adaptation strives to fill the interoperability gaps as devices and content formats evolve. Examples of content adaptation include: image format conversion and the resolution reduction of images and video. Content adaptation can also modify content modality to suit the consumer's environment. For instance, text-to-speech technologies can be useful for retrieving text messages while driving a car.

Operation Support Systems in MITA

The Operation Support Systems (OSS) specification deals with the Mobile Internet management and support architecture. The functionality of the OSS architecture is divided into management layers:

- o   Network management
- o   Service management
- o   Subscription management
- o   Subscriber accounting and identity management

By combining the management layers with the possible business players, two frameworks could be constructed:

- o   Management framework
- o   Policy framework

Privacy in MITA

Privacy is not an add-on feature. All aspects of a system have to be designed to fulfill the inherent need for privacy. Privacy in MITA covers all relevant principles, solutions and technologies related to consumer privacy on the Mobile Internet.

Nokia is concerned about the emerging privacy threats. As the awareness of the Mobile Internet and privacy issues increases among the consumers, the pressure to maintain and build trust rises among all players. Companies need to address the privacy concerns seriously. Trust, once lost, is very hard to regain. Privacy, once lost, cannot be repaired.

Architecture Frameworks

Part of the MITA work deals with issues which are complicated to solve without having common and unified models. These issues are structured under common frameworks, before a more detailed technical architecture is built. Current such frameworks include directories, security, Mobile Internet Interfaces and Quality of Service (QoS).

Directories in MITA

Directories are network repositories for information about people, places and things. There are numerous instances of directories in the Mobile domain. For example, in GSM technologies the Home Location Register (HLR) is a directory of information about subscribers. Directories consist of three basic components:

- o Schema
- o Access method
- o Directory store

Information in a directory, whether it represents data about a person or a device, is structured in a particular way. The characteristic defining the selection of information that is stored about people, places and things is called a schema. There is also a need to represent directory information when it is transferred around the Mobile domain. Exchange formats provide a standard exchange scheme for such purposes. Significant MITA schemes include those that represent the mobile person object and the mobile device object. Other auxiliary schemes define managed mobile services for these basic objects.

Information in a directory is accessed using a directory access protocol. Directory access protocols provide methods for creating, modifying, searching and deleting entries. The administrators of a directory service can also utilize restricted administrator access methods to create new directory object types. There are a number of directory access methods that are needed for Mobile domain directories.

Information in a directory is maintained and managed similarly to information within a database. In fact, a directory is a specialized database application. Within the Mobile domain, there are numerous directory stores, each being used to capture information about a particular feature of the mobile person object or the mobile device.

Additional components in the MITA directory specification include the use of directory synchronization to keep the numerous mobile directory stores updated with each other. Directories contain critical mobile assets including security information about consumers.

Security in MITA

Security is about safety and trust. Safety is considerably easier to manage than trust, and without safety, it is hard to create trust between unseen entities. It is important to understand that security addresses requirements on all entities of the Mobile Internet Technical Architecture. Security is required when calling someone or making an electronic monetary transaction with browser security functionality or features protecting consumer information from being revealed or being altered during transmission. Similar security requirements also apply to server-to-server transactions.

Introduction

MITA security objectives are:

- Integrity
- Confidentiality
- Availability
- Authentication
- Authorization

Integrity deals with the prevention of erroneous modification of information. Authorized users are probably the biggest cause of errors, omissions and the alteration of data. Storing incorrect data within the system can be as bad as losing data. Malicious attackers can also modify, delete or corrupt information that is vital to the correct operation of business functions.

Confidentiality deals with the prevention of unauthorized disclosure of third party information. This can happen as a result of poor security measures or information leaks by personnel. An example of poor security measures would be to allow anonymous access to sensitive information.

Availability deals with the prevention of unauthorized withholding of information or resources. This does not apply just to personnel withholding information. Information should be as freely available as possible to authorized users.

Authentication deals with the process of verifying that users are who they claim to be when logging onto a system. Generally, the use of user names and passwords accomplishes this. More sophisticated is the use of smart cards and biometric authentication. The process of authentication does not grant the user access rights to resources - this is achieved through the authorization process.

Authorization deals with the process of allowing only authorized users access to sensitive information, protected services or resources and controlled actions. An authorization process uses the appropriate security authority to determine whether a user should have access to resources.

Mobile Internet Interfaces in MITA

Mobile Internet Interfaces are fundamental elements of the Mobile Internet Technical Architecture. Elsewhere in this book, both the layered model of a MITA element and the fundamentals of the MITA subsystem architecture were presented. These models identified many interfaces. Some of them are software interfaces within a MITA element and others are protocol interfaces between interconnected MITA elements.

Mobile Internet Interfaces are a group of open interfaces provided by Nokia for both software developers and system integrators, so that they can implement MITA compliant applications and systems. It should be noted that Mobile Internet Interfaces live longer than product generations and have similar features over product and server categories. Open software interfaces in the Mobile Internet Technical Architecture are called Mobile Internet Software Interfaces (MSI) and open protocol interfaces are called Mobile Internet Protocol Interfaces (MPI).

MSIs are divided into Application Programming Interfaces and System Software Interfaces (SSI). In a similar way, MPIs are divided into two parts: Protocol Data Units (PDU), including protocol payload definitions, and content formats, describing presentation formats for the content.

In the Mobile Internet Technical Architecture, all MITA elements have Mobile Internet Interfaces. In terminals, most of the open interfaces are software interfaces for application developers, whereas in servers, the open interfaces are both software and protocol interfaces enabling fast and rich service implementation by system integrators and developers. In network elements, Mobile Internet Interfaces are mainly protocol interfaces enabling different MITA end-to-end views.

Quality of Service in MITA

The QoS is an important topic that requires a common and unified model. QoS in MITA is approached by defining the basic objectives for a QoS system on four levels:

- o   Support the mission of the service provider or carrier
- o   Serve the principal needs of consumers
- o   Meet the quantity and quality requirements of applications
- o   Transmit as many packets as possible

The fundamental assumption is that when defining the actions of a network, a hierarchical structure is formed where the first objective is the most important and the last objective is the least important. For instance, when the needs of different customers are conflicting, we have to look at the upper level in the hierarchy (i.e., the mission of service provider) instead of a lower level (i.e., the applications requirements) in order to solve the conflict. From this perspective, QoS is more about the mission of the service provider, that is, typically business, than the fulfillment of the special requirements of applications. In order to meet the objectives set for QoS, the carrier has to proceed in phases that include defining the fundamental mission, designing the service model, building an appropriate network, and using suitable QoS mechanisms.

The Mobile Internet will consist of several different networking technologies, network environments and services. In some parts of the Mobile Internet, the Quality of Service mechanisms and principles may be based on an approach that is not suitable for other parts of the network. For example, radio technologies often support a natural flow or connection-oriented QoS while the legacy Internet has been designed for packet-based approaches. This implies that the only reasonable way to provide real end-to-end QoS must be based on a common measure, and fulfilling the business goals of carriers is exactly this measure.

Introduction to the Systems Software Architecture

The mobile industry has experienced a period of exceptional growth during the past ten years. The next wave of growth is expected to come from the mobile services sector. New service enablers based on open global standards (e.g., Multimedia Messaging Service (MMS), Java™ and the Extensible HyperText Markup Language (XHTML)) will enable new compelling services for consumers and new sources of growth for the mobile industry. To ensure the successful take-off of mobile services, it is important to minimize the fragmentation of service platforms and to ensure seamless interoperability.

Lack of interoperability between service platforms from different vendors would lessen the attractiveness of many mobile services in the eyes of consumers. Fragmentation of service platforms would also force content developers to make costly and time consuming porting for each platform and hence limit the availability of attractive content. In a similar way, the economies of scale and the innovation pace of equipment vendors would be hurt by service platform fragmentation. For carriers, this kind of fragmented scenario would likely mean slower service deployment, higher costs and longer time to revenues.

A software architecture that enables open global standards and creates non-fragmented service platforms is needed for Mobile World services. The Systems Software Architecture is the Nokia solution for the implementation of Mobile World services. The key architecture design objective is to enable consumers to use interoperable mobile services across markets, carriers and mobile devices.

Design Principles of the Systems Software Architecture

The Systems Software Architecture rests on the following key design principles:

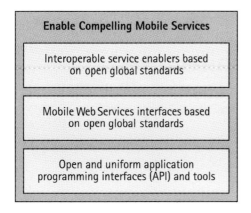

Key design principles of the Systems Software Architecture

Introduction

The success of any mobile service architecture rests on the ability to enable compelling applications for consumers and successful business for the industry's stakeholders. In order to increase the total service market, the objective of the Systems Software Architecture is to enable a flexible business architecture that makes a wide variety of business models for all stakeholders possible. These are the ultimate design requirements of the Systems Software Architecture.

By promoting interoperable service enablers based on open global standards, the Systems Software Architecture creates a foundation for mass-market services. The industry players who implement interoperable service enablers end-to-end in terminals, servers, networks and developer tools based on open global standards offer many new compelling services for consumers. The end-to-end view is critical for success, because one missing link disables the take off of mobile services.

Mobile Web Service Interfaces based on open global standards enable new ways for carriers to open their Mobile domain assets and provide their added value to applications and service providers in the Web domain. By building on technologies that are widely accepted in the Web domain, Mobile Web Service Interfaces will help to bridge the differences between the Mobile and Web domains. This will enable an easy method for non-telecom service providers to gain access to the mobile network and portal assets and enrich their services to consumers, with the added value of a unique Mobile domain. For carriers, this enables new revenue sources from third party service providers beyond revenues from only providing transport and access. In short, Mobile Web Service Interfaces will attract new innovation potential in services from the Web domain to the Mobile domain, thus increasing the total market for all players.

The ability to attract third party developers and content providers is crucial for creating a successful mobile services market. Currently, developers and content providers face a fragmented market where content and applications must be separately ported to service platforms from different carriers and vendors. This limits the availability of attractive content for mobile services. Therefore, uniform open Application Programming Interfaces (APIs) and service execution environments for developers covering both terminals and the service provisioning infrastructure are a central part of the Systems Software Architecture.

By offering uniform and open APIs and developer tools for terminals and servers, a non-fragmented global mass market for developers and content providers can be created. Also, by utilizing developer environments, developer languages and APIs already familiar to developers, the Systems Software Architecture will lower the barrier of entry for developers to make applications for the Mobile domain, and the mobile industry can benefit from an existing Web domain developer ecosystem.

Systems Software Architecture

This section briefly presents the key elements and interfaces of the Systems Software Architecture. The key components and interfaces built on the mobile network include:

- o Terminal
- o Service Provisioning Infrastructure (SPI) or application platform

Introduction to the Systems Software Architecture

- Application Programming Interfaces
- Applications and content
- Mobile Web Service Interfaces
- Mobile service brokers
- Network servers

The above key components compose the building blocks of the Systems Software Architecture as illustrated in the next figure.

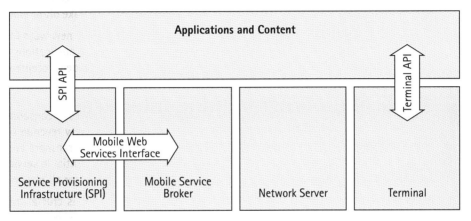

Simplified overview of the key technology components and interfaces of the Systems Software Architecture

Terminal

The terminal software is an essential part of the implementation of service enablers. It consists of two parts: consumer applications and the terminal middleware.

Typical examples of consumer applications include address books, calendars, messaging, browsers, music players and games. Applications may be either native or interpreted and they may be factory-installed or installed on-demand over the air. These options provide consumers with access to a large base of valuable applications and developers with a large volume of consumers.

The terminal middleware provides a comprehensive execution environment for consumer applications, including the necessary facilities (e.g., virtual machines, device drivers, networking, rich call and messaging). A solid and rich application environment speeds up application development by allowing developers to skip the basic plumbing and focus on the application itself. Applications that are portable across all compatible terminals increase the business potential for developers.

Providing developers with an easy access to the enabling services (e.g., location or presence information) is one key feature of terminal middleware.

Service Provisioning Infrastructure

The Service Provisioning Infrastructure includes the software for hosting applications and content. SPI covers the typical middleware products (e.g., application servers, virtual machines, Web servers, databases and run time libraries). It may also cover the necessary tools and adapters for integrating with various back-end systems.

SPI typically provides applications with a comprehensive application execution environment including all the basic services allowing the developers to focus on the essential. For example, both Java 2 Standard Edition (J2SE™) and Java 2 Enterprise Edition (J2EE™) provide developers with a virtual machine and an approach to write an application once and run it on practically any hardware platform. J2EE further allows developers to focus on the business logic while all the basic facilities (e.g., scalability and messaging) are provided off-the-shelf to all applications.

Providing developers with easy access to the enabling services (e.g., sending MMS messages or acquiring location information) is one important feature of SPI.

Application Programming Interfaces

According to the Systems Software Architecture, the end-to-end application environment includes both the terminal and server application environments.

On the terminal side, the application environment includes the Application Programming Interfaces provided by the terminal, called terminal APIs. The standard and easy to use APIs allow developers to write their application once and run it easily in a large variety of terminals compliant with the endorsed standards. Use of technologies, such as Java 2 Micro Edition (J2ME™) make it possible for a developer to write an application that automatically adapts to different types of mobile devices: screen sizes, colors and other device properties.

On the server side, the application environment includes the APIs provided by the SPI, called SPI APIs. Portability across multi-vendor SPIs and different hardware platforms is a crucial benefit, just as it is on the terminal side.

In addition to the regular application services, APIs on both the terminal and the server side provide applications with an easy way to benefit from service enablers and make applications more compelling and user friendly. These applications will leverage the potential and characteristics of the Mobile domain. For instance, easy access to a friend's location or presence information makes it possible for applications to provide consumers with a better user experience than before.

Applications and Content

Applications and content cover software programs and digital material run or consumed in either the terminal or the SPI. In the terminal, this includes for instance, downloadable Java games, XHTML content and music files. On the server side, it typically includes Java beans or applications.

The goal is that applications and content written according to standards are largely portable across different software and hardware platforms.

The availability of advanced development-time tools and emulators that speed up production is an important factor in boosting the development of compelling and valuable applications and content for consumers.

Device manufacturers and infrastructure providers will provide a large range of Software Development Kits (SDKs) for developers ranging from API libraries to emulators for devices and Web services. Having access to these tools will enable the developer to concentrate on the development of the applications rather than on developing their own utilities.

The Java APIs will enable the developer to use tools that are well known and support several SPI environments. The APIs will hide the underlying technology and complexity and make the applications portable between different SPIs and carrier environments with a minimum of work.

Testing an application is always time consuming. Mobile applications can be extremely difficult to test because this would require a complete test environment. The look and feel of an application in a device must be verified for different device types. Having emulators for devices and Web service interfaces as part of the SDKs will make testing easier and more efficient.

The ultimate criterion for success of any mobile services architecture rests on its ability to enable compelling applications for consumers.

Mobile Web Service Interfaces

Web services are gaining wide acceptance in the Web domain. They are built on existing and emerging technologies (e.g., Extensible Markup Language (XML), Simple Object Access Protocol (SOAP), Web Services Description Language (WSDL), Universal Description, Discovery and Integration (UDDI) and HyperText Transfer Protocol (HTTP)). One of the primary targets of Web service technologies is to enable the automation of business-to-business services.

Hence, Mobile Web Service Interfaces provide an easy and efficient method for carriers to increase their revenue by opening their assets (e.g., billing capabilities, identification/authentication, location, and presence information) to any service provider that could benefit from such services.

By building on technologies that are widely accepted in the Web domain, Mobile Web Service Interfaces will help to bridge the differences between the Mobile and Web domains. This will enable an easy method for non-telecom service providers to gain access to the carrier's network and portal assets and enrich their services to consumers. For carriers, this enables new revenue sources from third party service providers beyond transport and access.

In short, the new mobile enriched services enabled by open global Mobile Web Service Interfaces in the Systems Software Architecture will:

- o   Enable carriers to open their platforms in a new way that bridges the gap between the Mobile and Web domains
- o   Enable service providers to access a carrier's value added services (e.g., billing and location)
- o   Enable consumers to get enriched mobile services in a way that creates new business for carriers and service providers

Introduction

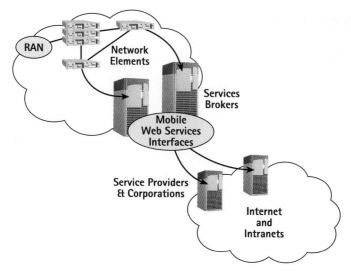

The concept of Mobile Web Service Interfaces

Mobile Service Brokers

A mobile service broker is a technology concept of how to implement functionality in Mobile Web Service Interfaces.

Mobile service brokers is a logical element that includes the software in the carrier domain that is specialized in providing valuable and meaningful mobile services to any service provider. Brokers may provide useful services (e.g., acquiring a customer's location information for parties within or outside the carrier domain, for example, on the Internet). Brokers can also hide the implementation details of different types of network technologies or devices from its users.

For carriers, the mobile service brokers concept provides an opportunity to increase their revenue beyond transport and access income. By selling these mobile services in a controlled and secure manner while boosting the creation and usage of a large volume of more attractive services makes it possible for carriers to fully leverage the assets of their mobile networks.

For service providers, the opportunity to use mobile services (e.g., location or presence) makes it possible to produce more attractive services. The possibility to benefit from the carrier's existing authentication and billing mechanisms opens new business models and opportunities.

For developers, the brokers provide mobile services in a logical, easy to use and familiar form that enables rapid adoption of mobile services and seamless interoperability between different carriers.

Brokers inside a carrier domain open mobile network services to any service provider through Web service interfaces.

Network Servers

Network servers is a logical element that covers the software on carrier-hosted servers that specialize in providing mobile- or telecom-oriented services to consumers. Examples of network servers are WAP gateways, the Short Message Service Center (SMSC) and the Multimedia Messaging Service Center (MMSC). Network servers can also communicate with applications directly (e.g., in the case of a WAP gateway).

Introduction to Service Enablers

The mobile industry has experienced a period of exceptional growth during the past ten years. Mobile voice and simple Short Message Service (SMS) text messaging have been the primary drivers of this growth. The next wave of growth is expected to come from new mobile services where content, not only voice, will go mobile.

Service enablers are the basic technology building blocks for creating mobile services. The implementation of service enablers can potentially take place at several locations along the end-to-end chain (e.g., terminal software, server software and development tools). Many new service enablers are needed for producing new and compelling mobile services and enabling the next growth wave for the mobile industry. The key criteria for service enablers are:

- o Enable new and better services for consumers
- o Provide developers with facilities which speed up the development of new services
- o Make new business possible for industry stakeholders
- o Be based on open global standards, which secures interoperability

To ensure successful take-up of mobile services, it is important to minimize the fragmentation of service enablers and to ensure seamless interoperability between different vendors. For these reasons, Nokia is strongly committed to open global standards. Therefore, Nokia is working together with other industry leaders in standardization (e.g., Open Mobile Alliance) to promote open global standards and the interoperability of service enablers. From the Nokia perspective, the key services enablers are:

- o Multimedia Messaging Service
- o Java technology
- o Browsing
- o Device management
- o Device profile
- o Delivery
- o Identification/Authentication
- o Payment
- o Mobile Digital Rights Management
- o Group Management
- o Location
- o Presence
- o User Profile Management

Introduction

- Instant Messaging
- Streaming
- Data Synchronization

Multimedia Messaging Service

The Multimedia Messaging Service (MMS) is a versatile messaging service which enables rich content (e.g., images, audio clips, video clips, data and text) to be sent between mobile devices, from a mobile device to e-mail, or from an online service to a mobile device. MMS is standardized globally and is one of the first services to be standardized by the Third Generation Partnership Project (3GPP).

MMS delivers a location-independent messaging experience combined with the ease of use guaranteed by a simple and logical extension of SMS. It builds on the well-established SMS model by adding new functions and new content types in steps that are understandable to a consumer. This step-by-step evolution will encourage the adoption of MMS, leading to rapid take-up and high penetration.

MMS can be used in various situations, in business or leisure, and meets the needs of many consumer segments. The possibility of taking a picture and immediately sending it via MMS allows the consumer to be in full creative control of the content and share it easily. This can be extended for business use where a photograph can be annotated with text or explanations and sent instantly back to the office. The versatility of content in MMS enables photographs, video clips, maps, graphs, layouts, plans and animations to be sent.

The key element in the MMS network architecture is the Multimedia Messaging Service Center (MMSC). The MMSC enables multimedia messages to be sent with various content types from device to device with instant delivery much like SMS messages today.

The MMS standard also defines messaging between Internet applications and mobile devices as well as support for flexible addressing of multimedia messages to both familiar phone numbers and e-mail. MMS messages can be created in either Internet applications or mobile devices equipped with an integrated or connected camera. Images can also be downloaded from various Internet applications, through which MMS messages can be created, stored and forwarded.

Java™ Technology

The commercial mobile device has until recently been a closed platform, without the possibility of adding new functionality after the mobile device has been purchased. The introduction of Java™ has changed this situation and has opened up the mobile platform. Java will enable the consumer to add new compelling features and applications to the mobile device anytime and anywhere.

Java technology includes both a programming language and a software platform. In order for Java to be successfully introduced into the global market, it needs to be globally standardized and optimized for the limitations of different devices. The core idea of the technology is that a standardized Java platform hides the complexity of the device from the applications. The applications see the standardized interfaces of a global Java platform, and they do not have to deal with the special characteristics of different devices. A standardized Java platform will allow content and service developers to focus on delivering quality services to a larger market. This in turn will bring more attractive services to consumers and grow the total Java market in the Mobile Industry.

Browsing

The essence of mobile browsing lies in its close alignment with widely accepted Internet standards. The WAP Forum and the World Wide Web Consortium (W3C) have successfully defined mobile Internet standards over the past several years. The WAP Forum has now adopted the Extensible HyperText Markup Language (XHTML) Basic standard from the W3C as the basis for WAP 2.0. The transition to the XHTML Mobile Profile will strengthen the position of the mobile browser in the mainstream Internet and allow for a far greater range of presentation and formatting than previously possible.

The essential elements of browsing content are a page description language, a content formatting language and a scripting language. These elements enable consumers to enjoy a wider array of services, more intuitive user interfaces and a generally more useful experience. At the same time, carriers will be able to exercise more control over the look and feel of the services they provide through their mobile portals.

Device Management

A key inhibitor in the uptake of new services has been that consumers are unable to correctly configure their devices to actually start using the service. The outcome has been a slower take-up of new services and features, considerable consumer costs for carriers and service providers, and poor consumer experiences with new services. In short, this dilemma has translated into considerable losses in revenues, increased customer care costs, lost opportunities, and an overall poor consumer experience.

Device Management is the generic term used for the technology which allows third parties (e.g., carriers, service providers, corporate IT departments) to remotely provision and configure mobile devices on behalf of subscribers. This includes remote provisioning of new services, configuration and management of consumer device parameters, setting remote device diagnostics and troubleshooting. Key use cases for the technology include:

- o   Device management for new device or service purchase
- o   Helpdesk problem identification and resolution
- o   Device backup/restore
- o   High volume configuration
- o   Advertising

Introduction

Device management is a critical enabler for making it easier for consumers to adopt mobile services and for carriers to cost-efficiently manage the terminals in their customer base. By alleviating the consumers' difficulties, carriers and service providers are able to dynamically differentiate their service offering, and customize their services for different consumer segments. Incorporation of device management technology will translate directly and indirectly to the bottom line, via revenue growth, cost reductions, and reduced churning due to the enhanced service level.

There are two main elements in device management: bootstrapping and continuous provisioning. The WAP Forum 2.0 provisioning specification for bootstrapping and the SyncML Device Management for continuous provisioning fulfill the criteria for openness and offer sufficient flexibility and extensibility to become industry standards.

Device Profile

The wealth of emerging wireless data applications and the continuing variety of form factors in handheld devices make it impossible to hide the differences in capabilities among mobile devices. In contrast to the legacy Internet, mobile devices exhibit striking differences in their display characteristics, their support for handling content types, their input and output modes, the bearers over which they communicate, and their memory capacity, among others. As a consequence, these different capabilities must be explicitly considered if one wants to avoid situations where, for instance, terminals receive content too large to be stored or in a format they cannot display properly. Alternatively, abstracting out these differences and designing applications based on the lowest common denominator does not result in appealing services.

Terminal profile management comprises the technologies required to provide information about terminal capabilities to all elements of an end-to-end service chain, so that they can be taken into account during application development, content selection adaptation, and content delivery.

Terminal profile management addresses the following issues:

- o   Representation of terminal profiles via a formal, structured notation amenable to automatic processing;
- o   Standard vocabularies - in other words, sets of universally agreed attributes and their allowable values that serve to define terminal profiles;
- o   Protocols enabling terminals to inform network elements about their profile;
- o   Rules to interpret and manipulate terminal profiles.

Knowledge of a terminal profile makes it easier for application developers to build applications for mobile devices. Among others, terminal profiles serve the following purposes:

- o   Automatically select the most suitable content to send to a terminal
- o   Adapt content automatically to the characteristics of the terminal
- o   Overcome limitations in the content accepted by a terminal

Terminal profile management does not prescribe whether applications can be optimized for specific terminal characteristics, nor how this should be done. Application providers and carriers are thus free to determine how best to take advantage of terminal profiles. Terminal profiles can be combined with subscriber profiles, providing further possibilities to optimize or tailor applications.

Delivery

Delivery is a family of technologies, which delivers digital content (e.g., entertainment and business applications) to mobile devices. Another important application area for delivery is terminal personalization. A user or a network application can initiate content delivery.

The download mechanisms presented are logical and complement each other: interactive download and non-interactive download. With interactive download, the discovery process and the actual download are an integrated process. This means that the discovery of content to be downloaded and the actual mechanics of downloading are part of a single session. This single session is often referred to as browsing.

In non-interactive downloading, the discovery and the delivery processes are typically separate. This happens when messaging is used as the main download mechanism. The discovery may occur using text, graphical or voice browsing, or it may happen by sending unsolicited trigger messages to a service. The actual download of the content uses messaging as the main transport mechanism.

These two mechanisms are also referred to as browsing-driven download and messaging-driven download. With interactive download, i.e., browsing-driven download, a user may discover content (e.g., images or Java games) from a service portal and download it within the same context to the browser. The ability to pull content into the browsing context enables seamless content discovery and download. The user does not need to switch between applications on the mobile device. Content download over browsing is very convenient for the user. For example, clicking on the "next image" will result in receiving another image within the same browsing context.

However, non-interactive downloads, in which the discovery and delivery processes are separate, are equally important. This method serves different uses and business models.

Identification/Authentication

Identification and authentication are critical functions for practically all services. Consumers and clients need to be identified and authenticated in order to get access to services ranging from mobile networks, Wireless Local Area Networks (WLANs) and corporate intranets to valuable Web and WAP sites. It is also equally important that the consumer can trust the service, to ensure that no sensitive information is given to unintended parties.

The system for identification and authentication shall enable consumers to:

- o   Identify themselves with an appropriate level of traceability and security
- o   Disclose some of their private data to various services in a controlled and confidential manner so that personal services can be provided

- Authorize services and other agents to use brokers
- Achieve single sign-on

A consumer may have a number of *virtual identities* (e.g., her credit cards, debit cards, loyalty cards). When the consumer accesses a service she may choose to present one or more of these identities. In its simplest form such an identity contains an account and password, specific for the service. It can be expected that some identities can be used at a growing number of services, analogous to a credit card that is widely accepted.

The next extension is that an *identity card* contains additional information, a part of the user profile. And instead of a simple account and password, the presented identity could include a signed certificate. While the consumer is in control of the content and use of the identities, the carrier plays an important role in enabling the use of such identities.

Identification support is expected to grow in phases. The first step includes identification towards third party web services. Consumers who access such services through a mobile network will enjoy seamless authentication based upon the Subscriber Identity Module (SIM) card in the mobile device. Later, identities can be used for authentication with intranet Virtual Private Network access gateways, and for authorizing and charging for network access, for example, to public WLAN network hotspots. Similarly, the same identities can be used to support local services where mobile devices communicate directly over local links (e.g., Bluetooth or infrared).

Mobile Payment

Mobile commerce is creating a whole new meaning for mobile devices and services. Mobile devices, being personal and always with the consumer, are evolving towards Personal Trusted Devices - the ultimate digital wallets - which will be used for a wide range of mobile transactions either remotely, over the mobile network, or locally at a point of sale.

In addition to paying for physical goods, a major part of mobile commerce consists of the purchase of different types of digital content that in most cases ends up being utilized in the mobile device. Consumers like to personalize their mobile devices with ringing tones, graphics and picture messages from content providers. Ticketing services, downloadable applications (e.g., games) and Musical Instrument Digital Interface (MIDI) ring tones as well as music and video feeds will soon follow.

With mobile devices, services are becoming far more accessible and can be tailored to your needs. The mobile device allows spontaneous purchases, as it is ready-to-use in seconds. Starting with General Packet Radio Service (GPRS) devices, consumers are always connected and can instantly access mobile Internet services and use the mobile device for convenient mobile shopping.

Providing the means and the security for transactions across the mobile network will be an important key function for carriers and service providers. Key characteristics for successful mobile commerce services will be immediacy, ease of use, personalization, location awareness, user authentication and security.

Mobile Digital Rights Management

Original content owners and content aggregators (e.g., branded service providers) and carriers will only be able to gain value from their mobile services by protecting downloaded content against misuse. Mobile Digital Rights Management (mobile DRM) provides the infrastructure for the usage and transaction control of content delivered Over the Air (OTA).

Mobile DRM is aimed primarily at OTA delivery of content formats (e.g., ringing tones, screen savers, games, Java applets and images). However, mobile DRM will evolve to meet the needs of more demanding media types (e.g., high-quality music). The mobile DRM architecture will be designed, standardized and supported in the spirit of openness, relying on the work of existing mobile industry self-governance bodies (e.g., WAP Forum, 3GPP) and Internet bodies (e.g., Internet Engineering Task Force (IETF), W3C, Moving Pictures Experts Group (MPEG)).

Presence

The emergence of presence technologies and services into the Mobile domain will be one of the most fundamental changes affecting communication behavior in the coming years. Today when communication is initiated, the consumer rarely knows if the other party is available or wishes to communicate. Making a call has been a game of chance. It is no surprise that a growing trend in mobile communication has been the consumer's need to be able to better control their own availability and communication.

Presence enables a new communication paradigm: "look before you communicate." A consumer is able to see prior to initiating communication if other parties are available, wish to communicate and by what means they would prefer to be contacted. Internet instant messaging services have popularized the concept of presence where users are able to see if their friends are online and available for messaging. Within mobile communications, the value of presence will not be limited to messaging. Presence will facilitate all communication. This is also the key for presence services to succeed in the mobile mass market: there must be utility in presence within existing mainstream mobile communication and communication behavior.

The carrier is in a key position to offer presence services. It already has a strong relationship with consumers. Some of the critical presence information is only available from the mobile network (e.g., location or terminal status). The carrier is also in a position to offer more added value through presence-enabled services on the network (e.g., smart call or communication routing).

Instant Messaging

Instant messaging services are now going mobile. Consumers will benefit from no longer being tied to their desktops. They will be able to engage in instant messaging or chat-sessions anywhere and anytime. With mobile devices, instant messaging will facilitate all communication and allow consumers to be connected to their friends from anywhere.

User Profile Management

A user profile is a set of attributes about a consumer, and may be physically distributed. Usually a party who holds a user profile may not disclose it to other parties without the consent of the consumer. A consumer profile may contain simple attributes (e.g., name, language and address). But it often also contains more sensitive data (e.g., credit card numbers). Often a user profile includes preferences and interests (e.g., hobbies).

The user profile is the key component needed to provide personalized services. Personalization is especially important for mobile services (e.g., to avoid cumbersome and unnecessary interaction, to increase value, to maximize the limited amount of display space). The user profile should be made available to various services, but in a way that respects the privacy of the consumer. Indeed, a user profile is a valuable asset and services will offer the consumer, the owner of the profile, value in exchange for the profile.

Carriers are in a good position to act as profile brokers. Although some parts of the user profile can be expected to be located in a virtual wallet or phonebook of a mobile device, the carrier has access to important presence and location information. Moreover, the mobile device is often not in a position to serve the profile data. A carrier can host a network copy of the profile data in the device and combine the data with the other consumer data it has to provide a unified view of a much larger chunk of the complete profile.

Data Synchronization

Data synchronization is a critical technology enabler in the Mobile Domain, since for practical reasons (e.g., network coverage) or due to the expenses of maintaining an always-on connection, most devices are only intermittently connected to the network. Also, as the number of local databases grows, and the operations on them increase, there is a constantly growing need for a convenient tool for sharing updated information. In short, there is a growing demand for data synchronization solutions for the Mobile Internet.

Streaming

Basic multimedia streaming services must use open standards to enable interoperability between devices and services, as well as to enable content and service creation. The key principles for the technology are: suitability for mobile implementation, interworking with other standard mobile multimedia services, and open standards. Mobile streaming services will be interoperable between different devices and carriers.

Prior to the transition to the Mobile Internet, in which all mobile services will be delivered over IP, radio access technology will evolve to allow streaming over packet switched GPRS and Wideband Code Division Multiple Access (WCDMA) bearers to mobile devices. The take-up of wireless streaming is vital for developing the market.

Location

Location service enablers are key differentiation levers for mobile services, which will give them an edge over legacy Internet services. There are nearly unlimited opportunities for radically enhancing services in areas, such as mobile commerce, advertising and entertainment, with location information. In short, location service enablers open up new revenue opportunities for carriers and service providers. In some countries, there will also be regulatory requirements for carriers to provide location information (e.g., the Federal Communication Commission (FCC) 911 mandate in the US).

Group Management

Group Management service enablers attempt to make it intuitive, easy and secure for consumers to manage their contacts and communications, and share presence information or personal content with their communities and groups.

Conclusions

Service enablers are the basic technology building blocks for creating mobile services. The implementation of service enablers can potentially take place at several locations along the end-to-end service chain. Many new service enablers are needed to produce new compelling mobile services and enable the next wave of growth for the mobile industry.

To ensure successful take-up of mobile services, it is important to minimize the fragmentation of service enablers and to ensure seamless interoperability. Therefore, service enablers should be based on open global standards.

Complying with open global standards in the underlying service enablers still provides many opportunities for Mobile World players to differentiate their offering to consumers. By playing with parameters on the service layer, an individual consumer's experience of services can be radically differentiated from the experiences of others, even though the underlying service enablers are based on open global standards.

Today, differentiation is too often based on separation in the technology of service enablers, hindering the overall take off of mobile services. Mobile World players should be enabled to direct their development resources appropriately; rather than focusing on common services and differentiations in the technologies for service enablers, they should be able to focus on truly innovative services which would differentiate their whole product offering.

52

III WEB DOMAIN TECHNOLOGIES

- o 3.1 Platform Layer
- o 3.2 Mobile Internet Layer
- o 3.3 Application Layer

Part 3.1

Platform Layer

o Linux

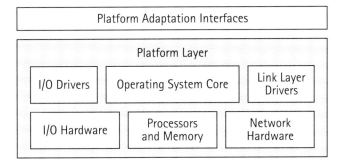

Linux

Linux ® is an operating system that is widely used in the Web domain for performing different functions. It is known as a robust, versatile, and efficient system that can be used for running a wide range of services. These include Web servers, e-mail servers, and other Internet services. Linux is also used extensively on the client side, especially in the research and academic sectors, and in those parts of organizations that are cost sensitive – Linux is free of charge.

Furthermore, Linux is not only free but it is also open source. This means that the operating system source code is available for anyone to view and modify, according to their needs. This is clearly an asset for the developer community, since there are no limitations to optimization or modification. A variety of tools are available, but you can also develop your own tools, if necessary.

The Linux operating system can be used in a wide range of system categories. Linux plays a role at least in the following three categories: servers, embedded systems, and desktop terminals. In the server area, which is clearly the focal point with respect to the Web domain, Linux is playing a significant role. Many products can use Linux as their operating system platform, e.g., the Apache™ Web server. According to IDC, currently Linux is used in 27% of new server installations, and its share is estimated to grow in the future. [1]

Linux is gaining popularity also in the embedded systems area. An embedded system refers to a device category that requires some tailoring in the actual operating systems distribution package. For example, from an operating system distribution package point of view, the Nokia Media Terminal™ is a product that falls into the embedded systems category. It is an excellent example of an open platform product designed for viewing and recording digital television. It uses Linux as a platform and includes the needed kernel functions (device drivers, memory management, networking and file system support) and some basic user services (windowing system XFree86 and a Mozilla™ browser). Developers are free to modify and add functionality and applications into this open platform, according to their needs. [2]

In the desktop environment, there are several commercial companies that package Linux with a variety of applications and an up-to-date windowing system with installation support. The aim is to allow consumers unskilled in the Unix world to be able to use and install Linux and the needed applications. These companies are in the business of making distribution packages of Linux by using the standard kernel and packaging it with the appropriate add-ons. Consumers can receive the distribution package on CD-ROM for a competitive price or free-of-charge from the Internet.

There are a number of technical reasons that make Linux an excellent choice for a server and services platform. Obviously, from the cost point of view, the initial investment in a server system based on Linux is modest when compared to many other systems, since there is no license fee for the operating system itself. Additionally, the cost of ownership is also typically less or at the same level when compared to typical proprietary operating systems. Furthermore, since Linux is not tied to any single company, you can choose the service provider according to your business principles.

Origins of Linux

Linux originated from the desire of Linus Torvalds to have a variant of Unix for the Intel 80386 microprocessor. A good starting point for this would have been Minix, a Unix variant written by Professor Andrew Tanenbaum for educational purposes. The Minix source code was available, but it was not open in the sense that users could modify it (Addison-Wesley had copyrighted the source code). There was port of Minix to 386, but it was not officially sanctioned by Tanenbaum. He was reluctant to accept additions to Minix because he used it as a classroom example of an operating system.

Since Torvalds could not get the additions he wanted to be included in Minix, he started developing his own operating system instead. Torvalds and Tanenbaum had an interesting discussion on Internet newsgroups, revealing some of their assumptions at the time:

> It had a monolithic kernel instead of a microkernel such as the one in Chorus ™, Mach, or Minix. Its design was too close to the Intel® architecture. [3]

Today, most operating systems are (still) monolithic, so history has shown that Linus was right on the first point. The benefits of microkernels have not yet overweighed their performance penalties.

As for the second point, the first version of Linux was in fact tied to the Intel architecture – after all, it started from the desire to utilize the 386 features to the fullest. Torvalds argued that writing the kernel-specific parts of the kernel was no big deal, and he was later able to prove his point when he ported Linux to other architectures.

Torvalds foresaw the importance of the Intel platform. Since Linux was originally designed for it, it had very good performance. In fact, since it did not have to be backward compatible to any earlier versions, it used the 386 features very efficiently.

Other developers were soon attracted to Linux. Unix was a real operating system, but there were no free variants of it available at the time. Intel 386 was becoming very popular. Tanenbaum had an operating system with source code available, but he was not eager to accept additional features to it. Berkeley Standard Distribution (BSD) Unix was troubled by lawsuits. GNU Hurd was far in the future.

One of the first objectives for Linux was to be able to use GNU tools. Many of these had been created by Richard Stallman or had been supported by his Free Software Foundation. Stallman had originally set out to create a free, Unix-like operating system. For this, he needed several programming tools. His first project had been Emacs, a free text editor that had become unexpectedly popular on Unix systems. Stallman next created the GNU C Compiler (GCC), which has been one of the best C compilers on any platform for a long time. The Free Software Foundation also sponsored bash, a Unix shell. Although Stallman has created many successful tools, his initial dream, the operating system, is still not complete.

Consequently, Linus Torvalds had the basic tools available – he only needed to port them to Linux and to make sure that Linux was able to run them. Getting more and more tools to run on Linux was an early goal for the operating system project and it also directed later work. This differentiated Linux from other free Unix-like operating systems.

Jon Hall was the person who orchestrated a Linux port to Digital Alpha. Working at Digital at the time, he invited Torvalds to speak at the Digital Equipment Corporation User Society. [4] During the meeting, Torvalds explained that he had been thinking about porting Linux to Alpha, but he did not have the equipment available. Jon Hall arranged for a Digital Alpha machine to be sent to the University of Helsinki, and Linus used the opportunity to port Linux to it.

The Gartner Group predicts that the annual growth rate in the number of shipped Linux workstations will average 56% in 2001-2005. [5] Torvalds himself has indicated that there are limits to the scalability of the kernel for larger machines with more Central Processing Units (CPUs). [6] There are also doubts about how well Linux will suit the average PC user.

Linux Architecture

Linus Torvalds released the Linux kernel, the core of the operating system. Since then, the work has continued as an effort of several key persons looking after specific areas of the operating system, i.e., networking support, process management, file systems, and device control. As the kernel itself is only visible to the machinery, Linux distribution packages cover the rest of the ground – the Graphical User Interface (GUI), the windowing system, applications such as Emacs and File Transfer Protocol (FTP) – packaging the collection of entities as one needs it. Naturally, the environment, whether desktop or router, will mandate the use of different applications and protocols.

Anyone can modify the kernel according to their needs, because Linux comes with a source code. Thus, it is possible to have a build that contains only the required parts of the kernel modules, and to modify or add completely new modules.

The Linux kernel has been ported to several different microprocessor architectures, e.g., PowerPC™, ARM™, and Sparc ®. Therefore, there are no limitations in the kernel design that restrict it to be used only in Intel-based processor architectures.

Features

Most of the features (e.g., multitasking, virtual memory, and networking) found in modern Unix operating systems can also be found in Linux. It supports kernel threads through the system call clone. Threads in Linux are identical to processes, except that all threads in a process share the same memory space. Threads have individual priorities and are scheduled as separate entities.

Monolithic

Linux has a *monolithic* operating system kernel: all subsystems of the kernel belong to the same memory space and use function calls to communicate with each other. Monolithic refers to the choice of the basic communication mechanism inside the kernel. The alternative is a microkernel architecture, in which message passing and process abstractions are implemented at the lowest level of the kernel. All other parts of the kernel are then in separate memory spaces and use the message passing system.

Modules

Linux also provides kernel modules, a dynamic mechanism to add functionality to the kernel. This is necessary, e.g., to support low-memory configurations or when two different devices would otherwise clash. It is also convenient for device driver development, since a device driver can be created as a kernel module and then loaded without recompiling the kernel and restarting the machine. Later, the kernel module can be statically linked to the kernel to speed up startup and execution.

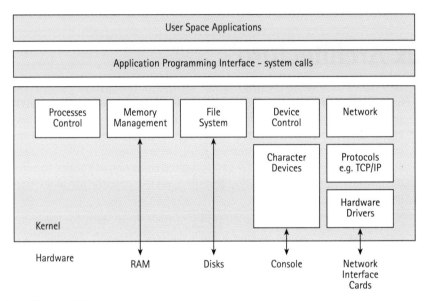

Kernel parts illustrated [7]

Kernel modules can also be used for real-time programming. Since they run in the kernel space, they have direct access to all hardware devices. They cannot be pre-empted by other processes unless they explicitly yield control. A Linux kernel patch (RTLinux™) that takes care of the execution of real-time tasks is guaranteed to function even if the kernel disables interrupts. Thus, the interrupts are only disabled from the kernel point of view, and in reality in the hardware.

However, there is a downside: the kernel module mechanism is not as flexible as a microkernel. A microkernel allows you to replace the scheduler and virtual memory policies. Linux has three different scheduling policies hard-coded in the scheduler, and if these are not sufficient, then the scheduler code must be modified. Still, kernel modules can be used, e.g., for creating device drivers and implementing new file systems.

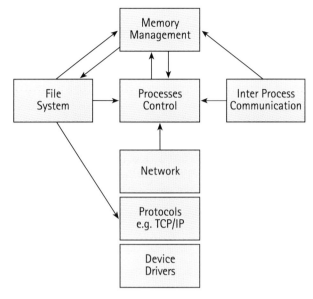

Conceptual presentation of kernel [8]

Kernel HTTPD Accelerator

The kernel httpd accelerator is an example of a module. Since Linux is so often used as a Web server, the kernel has some support for this function. The kernel httpd accelerator is a kernel module that interrupts Web page requests before they are passed to a Web server. For static Web pages, it is able to read the page from the file system and send it back to the requestor. Dynamic Web page requests are passed on to a real Web server. This speeds up the responses to static Web page requests and also reduces the load on the Web server. Even Web pages that display information customized to the reader usually consist of a number of large images and some dynamically generated text. These images can be served by khttpd.

Distribution Packages

Distribution packages are important because it would be too difficult to collect all individual pieces of software separately and to make sure that they run together. In a sense, distribution packages take the individual pieces of software and make them a complete system. Important distribution package functionalities include:

- o Installation Help
- o Upgrades
- o Automatic Settings

Installation Help

The installation program must be able to detect the available hardware and to tune each application to it. The installation program must also support other operating systems that may exist on a computer.

Upgrades

The most popular distribution packages (e.g., Red Hat™ and Debian™) have good package management systems. They can find newer versions of installed components from the Internet and manage the dependencies that a new piece of software has. Package management also helps in removing programs and their dependencies. The latest business idea that Red Hat and Ximia™ are implementing is automated, remote upgrades.

Automatic Settings

The installation programs in distributions must also be able to set default values for all applications. This is important especially for firewalls and other security-related parts.

A distribution package also provides a known reference point for other programs.

Most companies that provide Linux-based products have selected a third-party distribution package as a basis. Only some embedded devices that require heavy modification to the kernel need to have a proprietary distribution package in place.

Real-Time Support

Linux is not designed to run real-time tasks. This is because providing guaranteed execution times is often incompatible with making the system run normal tasks as efficiently as possible. One example of this is how often kernel tasks call the scheduler: it is more efficient for a kernel task to finish whatever it is doing than to try to do the same job in as small pieces as possible. However, as Linux is freely available, there have been efforts to make Linux more real-time.

The first approach to improve the responsiveness of the kernel is to insert scheduling points to the kernel and to reduce the time that the kernel blocks interrupts. This is known as a *low-latency patch*. In practice, it works well for tasks that do not require absolute control of the hardware. The benefit is that the applications can still use the normal Linux kernel, albeit a patched one. The trade-off is that patches are made for certain kernel versions and they may not be available for the latest ones. The patch is also not as well tested as the Linux kernel is in general, because they are not used as much.

The second approach (RTLinux) overrides the Linux scheduler completely. Real-time tasks always run first and the kernel is scheduled to run only when there are no real-time tasks to run. This makes it possible to run processes with very little overhead and no interference from the kernel. Normal Linux applications can still be run on top of the Linux kernel. Unfortunately, these real-time tasks cannot use normal Linux operating system services and must do everything by themselves (e.g., normal device drivers will not work with them unless modified). It is also a bit difficult to arrange communication between real-time and non-real time tasks.

```
┌─────────────┐
│   Normal    │
│   Linux     │
│ applications│
└─────────────┘

┌─────────────┐   ┌──────┐
│   Normal    │   │ Real-│
│   Linux     │   │ time │
│   kernel    │   │ tasks│
└─────────────┘   └──────┘

┌───────────────────────┐
│  Real-time scheduler  │
└───────────────────────┘
```

Real-time Linux approach

Security

Security is a crucial component in any computer or network system. It is well known that the system is as secure as its weakest link, and Linux is no exception. It needs to be managed and configured appropriately, just like any Unix system. However, one point could be mentioned. Linux is perhaps more secure because its source code is open and can be viewed by anyone competent enough in programming, also to verify that the security is not compromised.

Another aspect is that the precompiled kernels need to come from a trusted party. Otherwise, they might contain parts that are not in the consumer's best interest. Also, one should be careful with the modules that are loaded into the kernel, because their behavior will have an effect to the primary functions of the system.

Networking

It is widely recognized that networking support is excellent in Linux. The networking support needs to be handled in the kernel in order to be efficient and because networking is not a function of a specific application. Packet reception and forwarding are asynchronous events, which are well illustrated in today's bursty Internet traffic.

Two issues need to be dealt with: packet and process handling. Packets need to be identified and delivered to the correct interface, be it the network interface or the user programming interface. Thus, packet forwarding, routing, and address resolution functions need to be implemented in the kernel. In some cases, there are also functions related to mobility and security that are implemented in the networking part of the kernel. Processes that interface with the networking part need to be controlled accordingly: put to sleep when no packets are targeted for the process and woken up once there is a packet for the specific process.

Networking Functions

The versatility of Linux is best illustrated by naming the roles it can play on the Web. It can be found handling routing, bridging, and firewall functions, as well as being an e-mail, Web, Domain Name System (DNS), or Dynamic Host Configuration Protocol (DHCP) server.

Routing

The Linux kernel has built-in support for routing functions. A Linux device with proper network hardware interfaces can act as an IPv4 and IPv6 router for a fraction of the cost of a commercial router. Naturally, there may be a performance difference if the criterion is the number of delivered packets per second. However, it is more important to notice that the functionality exists and can well serve, e.g., small and medium size companies. Recent kernels include special options for machines that act primarily as routers (e.g., IP multicast routing and policy routing).

Firewall

Different firewall toolkits exist for Linux, in addition to the built-in support in the kernel. These firewall toolkits are very complete, and together with other tools allow the blocking and redirection of all kinds of traffic and protocols. Different policies can be implemented via configuration files or Graphical User Interface programs.

Development Cycle

Open source software is often claimed to be more reliable and to develop faster than proprietary software. This mainly results from the vast number of developers and ordinary users who participate in the process. Obviously, a large number of developers can produce new features faster than a small group, especially if the architecture supports this. As discussed earlier, all successful open source projects need such architectures anyway. An interesting question is whether such development is also efficient in terms of effort, or if it is simply the huge amount of effort that speeds the development.

The argument for reliability is "Linus's Law" as quoted by Eric S. Raymond: "Given enough eyeballs, all bugs are shallow." [9] In other words, given a sufficient number of beta testers and those who correct the bugs, it is possible to find essentially all defects. This means that the testing is extremely thorough, provided that the given feature or protocol is seriously needed.

Most open source projects follow the model of stable and experimental releases. Stable releases are well tested and known to work, while experimental releases are just snapshots of what currently exists.

Another distinction is between major and minor revisions that are reflected in version numbers. Major revisions are larger rewrites that can cause some incompatibilities to occur.

Licensing is one aspect to be considered when developing systems relying on Linux. Although Linux is free, it is copyrighted by Linus Torvalds. Its license is based on the Library GNU Public License (LGPL), which says that it is possible to create applications that run on top of Linux without revealing their source code. However, any modifications to the kernel are covered by the license and must be made public.

Kernel modules present an interesting case. Since they are dynamically loadable, they could be compared to normal application programs. However, they can potentially add new features to the kernel and for this reason they should be considered modifications and subsequently, they should be open source. Torvalds has said that as long as kernel modules are used "as intended," they are covered by the LGPL.

Conclusions

With all the incorporated features, Linux has evolved into a feature-rich Unix-like operating system with great importance to today's businesses in the Web domain. It provides an open development platform for building reliable Internet services for businesses and consumers.

References

[1] Tietoviikko magazine, 20 September 2001, p.10.

[2] Technical information, "Nokia Mediaterminal 511 S Technical information." http://www.nokia.com/multimedia/pdf/Mediaterminal.pdf, 2001.

[3] "The Tanenbaum-Torvalds Debate." In Chris DiBona, Sam Ockman, and Mark Stone: Open Sources. Voices from the Open Source Revolution. O'Reilly 1999, pp. 221-251.

[4] Peter Wayner, "Free for All. How Linux and the Free Software Movement Undercut the High-Tech Titans." HarperBusiness 2000.

[5] Pia Rieppo Adrian OConnell Suzie Low Tadaaki Mataga Jennifer Wu, "Workstations Worldwide Forecast, 2001-2005," Gartner Group, 8 August 2001.

[6] Linus Torvalds, "The Linux Edge." In In Chris DiBona, Sam Ockman, and Mark Stone: Open Sources. Voices from the Open Source Revolution. O'Reilly 1999, pp. 101-111.

[7] Alessandro Rubini. "Linux Device Drivers," O'Reilly 1998, p. 4.

[8] Ivan Bowman, Saheem Siddiqi, Meyer C. Tanyan, "Concrete architecture of Linux kernel," http://plg.uwaterloo.ca/~itbowman/CS746G/a2/, 1998.

[9] Eric S. Raymond, "The Cathedral and the Bazaar" at http://www.tuxedo.org/~esr/cathedral-bazaar/.

Part 3.2

Mobile Internet Layer

- o Basic Internet Protocols
- o Internet Protocol Version 6
- o Multimedia Sessions in the Web Domain
- o Internet Multicast and Services
- o Quality of Service in the Web Domain
- o Java in the Web Domain

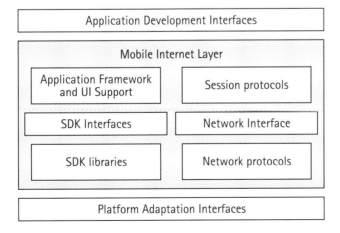

Basic Internet Protocols

This chapter describes the basic transport layer protocols as well as some application layer protocols.

The basic Internet Protocol (IP) architecture is shown in the next figure.

Application Layer
Transport Layer
IP Layer
Layer 2
Layer 1

Internet Protocol Architecture

Transport Protocols

The IP layer provides basic connectivity functions. Running on top of IP are a number of transport protocols such as the Transmission Control Protocol (TCP), User Datagram Protocol (UDP), and Stream Control Transmission Protocol (SCTP). These protocols provide basic bearer services for applications and application protocols.

Transmission Control Protocol

Many people are familiar with the TCP [RFC793] even if they do not realize it. TCP provides the basic transport for protocols such as the HyperText Transfer Protocol (HTTP). TCP provides connection services, reliable delivery, and data sequence preservation. Enhancements TCP has received over the years include these Requests For Comments (RFCs):

- o   RFC 896 Congestion Control in IP/TCP Networks
- o   RFC 2001, 2883 Slow Start, Congestion Avoidance, Fast Retransmit
- o   RFC 2018 TCP Selective Acknowledgement Options
- o   RFC 2581 TCP Congestion Control

User Datagram Protocol

UDP [RFC768] provides unreliable, connectionless transport for protocols such as the Real-time Transport Protocol (RTP). UDP is a lightweight protocol - it does not provide congestion control mechanisms. User plane protocols (e.g., RTP and the GPRS Tunneling Protocol (GTP)) and management protocols (e.g., Simple Network Management Protocol (SNMP), Domain Name System (DNS), Dynamic Host Configuration Protocol (DHCP), Lightweight Directory Access Protocol (LDAP) and Network Time Protocol (NTP)) run over UDP.

UDP provides 16-bit port numbers to let multiple processes use UDP services on the same host. A UDP address is the combination of a 32-bit IP address and the 16-bit port number.

The checksum in UDP over Internet Protocol version 4 (IPv4) either covers the entire datagram or is not used at all. In UDP over Internet Protocol version 6 (IPv6), the UDP checksum is mandatory and cannot be disabled. There is some interest in creating *UDP Lite* for IPv6. The UDP Lite header replaces the *Length* field with a *Checksum Coverage* field. The length of the UDP Lite packet can be found in the length field of the IP header, so packet length information in UDP is not required. Checksum Coverage is the number of bytes covered by the checksum, counting from the first byte of the UDP Lite header. Some payloads that are carried by UDP are tolerant of bit errors, so discarding damaged packets can cause problems, especially in burst error situations.

Stream Control Transmission Protocol

SCTP [RFC2960] is a relatively new protocol. It was originally designed to be a transport bearer for Public Switched Telephone Networks (PSTNs) and Signaling System No. 7 (SS7) signaling over IP. However, it has been enhanced to serve as a general reliable transport protocol, along the lines of a next-generation TCP. SCTP provides connection services, reliable delivery, and congestion control like TCP does, but SCTP has some additional features, such as support of multihoming, which allows a single SCTP association to use multiple interfaces/IP addresses. This allows for additional protection against network failures.

SCTP supports multiple streams to avoid head of line blocking. SCTP is also message-oriented, which improves upon TCP's byte alignment. This means that protocols running on top of SCTP need not be concerned with message boundaries - SCTP takes care of this. Additionally, the SCTP layer handles the fragmentation of the Maximum Transmission Unit (MTU). This should reduce the occurrence of message loss. SCTP provides for in-sequence delivery and order of arrival delivery. In-sequence delivery provides the preservation of data order to the user. Order of arrival delivery passes the received data immediately to the user, which avoids re-sequencing delays.

SCTP also improves upon TCP connection setup security by using a secure four-way handshaking procedure to prevent the flooding attacks that have been prevalent with TCP.

Initial deployments of SCTP will be for transporting control plane signaling in Radio Access Networks (RANs) as well as transporting SS7 over IP. Additionally, SCTP will transport Authentication, Authorization and Accounting (AAA) protocols such as Diameter. Other groups are looking at SCTP for applications such as IP Storage and Reliable Server Pooling.

The current status of SCTP is that a number of implementations already exist. Projects are underway to bring SCTP into the kernel of major operating systems such as Solaris®, Linux®, and FreeBSD®. These implementations of SCTP will include an enhanced socket API in order to take advantage of the new features provided by SCTP.

Related Protocols

Domain Name System

If IP is the glue that holds the Internet together, DNS is the workhorse that makes everything happen. The primary job of DNS is to map domain names or Uniform Resource Locators (URLs) to IP addresses and vice versa via DNS records. DNS is, at its simplest, a distributed database.

For example, when a consumer types www.nokia.com into a Web browser, the browser performs a DNS query on www.nokia.com and returns an IPv4 address of 192.100.104.200, which is used to contact the Web server.

In IPv4, DNS uses A Records, while IPv6 uses AAAA (also known as quad A) records - A6 records have been moved to experimental status.

New technologies, such as E.164 Number Mapping (ENUM) [RFC2916], are being deployed. ENUM is a technology that essentially maps telephone numbers (E.164 numbers) to DNS records. Under the domain name in DNS for the phone number, multiple records can exist, providing information such as a Web page, e-mail address, or Session Initiation Protocol (SIP) address for the phone number. Thus the user performing the ENUM query can select the manner in which to contact the party in question (e.g., e-mail, SIP).

Dynamic Host Configuration Protocol

The DHCP is a configuration protocol running over UDP; it provides a mechanism for dynamically assigning IP addresses and configuration parameters to other hosts and clients. DHCP is designed for large networks with complex software configurations. It allows servers to allocate reusable IP addresses and configuration parameters for clients. Clients can obtain IP addresses for a fixed time period, from 1 minute to 99 years, or permanently. This saves consumers from directly having to configure IP addresses and other parameters on their machines when they connect to the network.

Common Internet Application Protocols

This section concentrates on the basic and most commonly used application protocols and their history on the public Internet. The purpose is to give a short overview of the most common Internet application protocols as background for the most recent Web service protocols.

HyperText Transfer Protocol

The HyperText Transfer Protocol (HTTP) is without doubt the most used, best known and widest spread protocol in the Web domain. HTTP is an application-level protocol that is light and fast enough for interactions between distributed hypermedia information systems.

HTTP has a long history in the World Wide Web global information initiative; it has been in use since 1990.

The first version of the HTTP specification was HTTP 0.9. It was a very simple protocol implementation for data transfer across the Internet. The second improvement to the protocol specification was released in 1996, when HTTP version 1.0 was sent for comments to the Internet Engineering Task Force (IETF). Probably the most important improvement in version 1.0 was support for Multi-purpose Internet Mail Extensions (MIME) like formatted HTTP messages. Other major improvements included the meta-information about the transferred data in a message and several modifiers about request and response semantics. However, version 1.0 did not take into consideration the effects of hierarchical entities and functionality on the rapidly growing Internet.

There was still lack of support in the protocol for proxies, cache technology, and the needs of persistent connections or virtual hosts. In other words, there was still need for improvement in the protocol - applications required the functionality to properly determine each other's true capabilities for communication. Practical information systems require much more functionality than simple retrieval, search, front-end updates, and annotation. The current HTTP specification version 1.1 from 1999 tries to fulfill these requirements and solves the problematics of the blooming World Wide Web. Several additional specifications for HTTP concentrate on situations like authentication, state management, and content negotiation, and most of them are now included in specification version 1.1.

Like most of the Internet's application-level protocols, HTTP resides on top of the TCP layer in the TCP/IP stack. The default TCP port is 80. Of course, HTTP could also be used with some other reliable transport protocol in other networks. The difference from other application protocols is that in most cases HTTP is used to carry protocols inside its data segment. This is because HTTP is the most commonly used protocol on the Internet, so Internet and intranet infrastructures are built to support HTTP message pass-through (request/response) between the end points (server/client).

Secure HTTP (S-HTTP) provides secure communication mechanisms between an HTTP client and server in order to enable spontaneous commercial transactions for a wide range of applications. The intent is to provide a flexible protocol that supports multiple orthogonal operation modes, key management mechanisms, trust models, cryptographic algorithms, and encapsulation formats through option negotiation between parties for each transaction.

File Transfer Protocol

The File Transfer Protocol (FTP) is the primary Internet standard for file transfer. FTP runs on top of the TCP/IP stack and uses separate simultaneous TCP connections for control and for data transfer. FTP provides a connection-oriented service, and all data exchanged between the server and the client are guaranteed not only to be delivered but also to be delivered intact as originally sent. FTP uses TCP port 21 by default. FTP was initially defined in RFCs for the ARPANET network in the beginning of the '70s. Since then, it has been the main protocol for transferring files trough TCP/IP networks.

Simple Mail Transfer Protocol

The Simple Mail Transfer Protocol (SMTP) is the cornerstone of messaging interoperability across the Internet. SMTP runs on top of the TCP/IP stack. TCP port 25 is the default port for all SMTP operations.

In 1982, the IETF defined SMTP in RFCs 821 and 822. SMTP gets its name from the original protocol, which was quite simple and concentrated mainly on the task of sending plain text messages over an IP link between a client and a server.

Since 1982, SMTP has evolved to fulfill the requirements of modern messaging environments. The evolution of the SMTP protocol on the Internet has been very fast. Extended SMTP (ESMTP) and MIME have enabled SMTP to deliver highly functional messaging systems.

Although a decade old, SMTP has proven remarkably resilient. Nevertheless, the need for a number of protocol extensions has become evident. ESMTP enables vendors to add extensions to the basic SMTP service so that they can handle different classes of messages or other functionality.

Sending plain text messages is not always enough - modern messaging systems have to support graphics. Some early encoding technologies like UUencode let consumers send attachments and formatted text, but these technologies were not very well integrated in many cases. It is important to precisely configure the tags that different attachments use before it is possible to confidently send an understandable attachment to a recipient.

Multipurpose Internet Mail Extensions

MIME solves this SMTP attachment related configuration problem by defining the rules for the labeling and transmission of different data types within messages. MIME informs mail systems how to process parts of the message so that recipients see exactly what the sender intended. MIME is also the basis for the transmission of streaming data, such as audio and video messages.

Network News Transfer Protocol

The Network News Transfer Protocol (NNTP) protocol was initially defined in IETF RFC 971 and released in 1986. Since then, NNTP has become one of the most popular protocols on the Internet. Many implementations of the protocol have been created on several different platforms and operating systems. With the growth in use of the protocol, work began on a revision to

NNTP in 1991, but that work did not result in a new specification. However, many ideas from that working group found their way into subsequent implementations of NNTP. Additionally, many other extensions, often created by newsreader authors, are also in use. NNTP runs on top of the TCP/IP protocol. NNTP uses TCP port 119 by default.

NNTP is a good example of a protocol that has become an Internet standard even without an unambiguous specification.

Terminal Emulation Protocol

The Terminal Emulation Protocol (TELNET) was specified as a standard for the ARPA Internet community in 1983 (RFC 854).

The purpose of the Telnet protocol is to provide a general, bi-directional communications facility. Its primary goal is to allow a standard method of interfacing terminal devices and terminal-oriented processes with each other. It is envisioned that the protocol may also be used for terminal to terminal and process to process communication.

Applications use the telnet protocol to connect over the network with a remote computer that is running telnet server software. The telnet protocol provides two-way communication functionality that allows terminal devices and terminal-oriented processes to communicate with each other. Telnet uses TCP to transmit data and protocol-specific control information. The default port for telnet is TCP port 23.

RFC 854 states: "The Telnet Protocol is built upon three main ideas: first, the concept of a 'Network Virtual Terminal;' second, the principle of negotiated options; and third, a symmetric view of terminals and processes."

Telnet is probably the most commonly used protocol for enabling process-to-process communication over TCP/IP networks.

References

[RFC768]	Postel, J. (ed.), "User Datagram Protocol", STD 6, RFC 768, August 1980.
[RFC793]	Postel, J. (ed.), "Transmission Control Protocol", STD 7, RFC 793, September 1981.
[RFC2916]	RFC 2916, "E.164 Number and DNS", P. Faltstrom, September 2000.
[RFC2960]	RFC 2960 "Stream Control Transport Protocol" R. Stewart et al, November 2000.

Internet Protocol Version 6

The Internet Protocol (IP) has been the glue that has connected diverse networks together and forms the basis of the Internet. It allows hosts to communicate over diverse networks. The current version of the Internet Protocol, IPv4, is well over 20 years old and is showing its age. In late 1990, projections showed that the address space of the Internet would become an increasingly limited resource, so the Internet Engineering Task Force (IETF) undertook the task of specifying the next-generation IP protocol, which was at that time called Internet Protocol Next Generation (IPNG). [RFC1752] A number of proposals for IPNG were submitted, and much heated debate occurred.

IPv5 was an experimental streaming protocol (Experimental Internet Stream Protocol, Version 2) first described in RFC 1190 and updated in RFC 1819. This work has been abandoned.

IPv4 offers a 32-bit address, which translates into approximately four billion possible addresses; however, this number is greatly limited by address allocation policies. As a short-term fix, Network Address Translators (NAT) were proposed to provide private address space in order to provide additional addresses. Problems with NATs are well documented, however. IPv6 offers a 128-bit address, which translates to 340,282,366,920,938,463,463,374,607,431,768,211,456 possible addresses (including reserved address space), which should be sufficient for the foreseeable future.

IP Headers

The first figure shows the IPv4 header, while the second one presents the IPv6 header.

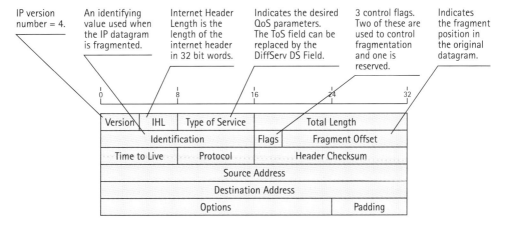

IPv4 Header

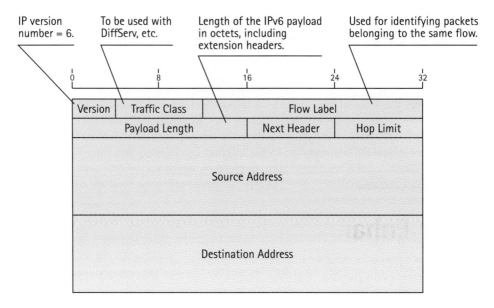

IPv6 Header

As one can easily see, the IPv6 header is streamlined in comparison to the IPv4 header. Unused or unnecessary fields in the IPv4 header were removed. The latter part of the IPv4 header contains extensions. These extensions are rarely used because they slow down routing considerably. With this in mind, IPv6 was designed to support extension headers. This allows the IPv6 header to remain a fixed length for simpler processing while still allowing extensibility. So far, six types of extension headers have been specified: hop-by-hop options header, routing (Type 0) header, fragment header, destination options header, authentication header, and encapsulating security payload.

IPv6 Header Handling

In IPv6, optional information may be encoded in separate headers, which are placed between the IPv6 header and the upper-layer protocol header contained in the IPv6 packet. The *Next Header* field indicates the location of the next header, whether the head of the payload protocol or an extension header. Each extension header uses the same mechanism to indicate the location of the next header. Thus the extension headers form a *daisy chain* of headers. A value of 59 in the Next Header field indicates that there is nothing following the header.

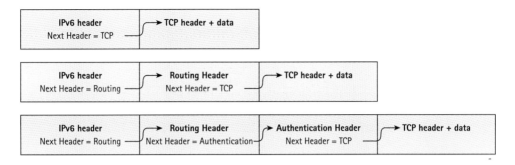

Extension Headers

Other Enhancements

Other enhancements were also built into IPv6. Mobility support is a basic feature in IPv6, while Mobile IPv4 is purely an add-on to IPv4. A Basic IPv4 node does not support mobility features at all, even as a Correspondent Node. Every IPv6 node should support Mobile IPv6 Route Optimization (as a Correspondent Node) and a Home Address option. IPv6 has a Stateless Address Autoconfiguration feature, which allows nodes to be automatically configured, even without the Dynamic Host Configuration Protocol (DHCP). IPv4 nodes need DHCP for automatic configuration. IPSec is a mandatory feature in IPv6. IPSec is an add-on to IPv4 and is not usually implemented. Additionally, IPv6 contains a flow label field which can be used to identify flows. Currently, the usage of the flow label is under discussion in the IETF.

IPv6 Address Types

With 2^{128} possible addresses, the astute reader would assume there would be different types of IPv6 addresses. The reader would be correct. "IP Version 6 Addressing Architecture" [RFC2372] describes IPv6 addresses in detail. Please note that this specification is currently undergoing revision.

There are three types of IPv6 addresses: unicast, anycast, and multicast. Unicast addresses are the most common type. They form the bulk of addresses on the Internet and generally define a single interface. IPv6 multicast addresses replace IPv4 broadcast messages, and they define a set of interfaces. Packets sent to multicast addresses are sent to all of the interfaces defined by the address. Anycast addresses are new and define a set of interfaces. Packets sent to an anycast address are sent to one of the addresses defined by the anycast address.

There are several ways to represent IPv6 addresses as text strings. The preferred form is

```
x:x:x:x:x:x:x:x
```

The "x"s are the hexadecimal values of the eight 16-bit pieces of the 128-bit address. It is not necessary to write any leading zeros. An example is:

```
FEDC:BA98:7654:3210:FEDC:BA98:7654:3210
```

Additionally, a shorthand method exists for writing IPv6 addresses which contain long strings of zeros. The use of "::" indicates multiple groups of zeros. The "::" may only be used once in an address but may be used to indicate leading or trailing zeros. Thus,

```
1080:0:0:0:8:800:200C:417A
```

may be represented as:

```
1080::8:800:200C:417A
```

IPv6 address prefixes are represented by the notation: ipv6-address/prefix-length, where the IPv6 address is an IPv6 address (as shown above) and the prefix-length is a decimal value specifying how many of the leftmost contiguous bits of the address comprise the prefix. An example, showing a 60-bit prefix, is:

```
12AB:0:0:CD30::/60
```

The Current Allocation of IPv6 Prefixes

Allocation	Prefix (binary)	Fraction of Address Space
Unassigned (see Note 1)	0000 0000	1/256
Unassigned	0000 0001	1/256
Reserved for NSAP Allocation	0000 001	1/128 [RFC 1888]
Unassigned	0000 01	1/64
Unassigned	0000 1	1/32
Unassigned	0001	1/16
Global Unicast	001	1/8 [RFC 2374]
Unassigned	010	1/8
Unassigned	011	1/8
Unassigned	100	1/8
Unassigned	101	1/8
Unassigned	110	1/8
Unassigned	1110	1/16
Unassigned	1111 0	1/32
Unassigned	1111 10	1/64
Unassigned	1111 110	1/128
Unassigned	1111 1110 0	1/512
Link-Local Unicast Addresses	1111 1110 10	1/1024
Site-Local Unicast Addresses	1111 1110 11	1/1024
Multicast Addresses	1111 1111	1/256

Note 1: The "unspecified address", the "loopback address", and the IPv6 Addresses with Embedded IPv4 Addresses are assigned from the 0000 0000 binary prefix space.

Note 2: For now, the Internet Assigned Numbers Authority (IANA) should limit its allocation of IPv6 unicast address space to the range of addresses that start with binary value 001. The rest of the global unicast address space.

Additionally, IPv6 provides for scoping of addresses [IPv6 Scope]. The term "scope" refers to the topological span addresses that may be used as unique identifiers for interfaces. For unicast and anycast addresses, there are three defined scopes: link local, site local, and global. Link local scope uniquely identifies an address on a single link only. Site local scope uniquely identifies an address within a site, such as an organization (e.g., an intranet). Global scope identifies an address uniquely on the global Internet. It is possible and likely that a single interface will have link local, site local, and global addresses. It is important to remember that addresses are not visible (or usable) outside of their scope.

Conclusions

IPv4 offers approximately four billion possible global IP addresses, which is clearly insufficient. IPv6 offers more than enough global IP addresses for now and the foreseeable future. Additionally, IPv6 builds upon the success of IPv4 and improves upon it. IPv6 is necessary and sufficient.

As can be expected, the applicability of IPv6 for many different devices has aroused the interest of many different bodies. For example, the Third Generation Partnership Project (3GPP) has mandated the use of IPv6 with the IP Multimedia Subsystem (IMS) in their architecture. Work is being conducted to clearly specify what the IPv6 requirements are for mobile devices. Additionally, there is ongoing work to specify general requirements for IPv6 nodes. This can help designers and implementers who are implementing IPv6 for a wide range of consumer devices.

References

[RFC1752]   S. Bradner & A. Mankin; "The Recommendation for the IP Next Generation Protocol," RFC 1752; January 1995.

[RFC2373]   R. Hinden & S. Deering; "IP Version 6 Addressing Architecture," RFC 2372, July 1998.

[RFC2460]   S. Deering & R. Hinden, "Internet Protocol, Version 6 (IPv6) Specification", RFC 2460, December 1998.

[RFC2993]   T. Hain, "Architectural Implications of NAT", RFC 2993, November 2000.

[IPv6SCOPE] S. Deering et al; "IPv6 Scoped Address Architecture", A Work in Progress.

Multimedia Sessions in the Web Domain

It is assumed that in the future almost all fixed and mobile communications networks will be based on Internet technology. The main benefits of running voice over the Internet Protocol (IP) are reduced costs and new services. In the beginning of the Voice over IP (VoIP) hype the emphasis was mostly on cost savings. While this is still considered to be true especially in the long run, the focus has recently been shifted to new services. After all, services are what consumers and carriers are interested in. Especially, services combining several communication types and modes will lead the way in future networks. Voice itself will be just one, although important, piece in the whole communication architecture.

The Session Initiation Protocol (SIP) [1] is an emerging Internet Engineering Task Force (IETF) standard for setting up multimedia sessions on the Internet. Its basic capabilities are setup, modification and teardown of any communication session, so it is a signaling protocol. SIP also provides personal mobility, meaning that a consumer is reachable via a single address regardless of his current point of attachment to the network. SIP is suitable for combined services, as it borrows many features from the HyperText Transfer Protocol (HTTP) and the Simple Mail Transfer Protocol (SMTP), which are currently widely used on the Internet for Web browsing and e-mail. SIP is also modular and extensible, so it can be used for other things than basic session establishment.

Besides session setup signaling, VoIP requires several other protocols and mechanisms. The actual voice is carried by the Real-time Transport Protocol (RTP) [2], which can also be used to carry video and other real-time media. In order to transport the RTP stream over the network, the Quality of Service (QoS) has to be guaranteed in some manner. Techniques such as Differentiated Services (DiffServ) and Multi-Protocol Label Switching (MPLS) can be used for this purpose.

This section presents an overview of the protocols used for voice and multimedia communications over the Internet. IETF's Internet multimedia architecture is then introduced and the role of each protocol is briefly described.

Internet Multimedia Architecture

The IETF has worked on the Internet Multimedia Architecture already for several years. The overall idea is that there should be a specific tool for each task and that a large system can be built by combining relatively simple building blocks. Layering is another guiding principle, as with all Internet protocol designs.

The following figure shows a simplified overview of the architecture as a layered protocol stack. At the bottom is the IP layer, which runs over every network technology and provides the basic connectionless packet delivery service for anything above it. This is the famous *IP over everything and everything over IP* paradigm of the IETF. In order to support multimedia services, seamless mobility and efficient multiparty conferencing, the IP layer needs to be enhanced compared to this simplistic view. Mobile IP allows terminals to move freely between different mobile networks. Differentiated Services brings different levels of service to the network, which allow, e.g., voice packets to be prioritized over e-mail traffic. IP Multicasting is an interesting technology that allows a sender to deliver a packet simultaneously to multiple receivers easing the scalability of large conferences or media streaming. Even security can be implemented on the IP layer by applying IP Security (IPSec), which provides confidentiality and integrity protection for all traffic.

	Application					
RSVP	SDP/SDPng					RTP/RTCP
	HTTP	SMTP	RTSP	SIP	SAP	
	TCP		TCP/UDP			UDP
IPv4/v6 with Mobility, DiffServ and Multicast						

Internet Multimedia Architecture

Above the IP layer reside the transport protocols, which operate end-to-end between hosts or terminals. The Transmission Control Protocol (TCP) offers connection-oriented reliable delivery with congestion control, while the User Datagram Protocol (UDP) provides a simple connectionless datagram service. Recently, a new protocol, Stream Control Transport Protocol (SCTP) has been designed to enhance TCP functionality. The Resource Reservation Protocol (RSVP) can be used to signal for QoS reservations on the IP layer through all the routers on the traffic path. RSVP runs directly on top of IP, and can be seen as a part of the IP layer functionality. RSVP suffers from scalability problems as it operates on a per packet flow level and its use for end-to-end reservations over the Internet is quite limited. At least in the core network, DiffServ and MPLS are enough to achieve guaranteed QoS.

The actual application layer protocols are situated on top of the transport and utilize it based on their needs. Voice and other multimedia content (e.g., video or animation) are transported by the Real-time Transport Protocol (RTP), which runs on top of connectionless UDP service. RTP offers synchronization of the media streams it carries by including a sequence number and timestamp headers. With voice data, the actual speech frame is inserted as payload for RTP and sent to the recipient. A specific payload format has been defined for each codec to be used within RTP. RTP is accompanied by the Real-time Transport Control Protocol (RTCP), which provides, e.g., statistics about each sender and recipient in an RTP session. An RTP session can be either unicast (point-to-point) or multicast. With multicast, the IP layer takes care of delivering the packets to each recipient so RTP does not need to take care of it.

It is important to understand that in IP networks signaling and media do not usually follow the same path. Media, such as voice, is sent over RTP directly between the communicating parties, while signaling often goes via servers. For media QoS is especially important; signaling does not usually have strict real-time requirements.

A group of signaling protocols exists for various purposes. The previous figure shows the transport that each of them utilizes, though in some cases there is more than one option. These protocols have the same look and feel:

o   Common addressing, user@domain.

o   Similar text-based headers and encoding according to the RFC 822 message format originally developed for SMTP Internet mail. Protocol parser implementations can utilize the same code base.

o   Same request-response model and response codes.

o   Utilization of Multipurpose Internet Mail Extension (MIME) for flexible payload formats.

o   Uniform Resource Identifiers (URIs) and Uniform Resource Locators (URLs) for addressing.

o   Domain Name System (DNS) for address mapping and inter-domain request routing.

o   Common security and authentication frameworks mechanisms can be used.

This similarity allows for easy combining of these protocols to perform services which need to use more than one of them. For example, an SIP message can carry a HyperText Transfer Protocol (HTTP) URL or even a HyperText Markup Language (HTML) document or it is easy to send e-mail from the network server if a callee was not reached to inform him about the missed call by using the common addressing format.

SIP is used to establish, modify and terminate sessions. It provides personal mobility by allowing a user to dynamically register to the network with his communication address (SIP URL). A session is usually a number of RTP streams to be exchanged. In a basic VoIP scenario a session is simply a combination of two unidirectional RTP streams between the calling and called party.

Session Announcement Protocol (SAP) [3] is used to advertise multicast sessions to larger audiences. It has been used for a long time in Mbone, the experimental multicast overlay network. The Real-Time Streaming Protocol (RTSP) [4] provides the possibility to control streamed media by sending commands (e.g., PLAY, RECORD and PAUSE) to the server where the media is located. In a VoIP environment RTSP can be used to control a voicemail server in order to record or listen to messages.

HTTP and SMTP are used for Web browsing and e-mail transport, respectively. As they are quite generic and flexible, especially HTTP is being used for various other purposes, even as a transport for remote procedure calls.

In order to describe the parameters and characteristics of the media sessions, each of the signaling protocols uses a common description language, Session Description Protocol (SDP) [5]. SDP is carried in SIP message payload or within an e-mail. SDP describes the codecs and addresses to be used to send audio and video. Recently, an effort to redesign SDP to include even more capabilities has begun in the IETF. It is referred to as SDP next generation (SDPng).

An application software can use all of these protocols to provide a rich, comprehensive and understandable service for a consumer. The application uses SIP for establishing voice and video sessions, RTSP for streaming video or voicemail announcements, SMTP for fetching and sending e-mail and HTTP for browsing Web content. It is up to the application designer, both on the terminal and server side, to best combine these protocols and to design an interface for the user. Certainly, the possibilities are much greater than with traditional telephony or even with second generation mobile systems.

An example of a simple user interface combining voice, messaging, Web and streaming is shown in the following figure.

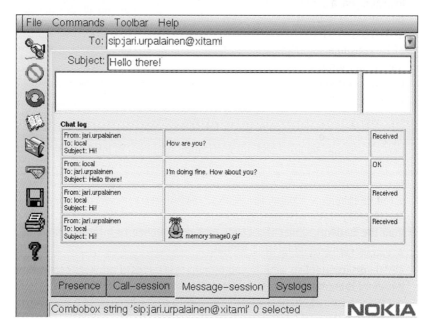

An example of a Rich Call user interface

Session Initiation Protocol

As explained in the previous chapters, SIP is used to setup, modify and tear down multimedia sessions. It is agnostic to the type of session to be established, and actually uses SDP to describe the characteristics of a session. Usually, the session is a combination of speech, audio and video streams, but it can also contain shared applications such as a whiteboard or text messages. Even network gaming sessions could be established with SIP as long as all the applications understand the required parameters for the game.

A basic *SIP network* is composed of four types of elements: User Agents (UA), Proxy servers, Redirect servers and Registrar servers. User Agents typically reside in end-points such as IP phones, personal computers or mobile devices. They initiate requests and provide responses. Usually UAs also have an interface to media handling and to the actual application software providing the user interface. For example, VoIP calls are carried out between UAs.

Proxy servers are intermediaries, which receive and forward requests providing them with, e.g., routing or other services. Redirect servers simply respond to a request by asking its originator to redirect it to a new address. Registrar servers keep track on the actual points of contact of the consumers by accepting registrations from the UAs. Registrar servers and the SIP registration procedure in general provide user mobility as the consumer is able to be reachable from any location via a single address. In this sense, they resemble Home Location Register (HLR) functionality in Global System for Mobile Communications (GSM) networks. Each consumer is part of a domain and each domain runs at least one registrar server, which knows the location of its consumers (if they are registered). In some sense, SIP registrations resemble Mobile IP functionality, but it must be understood that SIP is not capable of providing real-time handovers and not all communication is based on SIP sessions, so SIP and Mobile IP complement rather than contradict each other.

SIP uses an address format common to Internet mail, i.e., user@domain. An example is sip:john.doe@operator.com. The domain part is used to find the correct domain for the consumer: DNS can be queried for the address via which the SIP registrar within a domain is reachable. This is analogous to e-mail routing, which has been up and running for years already. The user part is to distinguish between individual consumers within a domain. In addition, traditional E.164 telephone numbers can be used within SIP as tel: URLs. An example is tel:+3581234567. The User Agent of a proxy can use a mechanism called E.164 number mapping (ENUM) [6] to map the E.164 number to a corresponding SIP URL.

A sample SIP network is shown in the following figure. It consists of User Agents (in this case mobile devices) and a number of domains operating SIP proxy, redirect and registrar servers. In addition, there can be servers dedicated to certain applications, such as presence. In order to connect to the telephony network, a gateway is required.

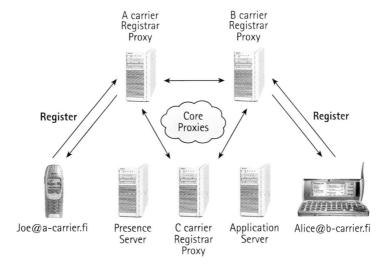

A sample SIP network with three domains

Basic SIP includes six requests or methods. These are INVITE, ACK, OPTIONS, CANCEL, BYE and REGISTER. Each request can be responded to with a number of response codes in a similar manner as in HTTP or SMTP. The basic response classes are:

- o 100 - Informational
- o 200 - OK
- o 300 - Redirection
- o 400 - Request failure
- o 500 - Server failure, and
- o 600 - Global failure.

Each class has a number of specific instances, such as 180 Ringing. Responses look very similar to requests.

The most important headers are *Request-URI* which defines where the request is to be sent next, *To* which contains the recipient (callee) address and *From* which contains the sender (caller) address. A SIP message may contain any payload conforming to MIME specification.

OPTIONS is used to query the capabilities of a UA. CANCEL can be sent to cancel a pending request. BYE is used to terminate an on-going session. REGISTER is sent by a subscriber to his registrar server to update his contact information.

INVITE is used to setup and modify sessions. It is a special request that is followed by an ACK after receiving the final response, while other request transactions end at the first final response. An example of basic session establishment is shown below.

Basic SIP session establishment and termination.

1. Joe sends an INVITE to Alice using his outbound SIP proxy server. Within INVITE, an SDP payload describes that Joe wants to establish a two-way voice call with AMR codec.
2. SIP proxies route the request to the address where Alice is currently registered. The second proxy is co-located with Alice's registrar server.
3. Alice is willing to accept the call and her device supports AMR, so the device starts to ring and 180 Ringing is sent to Joe.
4. When Alice picks up the device, a 200 OK response with the appropriate SDP is sent to Joe.
5. Joe responds with ACK.
6. The RTP sessions carrying voice coded with AMR are established. The packets flow directly between the two devices.
7. Alice hangs up and sends BYE to Bob to terminate the session.
8. Bob replies with 200 OK.

Note that the RTP streams do not follow the same path as the SIP messages did, but flow directly between the devices. It is possible to send the subsequent SIP requests directly between UAs. On the other hand, proxy servers in the middle may ask to remain on the signaling path for the duration of the call. This might be useful if the proxy offers some services to the call. The actual services can be implemented in various manners, such as with Java servlets or using the Call Processing Language (CPL).

It is possible to impose so-called Rich Call services in the call setup. SIP can carry small images, or icons in the INVITE payload and these can be shown to the callee. It is also possible to send the callee a URL which points to further information, such as the caller's Web page. The same can be applied in the reverse direction as well. For example, a shop can respond with a Web page to calls that come after opening hours to provide caller's with some information about their products and services.

SIP supports both parallel and sequential forking. A consumer can be registered from several contact points at the same time. In this case, the proxy service logic decides what to do with an incoming request. In case of parallel forking, the proxy forks the request simultaneously to each destination.

SIP can support the same basic and digest authentication mechanisms as HTTP. Proposals exist to embed Extensible Authentication Protocol (EAP) in SIP to offer a flexible way to introduce new authentication mechanisms and to use backend protocols, such as Diameter. Any SIP request can be authenticated.

It is possible to extend SIP in several ways. New headers and methods can be invented. The basic principle is that such additions should only require support in UAs, so proxies can remain untouched. A number of SIP extensions have been defined including the REFER method for call transfers and other referrals, the SIP Events framework for subscriptions and notification of events using SIP, and QoS interworking to allow QoS resource reservation within a call setup. Finally, even presence and instant messaging extensions for SIP are under definition in IETF.

Conclusions

The Internet provides an excellent platform to integrate all communication services. IETF has defined a multimedia architecture, which allows easy combination of different services including real-time/streaming voice, audio/video, text/multimedia messaging, shared applications, Web and e-mail. There are plenty of protocols to carry out the tasks of establishing and negotiating the required media sessions as well as transporting the actual media.

SIP is one of the key protocols in the Mobile Internet, as it provides the means to establish generic sessions, and it is already being extended to include various other capabilities. SIP also works well with the other Internet protocols.

References

[1] Session Initiation Protocol, IETF RFC 3261

[2] Real Time Transport Protocol, IETF RFC 1889

[3] Session Announcement Protocol, IETF RFC 2974

[4] Real Time Streaming Protocol, RFC IETF 2326

[5] Session Description Protocol, IETF RFC 2327

[6] E.164 Numbers and DNS, IETF RFC 2916

Internet Multicast and Services

Multicast is a group communication mechanism that allows the same information to be sent to a number of recipients. It is a technology, which allows services similar to traditional television and radio broadcasting to be provided over the Internet. However, some obstacles have to be overcome before the true potential of multicast can be realized. The wide deployment of multicast services has been held back by bandwidth constraints in the Internet Service Provider (ISP) backbone network, as well as the rather complex routing issues involved in delivering multicast services reliably to consumers. In the near future, there is an opportunity to bring multicast services to the mass consumer market by using datacast technology, which allows the easy deployment of the one-to-many multicast service scenario.

IP Multicast

From the network point of view, multicast is a mechanism that allows the same information (packet) to be sent to a group of clients. A group can be formed in different ways. In IP multicast, reception clients sign into a multicast group. A class D IP address in IPv4 and a multicast prefix in IPv6 identify a multicast group. The Internet Group Multicast Protocol (IGMP) in IPv4 and the Internet Control Message Protocol (ICMP) in IPv6 are used in joining a group.

For the core network, the main advantage of multicast is the bandwidth saved for other services. A single packet is delivered to a group of clients. The packet is duplicated as the path to different clients divides into two paths.

In the IP world, multicast is many-to-many communication. In essence, packets forwarded to a multicast group can be initiated by any host on the Internet. The host does not even have to be part of a multicast group to deliver a packet to the group. It is easy to see that any application or protocol, which handles hundreds of thousands or more hosts delivering and receiving packets, must be extremely complex. Therefore, in most cases multicasting is centralized in such a way that there is only one source of information. Hosts wishing to deliver packets to a group would have to have a unicast connection to the server in order to deliver packets. Only quite recently, the Internet Engineering Task Force (IETF) started to look more closely into one-to-many group communication mechanisms.

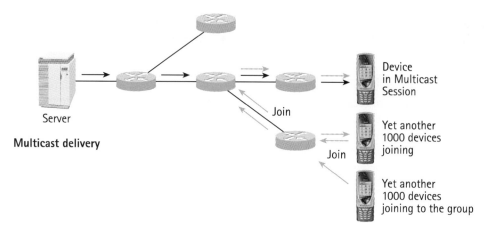

Multicast packet delivery in a group

Managing a Group

In multicast, group management is handled by using ICMP in IPv6 and IGMP in IPv4, as mentioned earlier. The mechanisms and messaging are essentially similar in both cases, so the following description focuses on the former case.

There are three group membership messages, namely, Query, Report, and Reduction. A host wishing to participate in a multicast session first needs to join a multicast by sending an ICMP message, and later by reporting to the router polling the membership status during the multicast session. To learn about ongoing multicasts, a host needs to be able to discover ongoing or soon-to-be-started multicast sessions and services. The Service Announcement Protocol (SAP) can be used as a mechanism for this purpose.

Group membership is dynamic and a host can leave the multicast group at any time by sending the appropriate ICMP Reduction message. In addition, a host can be a member of several multicast groups simultaneously.

When joining a multicast group, the host sends an ICMP Report message to the local network, since it might be the first host to join the given group. As the group is established in a specific local network, the router will thereafter periodically initiate ICMP Query messages to obtain information about the activity of the group. All hosts who are members of the group receive the messages, and after a random period, they will report their membership back to the multicast group. The random period is required to avoid all hosts reporting at the same time and making the network congested. Once the report has been received, there is no need to have more reports by any of the hosts as one report will make sure that the packet delivery from the router to that local network continues.

If there are no Reports to the Query, apparently there are no hosts listening to that group and the forwarding of that multicast to the given local network can be stopped. Alternatively, a host leaving the group initiates an ICMP Reduction message to the multicast group. When a host participating in the group hears the query, it initiates an ICMP Report to indicate that there still is activity in that group and that it should be continued. Reduction messaging makes it possible for route updates to take place promptly. Furthermore, in some cases, it would be beneficial for a service provider to see the Reduction messages in order to assess the popularity of the multicast session.

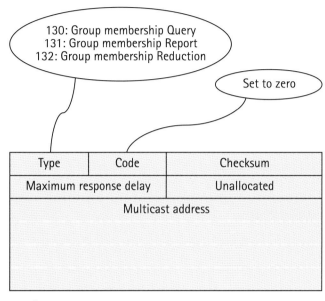

ICMP message structure for group management

Once they have learned about a host wanting to receive a specific multicast, routers propagate this information further so that it will be incorporated into the multicast delivery chain. Thus, the routes will be established as necessary, along the multicast capable routers using a multicast routing protocol.

Today, the routing protocols supporting multicast are somewhat limited. There is still research to be conducted in the area to make, e.g., reliable multicasting possible.

Layer 2 Broadcast and Multicast

The previous description dealt with IP multicast mechanisms, which are an abstraction of the actual Layer 2 broadcast and multicast. In essence, the efficiency of multicast is seen in this link layer as savings in scarce network resources. On the other hand, in IP abstraction layer resources, savings are made in having the routing table support multicasting and hence ease the routing table build up.

Depending on the actual link layer technology, multicast has various benefits. In a multiple access shared medium, multicast provides significant cost savings. On the other hand, in a star type topology, in which a single client is serviced in the last mail, multicast is not able to create such tremendous cost savings at the link layer. Still, considering the actual network layer, multicast offers benefits also in this scenario.

Service Discovery

The Service Announcement Protocol has a significant role in multicast service delivery systems. Together with the Service Description Protocol (SDP), it is used to create service information. In a uni-directional network, session announcement and description protocols are needed for notifying consumers about the services available currently or in the future.

SAP is a very simple protocol for announcing IP multicast sessions. These announcements are retransmitted after a period that grows in proportion to the number of session advertisers; on the Internet, the retransmission period is typically several minutes. Although the protocol scales well with respect to the number of recipients, it has scalability problems with a very large number of sessions. Another limitation is that it does not support the grouping of sessions.

An SDP message is a short, text-based description of the properties of one or more multimedia sessions. It carries information such as the session name, active session times, media types and formats, and the address/port combination needed to participate in the session. SDP is not really a protocol but a description format and needs some kind of transport protocol to be delivered over IP networks.

Services

There are many applications that could take advantage of this mass distribution system. For example, the results of a football game can be delivered online simultaneously to all recipients. Furthermore, it is interesting to note that one of the features characterizing broadcast channels is that the delivered information reaches all individuals at the exact same moment.

Numerous multicast services could interest the masses in a truly broadband network. The most interesting ones are probably still undiscovered. However, one can still identify streaming services, delayed file download, and Internet push services as areas to explore. Game applications also have great potential. One could foresee a massive multiplayer game that would allow several thousands of players to participate in the same virtual reality game.

Quality of Service in the Web Domain

The global Internet is based on Internet Protocol (IP) technology, which utilizes packet-based data transport. In general, any packet data network can suffer from two things. First, bit errors or network congestion can result in lost packets. Second, a packet may be delayed too long or the timing between successive packets can vary too much.

Consumers may encounter both of these problems. In general, all applications from Web browsing to network gaming suffer from lost packets. Delay-sensitive applications (e.g., voice or video telephony) also suffer from delayed packets. When the application is sensitive to both delay and loss, it poses the most difficult case, as it is not feasible to re-send lost or corrupted packets.

When these problems are addressed, we are talking about the Quality of Service (QoS). In practice, QoS in a network is the observable end-to-end behavior of the transport service that is seen by the network user (e.g., a person, an application, or a protocol layer). QoS mechanisms include all mechanisms that try to optimize the utilization of limited data transfer resources. They often consist of buffering and scheduling systems related to data transfer resources, and of a control system.

These definitions imply that QoS itself is more than simply the introduction of a QoS mechanism. It is essential to recognize that if a network does not suffer from a critical lack of resources, then there is no technical reason to introduce a QoS mechanism, either. On the other hand, QoS mechanisms are not limited to router output buffering. Any limited resource (e.g., the processing capacity of a network element performing a protocol function) is essential. Moreover, in addition to technical aspects, QoS often involves business and psychological aspects.

Technologies

There are basically two major variants of QoS mechanisms, connection-oriented QoS and connectionless QoS. The main idea with connection-oriented QoS mechanisms is to introduce flows or connections to the network with certain dedicated resources. For example, the simplest way is to allocate a single flow with some bandwidth through the network. In addition, it is possible to attach quality characteristics to the connection (e.g., concerning maximum delay and delay variation). This model appears attractive from the application viewpoint, because the application can declare its service requirements and assume that all requirements are met if the connection request is accepted. In contrast, the connection-oriented model is not as attractive from the network viewpoint, because the management and control of a huge number of connections is extremely laborious.

From the network viewpoint, the connectionless model may offer better scalability in large networks, since network nodes handle individual packets rather than connections with specific requirements. However, the combination of the connectionless service model and QoS support has turned out to be a challenging task.

There has been considerable effort to devise QoS support based both on connection-oriented and connectionless approaches. Integrated Services and Multi-Protocol Label Switching (MPLS) are examples of the connection-oriented model, while Differentiated Services is the main approach to introduce QoS in a pure connectionless network. In addition, we must remember that the Best Effort service model is still a potential solution, because practically all QoS issues can also be solved with sufficient link and forwarding capacity. The principles of these approaches are outlined in the following sections.

Best Effort

The original (and largely also the current) Internet is based on the so-called Best Effort service model: the network only makes its best effort to deliver data packets. This is a polite way to say that there is no QoS, in the sense that an application cannot know the characteristics of the transmission channel in advance. In particular, there are no guarantees about packet loss ratios or packet transfer delays, and no exact information about the available bit rate. Further, the Best Effort principle makes it difficult for the service provider to introduce any differentiation between consumers or between applications, although this is possible to some extent at the edge of the network by limiting the bit rate used by the consumer or by certain applications.

Regardless of the lack of specific QoS mechanisms, Best Effort does not mean that the QoS is poor. If an IP network has enough capacity everywhere, it will not loose or excessively delay any packets, and consequently the quality perceived by consumers and applications can be very high. This "big pipe" approach may work well within a network domain managed by one carrier, but it cannot work globally in all parts of the Internet. Therefore, the current Best Effort model cannot provide a universal solution, even though network capacity increases exponentially in some parts of the network. It seems that certain parts of the network, most notably access network and radio links to mobile devices, remain capacity-limited. Further, if and when certain applications require much higher bit rates than the majority of applications, it is very problematic from the carrier's viewpoint to treat all packets equally, because the share of network resources may become very uneven. Note that a video streaming flow may consume 100 times more resources than a voice call.

Integrated Services

The Integrated Services (IntServ) concept introduced a few years ago aimed to satisfy the demand for QoS. The original objective was to provide a good enough connection through the IP network for multicasting video applications. This required that the end system (such as a PC) first signaled a resource reservation request by using Resource Reservation Protocol (RSVP). The request was composed of a parameterized description of the IP call. Based on the parameters, the network checked the available capacity from every IP router along the connection path and made a decision on whether the call could be established.

This approach was very similar to that of traditional telecom networks, such as the Integrated Services Digital Network (ISDN). Unfortunately, IntServ did not meet the expectations and certainly did not become popular. The concept of well-defined calls, or flows, simply did not fit into the Internet model. Actually, if the only application for IntServ had been some rare multicasting connections, it might have been technically successful. In fact, at some point the expectation seemed to be that every internet should be able to make an IntServ connection everywhere. Unfortunately, that scenario was unfeasible both because of technical and business reasons. First, when the number of simultaneous flows in an IP core router goes into the millions, RSVP poses a huge scalability problem. Another problem is that a carrier cannot allow consumers to freely make reservations that require expensive and complicated systems to implement and manage.

Differentiated Services

More recently, the Internet Engineering Task Force (IETF) has created the Differentiated Services (DiffServ) approach that relies on relative guarantees. The starting point is that consumers have different needs, and consequently, they will be provided with QoS that they need and are willing to pay for.

In practical terms, at the first, ingress node of a DiffServ domain, IP packets are marked with a Differentiated Services codepoint that indicates how the core network should treat the packet. This treatment is called the Per Hop Behavior (PHB). The handling of IP packets in all nodes within the DiffServ domain is only based on the Differentiated Services codepoint and local conditions. This principle is illustrated in the following figure.

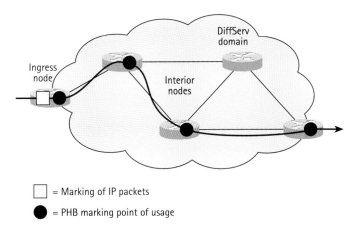

Packet marking in a DiffServ network

The number of possible Differentiated Services codepoints used in a single network is restricted to 64. Effectively, this means that a DiffServ core router only needs to know how to handle a small number of different aggregate flows (consisting of packets with the same Differentiated Services codepoint). This allows good scalability when compared to the IntServ approach that requires the router to handle all individual flows separately.

The task of the IETF has been to create a framework for Differentiated Services, but no detailed service models. There are different possibilities for choosing the Per Hop Behavior or Per Hop Behavior Group. The Differentiated Services Working Group has reserved only a small number of Differentiated Services codepoints to certain PHBs, such as Expedited Forwarding and Assured Forwarding PHB groups. The rest of the codepoints will be left to manufacturers and carriers to use as they wish. This leaves space for a lot of choices and possibilities to experiment and differentiate network services.

Some questions still remain open with respect to Differentiated Services, and they must be solved before its wide-scale deployment. One of the issues is the interoperation of various Differentiated Services networks. This is the result of a loose framework that allows a great amount of freedom in both the technical implementation and the business models used by the carrier. Some proposals now exist to provide smooth interoperation between Differentiated Services carriers.

It seems that Differentiated Services has good possibilities to introduce QoS in IP networks on a wide scale. It is scalable and relatively easy to introduce and implement. Since Differentiated Services is a framework instead of a strict standard, in the future we are likely to see a variety of Differentiated Services networks with individual technical and business solutions. For example, we could specify that our Voice over IP telephone call needs real-time service. Consequently, a DiffServ boundary router marks our IP packets with a stamp that tells the core network that those packets belong to a real-time flow. The core routers then handle our voice packets in such a way that the transfer delay is minimized.

On the other hand, some customers might be entitled to more bandwidth than others. In that case, the network has to measure the traffic flow sent by the customer and mark those packets to comply with the service agreement made between the customer and the service provider. In addition to the packet marking, the actual quality of the flow essentially depends on the load situation on the most congested link of the route. Because of the intrinsic characteristics of Internet traffic and network technology, the load situation throughout the network is unknown to boundary nodes. Thus, the result is not a set of fixed, predefined quality parameters for every flow, but a more relative situation.

Multiprotocol Label Switching

Multi-Protocol Label Switching (MPLS) is a technology aimed at introducing the connection-oriented paradigm to IP networks. Some of the major uses of MPLS are in the area of traffic engineering (e.g., load balancing, Virtual Private Networks (VPN), and traffic restoration).

The basic idea of MPLS is to provide an MPLS switching layer below the IP layer. With the MPLS layer, it is possible to create Label Switched Paths (LSPs) through an MPLS domain. For the IP layer, an LSP appears as a point-to-point link between two IP devices. Thus, the main function of MPLS is traffic engineering, as it can be used explicitly for setting up paths through MPLS-enabled networks. In addition, MPLS allows the setup of paths using a specific bandwidth. At least in theory, MPLS can work as an independent layer, supporting various protocols above, not just IP.

The technical starting point of MPLS is that IP packets are equipped with a label that is basically an identifier. MPLS-enabled routers, also called Label Switched Routers (LSR), only check the label and lookup the next destination from a table. This is done in the MPLS layer and the IP packet does not enter the IP protocol layer at all.

Naturally, before this can happen, the label lookup tables must be established in label switched routers. One part of MPLS is the label distribution protocol that is used for setting up these tables. This is needed since the labels are only valid locally and they may be changed in each LSR.

Java™ in the Web Domain

This chapter concentrates on the role of Java™ 2 Platform, Standard Edition (J2SE™) and Java 2 Platform, Enterprise Edition (J2EE™) in the Web domain.

Benefits of Java Technology

Java is an application platform originally developed by Sun Microsystems. It has grown continuously since it was first introduced to the world in late 1995. From the start, Java was designed to be: A simple, object-oriented, network-savvy, interpreted, robust, secure, architecture-independent, portable, high-performance, multithreaded, dynamic language.

The benefits of Java technology that have made it so popular are shown in the following figure.

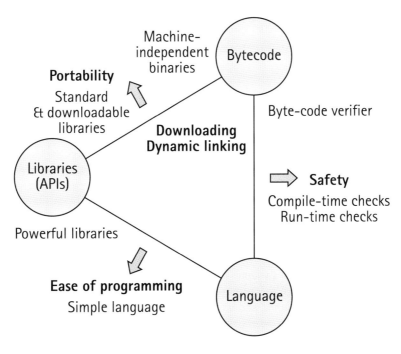

The cornerstones of Java technology

Platform Independence

Java was designed to be platform-independent, which results in a high degree of application portability. This is probably the most common reason that organizations choose to use Java. Its portability is based on standardized programming interfaces and the use of a device-independent format for representing executable programs. The programming interfaces are based on standardized libraries that are the same on different hardware platforms. Java applications are not distributed and installed in the form of machine-specific executable code. Instead, Java programs are compiled into *bytecode* that is encapsulated in a standardized "class file" format. The bytecode is executed by an appropriate Java Virtual Machine (JVM). The virtual machine has been ported to a wide variety of hardware and operating system platforms, enabling a high degree of portability for Java programs.

Safety

Compared to most other programming platforms, Java provides a certain degree of safety against programming mistakes and virus-like attacks.

Java helps programmers to avoid errors as it removes some elements that make C++ more error-prone, such as explicit pointers and multiple inheritance. It also uses strong typing and has several run-time checks such as array boundary checking.

At application start time, class files are run through a bytecode verifier to check that the bytecode has not been modified to bypass security mechanisms provided by the language.

Ease of Programming and Increased Productivity

Java is often compared with the C++ programming language and standard libraries since they are used for many similar types of programming tasks. Although the syntax of Java is based on C++, there are several important differences. One fundamental difference is that Java is a simpler language. The simplicity of Java makes programming easier and programmers more productive. Also, many of Java's compile-time and run-time checks reduce the time needed for testing and debugging in the initial stages of program development.

Dynamic Downloading and Late Binding

Java was designed for dynamic class loading via a network. This allows it to be used for *on-demand* applications. Although on-demand class loading is not always feasible in mobile networks, the dynamic class loading and late resolving of references still provides new types of dynamic behavior that have not been possible in embedded devices before.

Java Builds on an Existing Programmer Base

Java was designed to be a language that could be easily learned by the large pool of existing C and C++ programmers. The Java language and APIs were designed to be architecturally neutral, highly portable, network-aware, robust, and secure.

A Brief History

The development of Java started with the Green project at Sun Microsystems in 1991. Sun first developed a programming technology called *Oak* for interactive TV. The team, under the leadership of James Gosling, also developed a demonstrator including a real device. At that time, the players contacted in the interactive TV business were not interested. It is interesting to note however that Java-based technologies have now been standardized as a platform for interactive digital TV services.

After realizing that interactive TV would not provide enough business for Java, the team had to try something else. As the enormous growth in the use of the Internet was just emerging at that time, the team decided to try integrating Java with the Internet. In 1994, the project created a demo called *WebRunner* based on a Mosaic browser and Java. The project was finally published in 1995, and the decision to integrate Java with the Netscape browser ensured that Java would become widely known in the computing industry.

In the early days of commercialization, Java was sold in two flavors: a Java applet environment for computers and browsers, and a full operating system implemented in Java for embedded devices. The industry did not adopt the latter since most companies wished to keep their existing operating systems in place and use Java for downloadable applications. This led to Java being repackaged in a new way, and as a result Personal Java and Embedded Java were created.

At the same time, the Java platform for computers was expanded with new API libraries and new virtual machine technologies for speeding up the execution of Java bytecode. Only relatively minor changes have since been made to the specification of the virtual machine and language.

Core Java Platforms

The everyday world is for the most part divided into three main categories of computer-based systems: fault-tolerant and scalable enterprise computing systems and servers, desktop workstations and personal computers, and small mass-market consumer devices such as set-top boxes and mobile devices.

In a similar manner, Java has developed into three main types of platform *editions*. Each edition addresses one of the categories identified above:

o **Java 2 Platform, Enterprise Edition (J2EE):**

for Java-based Web, enterprise, and application servers

o **Java 2 Platform, Standard Edition (J2SE):**

for desktop workstations and personal computers

o **Java 2 Platform, Micro Edition (J2ME™):**

for resource-constrained consumer devices

Because a wide range of consumer devices exists for a variety of different types of uses, and because such devices can be very resource-constrained, further refinement of the J2ME platform is needed. For this reason, the concepts of *Configurations* and *Profiles* are used in the J2ME platform to define specific families of resource-limited devices and their capabilities.

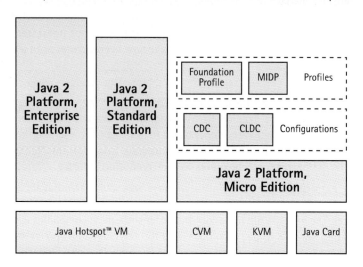

Java platform editions

Java 2 Platform, Standard Edition

The Java 2 Platform, Standard Edition (J2SE) was the first Java platform to gain widespread acceptance and use. By itself, it can be used to develop rich clients and some types of services. In addition, it is the foundation upon which the Java 2 Platform, Enterprise Edition is built.

Many technologies and application programming interfaces are included in the Java 2 Platform, Standard Edition. This section only presents a high-level overview of some of the main concepts and focuses on those that support application development in the Web domain. It should be noted that many good starting points [JAVAWEB4] and books exist that provide more detailed information.

Some technologies are specifically intended for use in graphical clients, such as user interface application programming interfaces, such as the Swing framework and the Abstract Windowing Toolkit (AWT). Support is also provided for localization and accessibility of applications, which are often important issues. Technologies such as Web Start and the Java Plug-in architecture help with deployment of applications or components.

Integration APIs play an important role in the J2SE. In some cases, they allow Java applications to leverage existing popular technologies. For example, the Java Database Connectivity (JDBC™) API is used for universal data access. Remote Method Invocation (RMI) allows developers to create distributed Java applications. RMI enables a Java object in one virtual machine to invoke the methods of remote Java objects in other Java virtual machines. The invoking and invoked objects may reside on different hosts. The Java Naming and Directory Interface (JNDI™) API provides a uniform interface to multiple naming and directory services, such as the commonly used Lightweight Directory Access Protocol (LDAP).

The Java API for XML Processing (JAXP) provides basic support for parsing and processing of Extensible Markup Language (XML) data or documents.

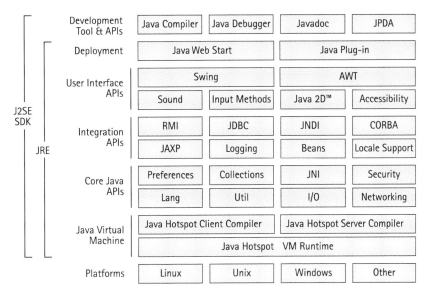

Java 2 Platform, Standard Edition

The J2SE also provides a developer with the core Java language classes and interfaces, and application programming interfaces for a variety of common problems (e.g., utility, basic input/output support, collections of objects, networked communication, and security). The Java Native Interface (JNI) defines a way to include native platform (i.e., non-Java) code in an application if needed.

The Java Hotspot™ virtual machine is used to execute J2SE applications. It is supported on a wide variety of hardware and operating system platforms, which gives a high degree of portability for Java applications.

Java 2 Platform, Enterprise Edition

The Java 2 Platform, Enterprise Edition (J2EE) enables a multi-tiered model for developing distributed applications. The design is based on the realization that nowadays server-side systems are specialized to support reliability and scalability for different tiers of an enterprise. The tiers commonly seen are: Web access servers, business logic and application servers, and enterprise information systems. The main tier divisions are shown in the following figure. Clients may exist that access any tier directly. However, because of the rapid growth of the Internet, the Web tier is increasingly the one used for business-to-business and business-to-consumer information flows.

Web Domain Technologies – Mobile Internet Layer

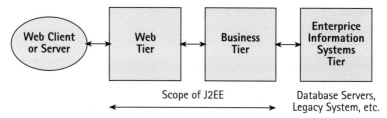

The Web and Business tiers

Many technologies and application programming interfaces are included in the J2EE. This section only presents a brief overview of some of the main concepts, and especially those that support application development in the Web domain. It should be noted that many good starting points [JAVAWEB5] and books exist that provide more detailed information.

In the Web tier, servlets provide a convenient model for developing HyperText Transfer Protocol (HTTP) based server-side components. Servlets may serve presentation content. Pages containing HyperText Markup Language (HTML) or Wireless Markup Language (WML) content are examples. Such content is often fairly static and similar to text in books or magazines. But it can also be dynamically generated. JavaServer Pages (JSP™) provides a convenient mechanism for mixing such static and dynamic content, which is to be served from the Web tier. Servlets are also used to support high-level messaging abstractions and protocols (e.g., XML-based) on top of HTTP. Servlets and JSPs run inside an appropriate container that is used to handle their scalability and reliability.

In the Business tier, Enterprise JavaBeans™ (EJB™) are used to wrap common database programming paradigms and to encapsulate business logic. They run inside an appropriate container that handles their scalability and reliability. An important part of database activities is the well-known concept of a transaction. J2EE supports a transaction model that permeates many parts of the J2EE model, including EJBs.

The Java Connector Architecture provides a means to integrate J2EE-based services with legacy enterprise information systems.

The Java Messaging Service (JMS) provides a reliable messaging service. Two types of messaging models are supported. One is a point-to-point messaging model. The other is a *Publish/Subscribe* model.

An important part of J2EE is the ability to easily deploy new (or change existing) enterprise components and applications that are run in appropriate containers. The deployment mechanisms for this are well defined. The management of J2EE-based Web and application servers is an important topic, which is addressed partly by the specification and partly by manufacturer tools.

Java™ in the Web Domain

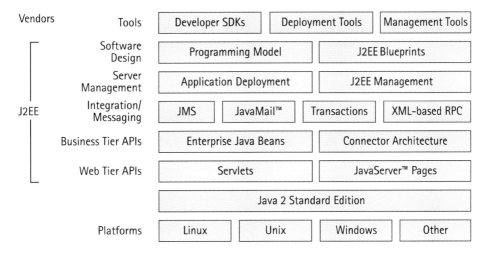

Java 2 Platform, Enterprise Edition

The first figure presented in this section gave an overview of the role of the Web and Business tiers. The following figure helps to get to the core of this by using some example components. It illustrates how Web or Mobile domain clients can use J2EE-based services.

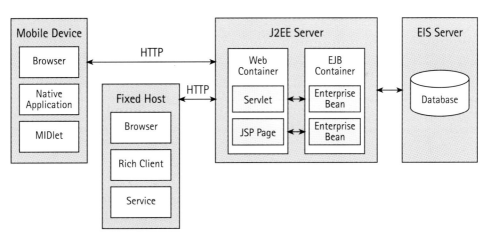

J2EE Server and Containers

A fixed or mobile device uses HTTP to access Web tier components, such as servlets and JSP pages. Those Web tier components use Business tier components, which encapsulate business logic such as EJBs. The components of the Business tier access enterprise information systems, such as databases or other legacy systems.

Separate containers are often used to run components of the Web and Business tiers. This allows for individual control of scaling and reliability for each tier.

Java Platform Support for Web Services

A detailed description of Web Services and related technologies is beyond the scope of this chapter. Only a brief description, as related to use of the Java platform, is provided here.

Two technologies commonly associated with the Web Services model are the Simple Object Access Protocol (SOAP) and the Web Services Description Language (WSDL). In addition, Service registries are needed to allow Web Services clients to perform lookups on appropriate services that they wish to use. Service registries can be based on Universal Description, Discovery and Integration (UDDI) or other appropriate mechanisms.

The following figure illustrates the basic processes of service registry, service discovery, and service use. Service use includes both two-way (request/response) and one-way (request) information transfer.

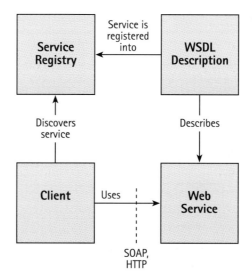

An overview of Web Services

The J2EE can be used as a basis for building Web Services systems of the type described in the previous figure. It includes the Java API for XML-based Remote Procedure Calls (JAX-RPC), which can be used as the communication mechanism between a client application and a Web Service. The Java API for XML Processing is part of the J2SE, and provides support for XML data and document processing.

The Java API for XML Messaging (JAXM) supports SOAP messaging and might be an alternative to JAX-RPC in some cases. JAXM differs from JAX-RPC in that the latter supports a specific remote procedure call-based model, whereas messaging is more general and can be used to support different types of communication models.

The Java API for XML Registries (JAXR) provides a uniform mechanism for accessing various types of XML-based registries.

Conclusions

This section gave a high-level overview of the role of Java technology in the Web domain. It concentrated on the design, benefits, and use of the J2SE and the J2EE. These are widely used in the industry to build powerful Web tier and Business tier applications as well as rich clients, due to their well-proven support for security, portability, reliability, robustness, and scalability. J2EE also provides a basis for building Web Services applications. In addition, high-quality J2EE servers and other support are available from a wide variety of vendors. This is due to the *open* approach of Java standardization.

All of the above reasons help explain the widespread popularity and use of J2EE-based Web and application servers. In summary, Java technology forms an excellent foundation for development of Web domain applications.

References

[JAVAWEB1]	A Brief History of the Green Project, http://java.sun.com/people/jag/green
[JAVAWEB2]	Java Language Specification, http://java.sun.com/docs/books/jls/
[JAVAWEB3]	Java Virtual Machine Specification, http://java.sun.com/docs/books/vmspec/
[JAVAWEB4]	The Java 2 Platform, Standard Edition, http://java.sun.com/j2se
[JAVAWEB5]	The Java 2 Platform, Enterprise Edition, http://java.sun.com/j2ee
[JAVAWEB6]	Java Technology and Web Services, http://java.sun.com/webservices
[JAVAWEB7]	Simple Object Access Protocol (SOAP) 1.1, W3C Note 8 May 2000, http://www.w3.org/TR/SOAP
[JAVAWEB8]	Web Services Description Language (WSDL) 1.1, W3C Note 15 March 2001, http://www.w3.org/TR/wsdl
[JAVAWEB9]	Universal Description, Discovery and Integration (UDDI) of Business for the Web, http://www.uddi.org/

Part 3.3

Application Layer

- o Web Services in the Web Domain
- o Semantic Web

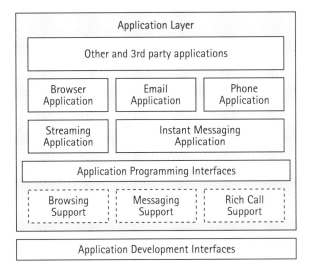

Part 2

Application Layer

Web Services in the Web Domain

The Internet today is evolving from a model where conventional services are provided via the HyperText Transport Protocol (HTTP) using human readable content, such as the HyperText Markup Language (HTML) and the Wireless Markup Language (WML). Services are developing toward a model of *mechanized service consumption* by autonomous computer systems that do not necessarily involve human interpretation. This is commonly referred to as the Web Services paradigm of computing. In the Web Services model, client systems interact with services using a model based on a rich set of meta-data made possible with interpretable Extensible Markup Language (XML) [2]. The transport is typically HTTP or based on the Simple Mail Transfer Protocol (SMTP). The domain where it has been applied most actively is the area of e-business and automation of business-to-business (B2B) processes and interactions. The goal is to improve efficiencies and economy of scale by adopting a standardized approach that reduces interoperability problems and the time to market. The Web Services paradigm is generally applicable to the Mobile domain as well.

Web Services is all about providing sets of services over the Internet to appropriate service consumers. A standard set of Web Services technologies (e.g., tools, platforms, XML-based service description, publication, discovery, inquiry and invocation) are used to build such services and their clients. A Web Services client may or may not involve human interaction. Human interaction would be more common in business-to-consumer (B2C) clients.

This section introduces Web Services and presents the currently available standardized Web Service Technologies (SWT) used to develop services for the Web. Advanced Web Service technologies and the need for a Web Services capable network are addressed in later sections.

By the end of this chapter, the reader should be able to:

o   Understand basic Web Services concepts
o   Distinguish between a Web Service and a non Web Service
o   Understand the relevant issues in the area of Web Service technologies
o   Better appreciate the needs for a Web Services enabled infrastructure

Overview

Web Services are services provided over a session layer (e.g., HTTP, SMTP, File Transfer Protocol (FTP), or another, similar Internet technology) and using certain, industry-standard software technologies, such as XML, XML Protocol (XMLP), Simple Object Access Protocol (SOAP), Web Services Description Language (WSDL), and Universal Description, Discovery and Integration (UDDI). Web Services should not be specific to any particular platform or a specific vendor's development environment. Web Services are described and classified in such a manner that clients (implemented using heterogeneous Web Services tools and platforms) can do the following:

- o Discover or find desired services for invocation
- o Interpret service offerings available from a registry of services
- o Invoke service requests with the appropriate request parameters and interpret responses correctly

Key Participants

The key participants in a Web Service are as depicted in the following figure. The participants communicate with each other utilizing SWT, which all the participants comply with. This enables seamless interoperability and integration. This should allow for rapid application development, reduced integration costs, and promotion of applications across various platforms, promoting the main goal of the Internet.

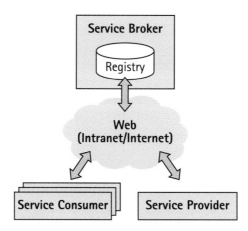

Key participants in Web Services

Service providers are participants who desire to publish a service as a Web Service. Service brokers are participants who provide a registry and discovery mechanism for such services provided by all Service providers. This is done with the help of either directly maintaining a registry of such service descriptions or by then providing value-added services over an existing Web Service registry. Service consumers are the immediate consumers of the Web Services offered by Service providers. The above figure presents a run-time view of the Web Services world. In addition, there are also development and deployment phases that involve service integrators, network and infrastructure providers, and toolkit and platform providers. However, the focus of this section is on the infrastructure requirements.

Basic Web Service Process

The basic process involved in the utilization of a Web Service is as depicted in the following figure. The Service Provider provides a Web Service. This service may be built from scratch or could wrap an existing legacy system (e.g., a purchase order request system). A Web Service interface is developed which wraps the service, enabling an appropriate client to invoke the service interfaces with SWT. The service is thus hosted and ready for access by the service consumers. The next step is for the service provider to describe the service, its interfaces, and the mechanism to invoke the service interfaces in order to utilize their functionalities. The description utilizes a standard service description mechanism, WSDL. This description is then registered with the service brokers so as to be discovered by the service consumers. The service providers use the interface provided by the service broker to manage the published service description.

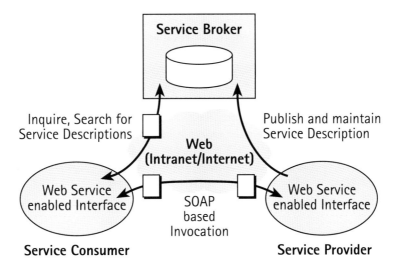

Basic Web Service process

Service consumers utilize the service interfaces provided by the service broker to discover and identify the correct Web Service. On finding the relevant Web Service description, the consumer will interpret the invocation mechanism described and dynamically invoke and communicate with the Web Service found. The communication between the Service Consumer, Producer and Broker is established by a SOAP based interaction.

Web Service Technologies

This section introduces the reader to the current technologies utilized in the Web domain. The commonly agreed upon industry-wide set of Web Service technologies consists of: XML [2], SOAP [3], WSDL [1], and UDDI [4]. These technologies are discussed in this section to introduce the reader to their relevance in the Web Services environment.

Data Encapsulation

Participants in Web Services need to exchange data and messages so they can request services and provide them accordingly. The data and message need to be in a platform-neutral format, yet interpretable by each of the participants. XML [2] provides a data encapsulation framework, which can be used to deliver a platform and vendor agnostic data representation and interpretation.

XML provides a method to represent structured, typed data and meta-data in a platform- and vendor-neutral representation. This makes it very suitable for applications intended to reach a wide set of consumer clients over the Web. The data exchanged between two participants can be efficiently interpreted if it follows a valid structure. This is where the XML Schema Definition (XSD) [10] is utilized. XSD defines a mechanism to validate the structure of the XML document. An XML document can comply with multiple such XML schemas, hence providing modularity in defining the document structure. The association of a single XML document structure with a specific schema can be referred to using the XML namespace [11]. XML forms the core of the Web Service technologies, as XML through the aid of schemas and namespaces provides a framework that can be used to define a platform-independent protocol. The Web Services are hence based on XML. The XML Core Working Group [9] of World Wide Web Consortium (W3C) drives the XML efforts.

Communication Framework

This section discusses the framework necessary for the exchange of information among the various participants in Web Services.

Consumers request the services offered by a Web Service by using an application, which communicates to a single Internet-accessible communication port. The services accordingly meet the needs of the requester. In Web Services terms, these are identified as the end-points of a Web Service. The end-points need a binding mechanism to link the needs of the requester and a mechanism for the service provider to deliver what is needed. The communication mechanism in the Web domain is primarily based on Remote Procedure Calls (RPC) or on message exchange schemes. These are the minimal required mechanisms to be supported by any Web-based solution.

The basic functionality is very similar to historic distributed computing systems, such as Remote Method Invocation (RMI), the Internet Inter-Object Request Broker Protocol (IIOP), and the Distributed Component Object Model (DCOM). A communication framework for Web Services should meet the distributed communication needs while being vendor and platform agnostic.

The framework should also provide sufficient flexibility to extend the communication protocol. The Simple Object Access Protocol is one such framework that is utilized in the Web Services world [3].

Simple Object Access Protocol

SOAP provides a message framework, which defines the message structure and processing mechanism; a set of encoding rules to represent service-specific data types; and, finally, a convention to enable remote procedure calls or the exchange of messages. SOAP does not provide any transport capabilities for the exchange of information between end-points, but it can be utilized to bind it with HTTP, SMTP, or other Internet transport technologies. The figure below shows the main parts of the SOAP message, which consists of three main elements: the root Envelope element, which is mandatory; an optional Header element; and a Body element, which is a mandatory part of a SOAP message.

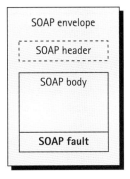

Parts of a SOAP message

The root element of the SOAP message is the Envelope, which must comply with and refer to the SOAP XML namespace [30]. The encoding style can be specified to comply with the standard SOAP encoding convention [31] by referring to a specific namespace.

The Header is an optional element of the SOAP envelope. It can contain the required elements following a specific namespace within it. The Header elements are for the benefit of the intermediary systems, so as to interpret the properties related to the message, i.e., the payload. The elements provide the necessary additional information related to delivering the message to the final recipient. The elements could be entries, such as message correlation ID, transaction ID, client identity, authorization, payment related identifiers, or other required attributes.

The SOAP Actor attribute can be used to define the intermediary recipient of the Header element of the message. This enables the message to be delivered in a hop-by-hop manner to the end recipient, enabling message processing at each hop.

The Body is a mandatory element of a SOAP message. The element contains information that is meant to be consumed and interpreted by the final recipient. The recipient, be it a message- or procedure-oriented system, will need to perform some activity based on interpretation of this element. Similarly to the Header, the Body can contain multiple element entries. The elements can be application-specific and need to be namespace qualified and referenced by a Uniform Resource Identifier (URI).

```
<SOAP-ENV:Envelope
     xmlns:SOAP-ENV="http://schemas.xmlsoap.org/soap/envelope/"
     SOAP-ENV:encodingStyle="http://schemas.xmlsoap.org/soap/encoding/">
<SOAP_ENV:Header> -- </SOAP_ENV:Header>
<SOAP_ENV:Body> -- </SOAP_ENV:Body>
</SOAP-ENV:Envelope>

-- = <schema-prefix:elementName xmlns: schema-prefix ="Some-URI">
       Schema Specific elementStructure Here..
     </schema-prefix:elementName>
```

SOAP message structure

Error or status information is carried between the requester and the recipient of the message using the SOAP Fault element. The SOAP Fault element is a part of the Body and can occur only once. The only sub-elements that a Fault can contain are:

- o **faultcode**, intended for the software to interpret,
- o **faultstring**, a human-readable fault message,
- o **faultactor**, the source of the error/status, and,
- o **the details**, the sub-element that is intended to carry the application-specific information.

XML Protocol

The World Wide Web Consortium's XML Protocol Working Group (XMLP-WG) specifies an XML-based protocol, XMLP [39], that can be utilized by applications to deploy simple service interfaces. This can be incrementally extended to support security, scalability, and robustness. The XMLP is specified from the standpoint of delivering a protocol that contains an envelope to transfer XML encapsulated data, which can be extensible; a convention for utilizing the envelope for RPC; a mechanism to marshal and de-marshal XML data types defined by XML schema; binding with HTTP to depict the transportation binding with the XMLP, and a framework for binding with other Internet-based transportation technologies (e.g., SMTP). While defining the XMLP, the XMLP-WG will not assume any programming model or particular model of communication between the peers. The corresponding SOAP1.1 [3] technology will be utilized; any extensions to it will be proposed by XMLP-WG. SOAP1.2 [40] is already one such result produced by the XMLP-WG [39].

Service Description

When a Web Service is properly hosted, it is ready to be invoked and utilized by the client, which complies and understands the service-specific invocation mechanism. This calls for a client implementation with knowledge of the service-specific invocation interfaces and message structures. This is achievable in a closed service-to-customer relationship, where the customers are known to be few and the client systems are developed and deployed beforehand based on the common conventions followed directly between the two.

The Web promotes dynamism, which means in this case that the consumers should be able to dynamically bind to a service and start utilizing the offered services. The dynamic binding can be crystallized only if the services can describe their service interfaces and the required message structures in a standardized manner. This will promote a wider set of consumers being able to utilize the services offered over the Web and hence utilizing the Web to its potential. The standardized description will fuel wider service integration and reduce the cost of integration.

The description can be utilized for multiple purposes, including:

o   To inspect the service end-points; which are the access points for invoking a service functionality.

o   To inspect the invocation mechanism

o   To understand if the service follows a procedure- or message-oriented communication convention

o   To interpret and understand the message structure utilized in the invocation mechanism

o   To understand the concrete binding to certain network protocols whose utilization is required for communication with the service.

Web Services utilize Web Services Description Language (WSDL) [1], a standard description language, to describe the service and its corresponding interfaces. WSDL is brought forward as a W3C Note, originally proposed by Ariba, Microsoft, and IBM. WSDL defines an XML-based grammar to define Web services as a collection of the communication end-points capable of exchanging messages that follow specified message structures. The description is modeled into abstract and concrete parts, promoting reusability of such descriptions. The abstract part describes the end-points and messages, while the concrete part of the description binds these end-points and messages to a concrete network deployment and data format bindings. This split in the service description promotes multiple deployment models of the baseline service definitions.

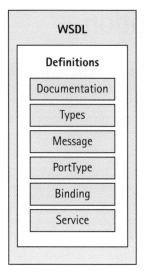

Web Service Description Language: Document structure

The previous figure depicts the essential parts of the WSDL document. Definitions provide the base of the WSDL document, containing all the service definitions. After the definitions are explored in brief, the need for them will become clear:

- o **Documentation** - This contains a human-interpretable description of the Web Services, meant for the service integrators or to serve developer needs.

- o **Types** - This contains the data type definitions used in the exchange of messages. WSDL supports the usage of the XSD [10] in an intrinsic manner. It also allows the utilization of other type definition systems through its extensibility elements.

- o **Message** - This definition consists of at least one part, and each part is associated to a specific type. The set of parts within messages represents the logical contents of a message. Message definitions are still considered to be the abstract content of a message, and the concrete portion of message definition only occurs at the binding phase of the messages.

- o **PortTypes** - This gives the possibility to represent an end-point operation containing messages in an abstract manner. The different types of operations that an end-point of a service or a customer can support are: One-Way, Request-Response, Solicit-Response, and Notification. The abstract operation can contain up to three types of messages, depending on the type of operation it belongs to. These are input, output, and fault. The service end-point operation receives an *input* message and, based on the type of operation, will respond back to the requester with the *output* message. The operation can report any errors to the requester through the use of the *fault* message.

- o **Binding** - This provides a concrete representation of the portTypes by binding the abstract definitions of the operations and messages defined in the portType to a specific network protocol. There could be more than one concrete binding to the abstract portType definition, hence promoting reusability of the basic abstract definition. The binding is done through the use of the extensibility framework of WSDL.

o   **Service** - A service groups a relevant set of ports, which represents an individual portType binding to a specific end-point address.

The WSDL description allows flexibility by importing definitions from other modular WSDL definition documents. The importing of these definitions enables modular description of the different parts of the Web Service definition and hence promotes reusability of these modular descriptions. The WSDL definitions can be bound to a specific communication framework through the means of the extension framework. For example, using the extension framework, the WSDL definitions can be bound to the SOAP, HTTP, or another relevant Web technology.

Service Brokers

The development of Web Services brings new business potential to the Web community, where businesses can trade with other new business partners in a dynamic fashion, as the service interfaces follow a standard approach to specification. The Web Services provide a standardized framework for such dynamic demands of businesses.

The Web Services bring a point where a standardized approach to communicating and describing services will benefit businesses by speeding and reducing the cost of integration. These services, though, still need to provide a solution for businesses to publish their service offerings and find those of other business partners. This is the point at which the service brokers play a major role in enabling an appropriate infrastructure. The core of the service brokers is a service registry, which will facilitate the publishing and discovery of such services. The model can be such that the Web Services, once available, could publish their descriptions and interfaces to a central or distributed registry. Business partners can further be helped to find each other by utilizing the services of the registries. The registries provide a service to business partners for publishing, managing, and finding the desired service details.

The Universal Description, Discovery and Integration

The Universal Description, Discovery and Integration (UDDI) [4] specification provides a global, platform-independent, open framework to enable businesses to discover each other, share information in a registry [32], and define the communication framework they utilize over the Internet. This will enable other businesses to communicate and utilize the services in a more standardized and open way. The UDDI registry consists of three main components: White Pages, listing address, contact information, and identifiers; Yellow Pages, industrial classification based on the standardized taxonomies; and the Green Pages, which contain technical information [33] about the services, such as service interface descriptions.

UDDI provides a programmatic service interfaces which is utilized by both the publisher and the consumer of services; The publisher utilizing these interfaces registers information about the service interfaces, while the consumer uses the interfaces to find and get the relevant business profiles. On inspection of the relevant technical details, the consumer will develop the necessary system to invoke the service offerings of a *Web Service*. The registry provides a *register once, published everywhere* [33] feature to a service publisher, hence making the service information accessible ubiquitously.

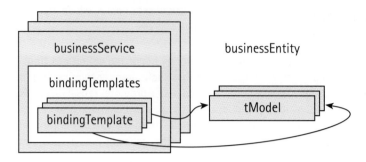

UDDI business entry

UDDI uses XML schema based business description to register the business information. The information registered is classified into business information, service information, binding information, and service specification information. All the information about a particular business is registered in a UDDI registry as a businessEntity. This enables discovery of information about a service based on, say, categorization (e.g., high-level information about the business such as address or contact information). This provides the features of the Yellow Pages for the discovery service. The businessEntity contains businessService and bindTemplates, which together provide the features of the Green Pages. The businessService groups a set of Web Services related to a specific business category. The bindingTemplate contains the actual technical information on the connectivity (e.g., the Internet address and the port), needed in order to communicate with the Web Service. It could also contain information related to routing or additional steps to be taken prior to actually connecting to the Web Service. The next requirement prior to actually invoking Web Services is getting the details of the operations and the format of the message communication between the services and its invokers. In UDDI terms, these are referred to as technical fingerprints. The technical fingerprints are used to identify the particular interface specification a Web Service implements. The tModel provides the details of such a technical fingerprint description. The information contains the name, publishing organization, description, and URL of the actual specification (e.g., WSDL). This property of the tModel enables services to indicate that it is in compliance with a particular specification.

Once the businessEntity is prepared by a business, it utilizes the UDDI service interface to publish the XML description. UDDI makes this description available to other business partners through its interfaces. The business partners using UDDI APIs will discover the required service descriptions and then accordingly proceed with service-specific invocations.

UDDI Service APIs

UDDI provides two classes of interfaces for services, the Inquiry API and Publisher's API. The Inquiry API is meant for businesses to find another suitable business in the UDDI registry by either browsing or specifically requesting particular business information (e.g., businessEntity, businessService, bindingTemplate or tModel). The Publisher's API is meant for the business developing Web Services to register its offering in the UDDI registry, so as to be located by potential business partners. The API also provides the possibility for the publisher to update or de-register services. UDDI service APIs are based on the SOAP communication framework. WSDL can be utilized in conjunction with UDDI's business description to complement the overall Web Service discovery, invocation, and implementation [34] process.

Publishing Web Services Locally

The Web Services Inspection Language (WSIL) is a draft made available by IBM and Microsoft. This solution is used by consumers who already know the business partner's site. The consumer can find the set of Web Services offered by a specific business partner's [6] site using WSIL. WSIL is an XML-based grammar used to aggregate the set of Web Service interface descriptions. WSIL documents are located at the business Web site offering the specific Web Services. WSIL provides a framework which can be used to aggregate a set of references to service description documents (e.g., as WSDLs), which can be accessed by a URI, acquired via FTP, or even located in an external container such as the UDDI registry.

Electronic Document Interchange Based Web Services

Electronic Business XML (ebXML) defines the basic Web Services mechanisms for registries, business process modeling, service description, and the communication framework. This is an initiative sponsored by the Organization for the Advancement of Structured Information Standards and UN/CEFACT, a United Nations body [7]. This initiative takes a view of Web Services from the Electronic Data Interchange (EDI) environment. The main goals of ebXML are to provide a solution for electronic business trading using a standardized global electronic business environment. The environment follows a standardized XML-based definition of businesses, processes, and the models that are commonly applicable to most business transactions [37].

ebXML proposes two views when it comes to modeling e-businesses. The Business Operational View (BOV) provides the Core Libraries. The libraries contain the data and process definitions representing business collaboration knowledge. The knowledge is represented by models in a more context-neutral language with sufficient abstractness to be applied across a specific business or industry.

In the Functional Service View (FSV), the trading partner who wishes to participate in the ebXML complaint transaction will need to utilize the Core Library and the Business Library defined in the BOV to implement the ebXML business service interface. The trading service can also register its own Collaboration Protocol Profile (CPP), which describes all the supported business processes, service interfaces, and its requirements in a manner that other ebXML complaint trading partners can understand. The CPP can also contain information such as contact information, industry classification, supported business processes, requirements for interfaces, messaging service requirements, and security information. The trading partner can now look up a trading partner's CPP. The intersection of two or more CPPs produces a Collaboration Protocol Agreement (CPA), which enables each trading partner to do business using ebXML governed commonly accepted conditions and agreement. The CPA governs the messaging services and business process requirements that are agreed upon between the trading partners.

The businesses look up the ebXML registries for the business process and the information meta-models relevant to the business they plan to conduct. On inspecting the information set, the business produces the implementation of the service and publishes the CPP profile via a registry. The trading partner that has already published its CPP to the ebXML registry looks up the CPPs of its business partners. The two trading partners' CPPs are used by the system to derive a CPA. Once the CPA is derived, the trading partners are ready to do business under the commonly accepted agreement. This is a complementary technology development in the area of Web Services.

Advanced Web Services

Transactions

For Web Service enthusiasts who are oriented toward business processes, individual Web Services do not make much sense as separate islands. One might actually want to build some dynamic applications on top of a set of selected services on the corporate intranet. One might also want to implement business processes on top of various loosely coupled services that are provided by business partners on the Internet. It is often very important that a set of related activities be treated as a single unit of work, producing consistent results on an all-or-nothing basis. Although Web Service transactions derive many of their attributes from traditional distributed object transactions, they have one very distinctive characteristic: these often business-level transactions may take an arbitrarily long time to complete.

The most popular transactional approach, one that has been traditionally used in application server architectures, is the two-phase commit combined with the traditional Atomicity, Consistency, Isolation and Durability (ACID) properties for distributed transactions. The two-phase commit relies on the concept of a transaction coordinator that controls the transaction outcome based on the votes made by constituent transaction participants.

However, having a central transaction coordinator in a Web Services environment implies mutual trust, which is not a given thing in the context of loosely coupled Web Services. Also, the two-phase commit protocol involves resource locking in one form or another, which is not very desirable for long-running transactions. Even for shorter-lived transactions, it may be tedious to implement the two-phase commit because the resources belong to systems within different organizations.

To better satisfy the demands of long-running transactions, there exists an alternative approach called compensating transactions. With the compensation approach, the transactions are always allowed to commit immediately, but the effects can be compensated (i.e., cancelled or otherwise modified) later. In compensating transactions, there is always a compensating operation associated with each normal operation. Therefore, the application logic is actually divided into two separate paths, the normal successful path and the erroneous path triggered for compensation. The actual implementation of compensations may vary a lot, all the way from simple undo operations to complex compensatory activities dealing with cancellation fees, for instance. Compensation does not necessarily require any specific transaction protocol or any other plumbing support, but, unfortunately, it does not come without a price either. Unlike the two-phase commit, compensation by definition may not satisfy the isolation property of ACID. This means that there is always the potential for a failure by reading data that is subject to be rolled back later unless some preventive measures are taken at the application level.

Despite this apparent controversy, there is one notable specification effort that attempts to define a suitable transaction protocol for Web Services. The OASIS Business Transactions technical committee has released the Business Transaction Protocol (BTP) [1]. The BTP extends the two-phase commit approach to Web Services by making the coordinator role flexible and negotiable during the transaction initiation. In addition to normal ACID transactions, BTP also covers non-ACID transactions (i.e., cohesion) for long-running transactions.

Workflow and Conversations

The transactional behavior defines a mechanism to determine the outcome of multiple Web Service invocations when treated as a single unit of work. The main challenges faced are definition of such a sequence and the execution of the operation sequences at runtime. This is where workflows come in: a workflow usually means a well-defined flow of operations that can be modeled as a directed graph with activities and transitions between them. Another commonly used term for the flow is *conversation*, as traditional workflows may have additional connotations in the context of GroupWare. The flow provides seamless integration across business processes and transaction life cycles that make use of multiple services. The implementation of a workflow often relies on specific workflow engines that execute activity via proxies between two or more services, pass data between activities, monitor workflow state, and evaluate dependencies between individual activities. A workflow specification may usually include the following aspects:

o   Messages to be exchanged when flowing from one activity to another

o   Actual transport and messaging protocols and their end-points

o   Message exchange behaviors such as operation sequencing and parallelizing, message correlation patterns, and long-running transactional behavior with compensation

o   QoS aspects such as reliable message delivery, latency constraints, and security requirements

In the context of Web Services, one of the existing workflow description languages is the Web Services Flow Language (WSFL) [16] proposed by IBM. The power of WSFL lies in its rigorous meta-model that makes it able to model business processes that may span technology and enterprise boundaries. WSFL consists of two essential models: a flow model and a global model that together make a complete workflow description. The flow model represents the XML representation of a directed, acyclic graph that models the structure of a business process with activities and processing steps, connected with control and data links. The global model produces the realization of one or more flow models by describing the interactions between appropriate service providers and requestors. As an additional outcome for a workflow specification, a new WSDL-based interface is defined to represent the exposed business process as a composite Web Service for the caller.

Hewlett-Packard has specified its own variant called Web Services Conversation Language (WSCL) [17]. The WSCL approach is significantly lighter than the WSFL approach, but the target is the same: dynamically defining the abstract interfaces of business level conversations or public processes supported by individual Web Services. Microsoft joins the table with the XLANG [18], or Web Services for Business Process Design by its subtitle, a specification that is essentially an extension to WSDL providing means to formally specify business processes as stateful, long-running interactions among participating Web Services. One of the distinctive features of XLANG is the notion of long-running transactions with compensatory actions.

Workflow specification mechanisms are becoming a crucial issue in the Web Services community today, and many expect existing proposals from both commercial and non-profit organizations to converge sooner or later into a commonly accepted standard. At the end of the day, all of them are reaching out for the same target: dynamically defined compositions of loosely coupled Web Services as applications or business processes. In an ideal world, these compositions carry out arbitrarily complicated interactions between individual services that do not have to explicitly support any of this conversational logic by themselves.

Security Issues

The very idea of Web Services is to seamlessly integrate systems across the Internet, and this may often imply the exposure of sensitive information to parties that are not explicitly known in advance. However, the core Web Service technologies do not make any statements about security; rather, they deliberately delegate the issue to other protocol layers below and above them. To be able to trustfully use or implement a Web Service, there must exist a security infrastructure covering the following issues:

- o Authentication of requestor (application) and the consumer (person) identities
- o Authorization based on authenticated identities
- o Maintaining information confidentiality and integrity with encryption
- o Protecting the transaction integrity with non-repudiation
- o Applying negotiable *privacy* rules for any information revealed

Traditional solutions for Web security are usually rather transport-oriented and therefore often inadequate for general Web Service purposes. For instance, the Secure Sockets Layer (SSL) [19] can be used for authenticating network end-points, but from the Web Services point of view it is not able to provide end-to-end security. However, taking one extra step and combining several existing security solutions makes it possible to put together basic security architecture for Web Services: SSL encryption on the transport level, XML message digests for integrity, customized access control lists for authorization, and, finally, user IDs and passwords for simple authentication.

Although these solutions could be easily implemented today with existing technologies, they might not be strong enough to establish a common security infrastructure for Web Services. To fill this gap there are many ongoing developments targeting better security frameworks for Web Services. On the identity management front, Microsoft is implementing the Kerberos-based Passport [20], while the Liberty Alliance [21] is working on an alternative solution but a less centralized model. Also, a great deal of specification work is taking place in the XML arena, producing solutions that are effectively decoupled from the actual network layer. An XML Digital Signature (XML-DSIG) [22] defines how to digitally sign complete or partial XML documents to ensure integrity and non-repudiation, and XML Encryption [23] describes similar but more generic encryption and decryption mechanism for arbitrary XML data. The XML Key Management Specification (XKMS) [24] defines the protocols for distributing and registering public keys. This as such is meant to work in concert with XML-DSIG. Extensible Access Control Markup Language (XACML) [25] can be used to specify access policies, including digital rights management. The Security Assertion Markup Language (SAML) [26] is meant to provide a mechanism for authentication and authorization information exchange with single sign-on capabilities.

IBM has released the SOAP Digital Signature (SOAP-DSIG) [27], a specialization of XML-DSIG for use in SOAP envelopes. Microsoft has also released two security-related specifications as SOAP extensions. Web Service License Language (WS-License) [28] defines how to encode commonly used license types, such as X.509 certificates, Kerberos tickets, or arbitrary binary credentials, in SOAP envelopes. Web Service Security Language (WS-Security) [29] is an umbrella specification that enables secure Web Service interactions by defining mechanisms for credential exchange (WS-License), message integrity (XML-DSIG), and message confidentiality (XML Encryption).

All these technologies are capable of working side by side or completely separately from each another, depending on the needs of the application. However, a good security framework is inevitably a crucial aspect of Web Services, as consumers' confidence is directly driven by it. The technologies listed here are meant to solve certain problems specific to Web Services, but other issues must be taken into account as well. These include server-side solutions for 24x7 availability, denial-of-service attacks, and so on. The bottom line is that for any decent security infrastructure there must be a well-defined security policy properly implemented and administered – Web Service security technologies are just building blocks for that.

Conclusions

This section has presented an overview of the Web Services paradigm and related platforms and technologies. Support for business processes, transactions, and workflows was also discussed, as were security aspects. Gaps that still exist in the industry today were presented in brief.

Web Services are already available today from major platform vendors. These technologies will still mature in the near future, after which Web Services will be widely applied to Web-based businesses. There are gaps, which need to be filled in, such as scalability, guaranteed reliability, and a very strong trust model, before Web Services can be applied to business over the Web in an open way. Meanwhile, the technologies mature, and the adaptation of Web Services will be visible within the corporate intranets as a method of integrating various solutions between variant platforms. In any case, the Web Services model has already begun to be applied in the real world.

However, Web Services technology will face performance problems and hence will depend on an infrastructure that will simplify and provide additional support on top of the standards available today. There is caution of the application of Web Services to mainstream, business-critical solutions, as the full range of wider applications of this technology still need to wait for maturity of the technologies. The Web Services model should be applied to business systems where resource consumption is not too critical an issue. These issues will eventually fade away as the technology is utilized more fully with its maturation in the near future.

References

[1] WSDL 1.1, "*Web Services Description Language*", W3C Note, March 15, 2001, http://www.w3.org/TR/wsdl.

[2] XML 1.0, "*Extensible Markup Language - Second Edition*", W3C Recommendation, October 6, 2000, http://www.w3.org/TR/REC-xml.

[3] SOAP 1.1, "*Simple Object Access Protocol 1.1*", W3C Note, May 8, 2000, http://www.w3.org/TR/SOAP/.

[4] UDDI, "*Universal Description, Discovery, and Integration*", http://www.uddi.org/.

[5] WSFL, "*Web Services Flow Language 1.0*", May 2001, By Prof. Dr. Frankleymann, IBM, http://www-4.ibm.com/software/solutions/webservices/pdf/WSFL.pdf.

[6] WSIL, "*Web Services Inspection Language 1.0*", November 21, 2001, http://www-106.ibm.com/developerworks/webservices/library/ws-wsilspec.html or http://msdn.microsoft.com/library/default.asp?url=/library/en-us/dnglobspec/html/ws-inspection.asp.

[7] ebXML, "*Electronic Business XML*", a global standard sponsored by UN/CEFACT and OASIS, http://www.ebxml.org/.

[8] XLANG, "*Web Services for Business Process Design*", http://www.gotdotnet.com/team/xml_wsspecs/xlang-c/default.htm.

[9] XML Core Working Group, "*W3C's XML Working Group*", http://www.w3.org/XML/Group/Core.

[10] XML Schema "*W3C's recommendation - XML Schema Part 0/1/2: Primer, Structures, Datatypes*", May 2, 2001, http://www.w3.org/TR/xmlschema-0/.

[11] XML Namespace, "*W3C's recommendation - Namespaces in XML*", January 14, 1999, http://www.w3.org/TR/REC-xml-names/.

[12] Net Know How, "*Web Services Model Introduces New Infrastructure Requirements*", October 22, 2001, http://www.internetweek.com/graymatter/lewis102201.htm.

[13] White Paper "*Web Services and the need for Web Services Networks*", a White Paper by Flamenco Networks, http://www.flamenconetworks.com/resource.asp.

[14] White Paper "*Implementing Web Services Enterprise Class web Services*", White Paper by Flamenco Networks, http://www.flamenconetworks.com/resource.asp.

[15] BTP 1.0, "*Business Transaction Protocol Specification*", http://www.oasis-open.org/committees/business-transactions/#documents.

[16] WSFL 1.0, "*Web Services Flow Language*", May 2001, By Prof. Dr. Frank Leymann, http://www-4.ibm.com/software/solutions/webservices/pdf/WSFL.pdf.

[17] WSCL 2.0, "*Web Services Conversation Language 1.0*", W3C Note, March 14, 2002, http://www.w3.org/TR/wscl10/.

[18/8] XLANG initial draft: http://www.gotdotnet.com/team/xml_wsspecs/xlang-c/default.htm.

[19] SSL3.0, "Secure Sockets Layer - 3.0 ", November 18, 1996, http://home.netscape.com/eng/ssl3/.

[20] Microsoft Passport, http://www.passport.com/.

[21] Liberty Alliance, http://www.projectliberty.org/.

[22] XML-DSIG, "*XML Signature WG*", http://www.w3.org/Signature/.

[23] XML Encryption, " *XML Encryption WG*", http://www.w3.org/Encryption/2001/.

[24] XKMS, "*XML Key Management WG*", http://www.w3.org/2001/XKMS/.

[25] XACML, "eXtensible Access Control Markup Language", http://www.oasis-open.org/committees/xacml/.

[26] SAML 1.0," *Security Assertion Markup Language*", February 20, 2002, http://www.oasis-open.org/committees/security/.

[27] SOAP-DSIG 1.0, "*SOAP Security Extensions: Digital Signature*", W3C Note, February 6, 2001, http://www.w3.org/TR/SOAP-dsig/.

[28] WS-License, "*Web Services License Language (WS-License)*", Draft, January 15, 2002, http://msdn.microsoft.com/ws/2002/01/License/.

[29] WS-Security, "Web Services Security Language (WS-Security)", Draft, January 17, 2002, http://msdn.microsoft.com/ws/2002/01/Security/.

[30] XML SOAP Schema, "*SOAP 1.1 Schema Document*", http://schemas.xmlsoap.org/soap/envelope/.

[31] XML SOAP Encoding, "*SOAP 1.1 encoding schema Document* ", http://schemas.xmlsoap.org/soap/encoding/.

[32] "*UDDI Executive White Paper*", November 14, 2001, http://www.uddi.org/pubs/UDDI_Executive_White_Paper.pdf.

[33] "*UDDI technical White Paper*", September 6, 2000, http://www.uddi.org/pubs/Iru_UDDI_Technical_White_Paper.pdf.

[34] "*Using WSDL in a UDDI Registry 1.05, UDDI working Draft Best Practices Document*", June 25, 2001, http://uddi.org/pubs/wsdlbestpractices-V1.05-Open-20010625.pdf.

[37] ebXML, "Technical Architecture Specification v1.0.4", February 16, 2001, http://www.ebxml.org/specs/ebTA_print.pdf.

[38] "*How can we build Trust between a consumer of a Web Service and Developer?*" - A White Paper by Value Added Web Services Providers, VAWWS, February 19, 2002, http://www.vawss.org/ca/vawss_001.pdf.

[39] XMLP and XMLP Working Group, http://www.w3.org/2000/xp/Group/.

[40] SOAP 1.2 "*SOAP Version 1.2 Part 0: Primer, Part 1: Messaging Framework, Part 2: Adjuncts*", W3C Working Draft, December 17, 2001, http://www.w3.org/TR/2001/WD-soap12-part0-20011217/, http://www.w3.org/TR/soap12-part1/, and http://www.w3.org/TR/soap12-part2/.

Semantic Web

The World Wide Web has developed primarily as a medium of content for *human* consumption. Automating anything on the Web (e.g., information retrieval, and synthesis of information) is difficult because human interpretation, in one form or another, is required for making Web content useful. A new Web architecture, known as the *Semantic Web*, is emerging to offer some relief. In broad terms, it encompasses efforts to create mechanisms that augment content with *formal semantics*, thereby producing content suitable for the consumption of automated systems [Berners-Lee et al 2001]. The Semantic Web will allow us to use more automated functions on the Web (such as reasoning, information and service discovery, and autonomous agents), easing the workload of humans. The Semantic Web will also pave the way for true *device independence* and customization of information content for consumers. Information on the Web can now exist in *raw form* and any context-dependent presentation can be rendered on demand (more generally, the Semantic Web represents a departure from the current *rendering-oriented* Web). It is important to note that the Semantic Web is not a separate Web but an extension of the current one, in which information, when given this well-defined meaning, better enables computers and people to work in tandem.

The World Wide Web essentially gives us an infrastructure for *pointing* (or, in other words, an infrastructure and mechanisms for linkages between documents). This infrastructure is quite general and consequently very powerful, but its problem is that the linking is indiscriminate: human interpretation (again) is required for understanding what any particular link means. The Semantic Web aspires to change this by associating formal meaning to the linkages. This brings us to a key aspect of the Semantic Web, something we will refer to as the *ontological approach*: Instead of merely introducing new markup tags (e.g., via the use of XML), the Semantic Web will focus on the meaning of the new markup, and on mechanisms that allow us to introduce, coordinate, and *share* the formal semantics of our data, as well as to *reason* about (i.e., draw inferences from) the semantic data. In practice, computers will find the meaning of semantic data by following (semantic) hyperlinks to definitions of key terms and rules for reasoning about them logically. This aspect of the resulting infrastructure will enable the development of fully automated Web Services, such as highly functional *intelligent* agents.

Role of Ontologies

The ontological approach is predicated on the existence and use of *ontologies*: by this, we mean a document or file that formally defines the relationships between terms in any particular domain of discourse. In this context (especially since the Semantic Web community draws heavily on a branch of artificial intelligence known as *knowledge representation*), a widely cited definition of ontology is "a specification of a conceptualization" [Gruber 1993]. In that sense, we depart from the abstract philosophical notion of ontology defined as "a branch of metaphysics concerned with the nature and relations of being" [Merriam-Webster 1998]. The most typical kind of ontology for the Web has a *concept taxonomy* and a set of *inference rules*.

Ontologies can enhance the functioning of the Web in many ways. They can be used in a simple fashion to improve the accuracy of Web searches by making the search program use precisely defined concepts (as opposed to ambiguous keywords *possibly* contained by a page). More advanced applications will use ontologies to reason the relationships between concepts, and will be able to relate the information on a page to any associated knowledge and inference rules. This will enable conceptual reasoning across independently created Web sites and their content.

People, as well as computational agents, typically have some notion or conceptualization of the meaning of terms. The specification a software program sometimes provides about its inputs and outputs could be used as a specification of the program itself. Similarly, ontologies can be used to provide a concrete specification of term names and meanings. Although, following this line of thought – in which an ontology is a specification of the conceptualization of a term – there is much room for variation, and the spectrum of Web ontologies typically range from simple controlled vocabularies through informal concept hierarchies to something in which arbitrarily complex logical relationships can be specified between defined concepts.

In practical terms, ontologies can be defined as finite yet *extensible controlled vocabularies*, as observing *unambiguous interpretation of classes and term relationships*, and as having *strict hierarchical subclass relationships* between classes. In addition, ontologies typically allow *property specification and value restriction on a per-class basis*, as well as the *inclusion of individuals* (class instances) in the ontology. Other features of ontologies, albeit not necessarily typical, are the specification of *disjoint classes*, *inverse relationships*, and *part-whole relationships*, as well as the specification of *arbitrary logical relationships* between terms [Lassila & McGuinness 2001].

Languages of the Semantic Web

In order to take the ontological approach, we need representation languages not only suited for expressing ontological information but also well matched with other Internet and World Wide Web standards. A layered approach is adopted.

The Extensible Markup Language (XML) is used for the purpose of serializing data and transporting it from one system to another. XML is a very powerful formalism: it allows *customized* languages to be created for almost any purpose. In practical terms, XML lets everyone create their own markup tags that, e.g., annotate Web pages or sections of text on a page. Programs could (and currently do) make use of these tags in sophisticated ways, but the programmer has to know the intended meaning attributed to the tag by the author. In short, XML allows designers to add structure to their documents but says nothing about what these structures mean.

Resource Description Framework

Another language, called the Resource Description Framework (RDF) uses XML for serialization, but introduces a powerful abstract data model that allows systems to share the meaning of the aforementioned structures [Lassila 1998, Lassila & Swick 1999, Lassila 2000]. RDF was originally a standard for Web metadata developed by the World Wide Web Consortium (W3C) and some of its member organizations (e.g., Nokia, IBM, and OCLC). Expanding the traditional notion of document metadata, such as library catalog information, RDF is suitable for describing any Web resource, and as such, provides interoperability between applications that exchange machine-understandable information on the Web.

Items that RDF expressions describe are called *resources*; broadly speaking, anything a Uniform Resource Identifier (URI) can name is a resource [Berners-Lee et al 1998]. Consequently, RDF can describe not just things on the Web (e.g., pages, parts of pages, or collections of pages) but also things *not* on the Web – as long as they can be named by using some URI scheme. The RDF description model uses object/attribute/value *triples*: we can view instances of the model as *directed labeled graphs* (resembling semantic networks), or we can take a more object-oriented view and think of RDF as a *frame-based representation system* [Fikes & Kehler 1985, Lassila & McGuinness 2001]. In RDF, these triples are known as *statements*.

Descriptions in RDF can be limited to one resource (e.g., a library catalog card that names a document's author and publisher), in which case the values of resource attributes (i.e., *properties)*, are typically just strings. Descriptions can also span multiple resources: values of properties can be other resources – this way we can describe arbitrary relationships between multiple resources. URIs name properties, which are also resources, and can be described by asking, "What are a particular property's permitted values, which types of resources can it describe, and what is its relationship with other properties?" Meaning in RDF is bound to specific terms and concepts that URIs define and name. Since URIs can be unique, two systems can define a concept (e.g., "person") and each can use a different URI to name it to avoid clashes. However, two systems agreeing on a common concept will use the same URI and will effectively *share semantics*.

The RDF model also defines some *meta-level constructs,* such as container types for describing collections of resources and *higher-order statements* (i.e., statements about other statements). Higher-order statements are modeled in RDF itself, and allow the representation of *modalities,* such as beliefs. Furthermore, an extensible, object-oriented *type system* known as RDF Schema is introduced as a layer on top of the basic RDF model [Brickley & Guha 2000]. Class definitions can be derived from multiple superclasses, and property definitions can specify *domain* and *range* constraints. We can think of RDF Schema as a set of ontological modeling primitives layered on top of RDF.

As mentioned earlier, RDF uses XML for the syntactic expression of model instances. This is a source of much controversy and confusion. RDF is essentially a *data model* and does not strive to replace XML. Instead, it builds a layer on top of XML, making interoperable exchange of semantic information possible (e.g., object-oriented extensibility is intended to allow a *partial understanding* of data). The current version of RDF still lacks primitive data types (e.g., integer and float), so text strings are essentially the only literals available. Languages that build on top of RDF, such as the DARPA Agent Markup Language may expand on this, typically adopting the atomic data types from XML Schema [Biron & Malhotra 2001].

RDF's object-oriented extensibility allows developers to take pieces of existing RDF schemata and extend them as needed. This is important, since it may lead to a type of *Darwinian evolution* of ontologies (and metadata), in which strong solutions survive and evolve further (as opposed to the all-or-nothing situation that of XML DTDs).

DARPA Agent Markup Language

The DARPA Agent Markup Language (DAML) program, launched in 2000, is a US Government sponsored effort aimed at producing foundation technology for the Semantic Web. The program funds critical research to develop languages, tools, and techniques that will allow the creation of more machine-understandable Web content. One of the program's results is DAML+OIL, a semantic markup language for constructing Web-based ontologies.

DAML+OIL is built on earlier RDF and RDF Schema, and extends them with richer modeling primitives (commonly found in frame-based representation languages). It also uses XML Schema to provide, e.g., numeric data types. The language has clean and well-defined semantics.

In contrast to RDF, DAML+OIL not only supports explicitly named classes, but also *class expressions*. These include classes whose *instances are enumerated*, as well as *boolean combinations* of other class expressions. The notion of *class equivalence* is supported (i.e., it is possible to specify that two classes are equivalent and thus must contain the same instances). Furthermore, DAML+OIL also supports *property restrictions*, which allow properties to be tailored on a class-by-class basis (in RDF, characteristics of properties are global). DAML+OIL further clarifies the nature of RDF Schema *domain* and *range* constraints of properties (making them conjunctive, not disjunctive). It also supports the notions of *property equivalence* and *inverse relations*.

Finally, DAML+OIL introduces a new container type that allows the definition of a collection in such a way that all its members are known (i.e., the collection cannot contain any fewer or more members than the ones explicitly specified). This quality is important, since ontological processing on the World Wide Web cannot typically rely on the so-called *closed world assumption*.

W3C Web Ontology

The W3C Web Ontology Working Group (Web-Ont) is working on a new ontology language for the Web, as a successor to DAML+OIL [Web-Ont]. This language, currently dubbed as the Ontology Web Language (OWL), will improve on DAML+OIL in several respects (it will be simpler, making it easier to deploy the language in the general Web developer community). It will also reflect input from a larger consumer and developer community. The Web-Ont group has published a requirements document for the new language [Heflin et al 2002].

Semantic Web as a Web of Services

"Web Services" – remote functionality invoked over the Internet – are a step towards automating tasks on the Web. The Semantic Web, representing a more far-reaching form of Web automation, is closely related (and highly complementary) to Web Services.

Web Services Using DAML

On the Semantic Web, to make use of a Web Service, a software agent needs a computer-interpretable description of both the service and the means by which it is accessed. An important goal within the DAML program is to establish a framework for creating and sharing these descriptions. Web sites will then be able to employ a set of basic classes and properties for declaring and describing services.

Part of the ongoing work of the DAML program is DAML-S, a Web Service ontology expressed in DAML+OIL [Ankolenkar et al 2001, Burstein et al 2001]. It gives Web Service providers a core set of markup language constructs for describing the properties and capabilities of their Web Services in unambiguous, computer-interpretable form. DAML-S markup of Web Services will facilitate the automation of the following Web Service tasks:

1. **Automatic discovery** involves the automatic location of Web Services that provide a particular service and that adhere to requested constraints.

2. **Automatic invocation** involves the execution of a discovered Web Service by a computer program or agent. The execution of a Web Service can be thought of as a collection of function calls. DAML-S markup of Web Services provides a declarative, computer-interpretable interface for executing these function calls. A software agent should be able to interpret the markup in order to understand what input is necessary for the service call, what information will be returned, and how to execute the service automatically.

3. **Automatic composition and interoperation** involves the automatic construction of a plan to use a number of Web Services to perform a task, with a high-level description of an objective. With DAML-S markup of Web Services, the information necessary for selecting and composing services will be encoded on the service Web sites. Software can be written to manipulate these representations, together with a specification of the task objectives, in order to execute the task automatically.

4. **Automatic execution monitoring**: Individual services, and especially compositions of services, will often require some time to be executed completely. During this period, users (human or artificial ones) may want to know the status of the request. Their plans may also have changed, and may require changes in the actions of the Web Service provider.

In contrast to *service profiles* (the DAML-S characterization of services), the other industry standards are limited primarily in that they cannot express logical statements (e.g., pre- and postconditions, or rules that describe dependencies between profile elements). Input and output types are supported to a varying extent. From the DAML-S viewpoint, services can be simple (or primitive): they invoke only one Web-accessible computer program, sensor, or device that does not rely on another Web Service, and beyond a simple response, there is no ongoing interaction between the consumer and the service. Alternately, services can be complex, composed of multiple primitive services, often requiring interaction or a dialogue between the consumer and the services, allowing the consumer to make choices and provide information conditionally. DAML-S is meant to support both categories of services, but complex services have provided the primary motivation for the features of the language.

DAML-S provides an upper ontology for services, including the properties normally associated with all kinds of services. The upper ontology does not address what the particular subclasses of the base service class should be, or even the conceptual basis for structuring this taxonomy. This structuring may take place according to functional and domain differences, or market needs. A service profile provides a high-level description of a service and its provider, and it is used as a request or an advertisement in a service discovery process.

Service profiles consist of three types of information: a human-readable *description* of the service, a specification of the *functionality* provided by the service (represented as a transformation of the inputs required by the service into the outputs produced), and a host of *functional attributes*, which provide additional information and requirements about the service that provide assistance when reasoning about several services with similar capabilities (these include guarantees of response time or accuracy, as well as the cost of the service).

Furthermore, a more detailed perspective on services is to view them as *processes*. The DAML-S representation of processes draws upon established work in a variety of fields, such as automated planning and workflow automation, and will support the representational needs of a broad array of services on the Web.

Ultimate Form of Interoperability

Given the automation aspect, the Semantic Web enables a radical departure from the current browsing paradigm of the Web. Indeed, we can foresee applications and usage situations in which browsing is not a viable alternative. On the other hand, one might argue that browsing still takes place, but by automated agents and not by human users. If we can represent information about Web-based services in a formal manner, the discovery of additional functionality needed by agents becomes easier. Generally, we call the process of matching one agent's description of *missing* functionality with descriptions of services offered by other agents *service discovery*, which resembles the act of advertising and querying. It is becoming an increasingly important aspect of distributed information systems.

In addition to the Web Service standards described earlier, a number of mechanisms for *low-level* service discovery are emerging – examples include Sun's Jini and Microsoft's Universal Plug and Play [e.g., Richard 2000] – but these mechanisms tackle the problem at a *syntactic level*, and like their higher-level relatives, they rely heavily on the standardization of a predetermined set of functionality descriptions.

Standardization, unfortunately, can only take us half way towards the goal of automated agents on the Web, as our ability to anticipate all possible future needs is limited. The Semantic Web offers a possibility to elevate the mechanisms of service discovery to a *semantic level*. Here, a more sophisticated description of functionality is possible, and a shared understanding between consumer and producer can be reached via the exchange of ontologies that provide the necessary vocabulary for discussion.

More importantly, it will also be possible to reach a "partial match" between a request and an advertisement, and still be able to take advantage of the discovered service [e.g., Sycara et al 1999]. An agent could enhance its functionality in the following manner:

1. Exchange ontologies to establish partial understanding,
2. Find a partial match of needs with services offered, and
3. Compose the exact required functionality from services discovered (using, e.g., automated planning or configuration techniques).

This approach takes information system interoperability well beyond what mere standardization or simple interface sharing enables, since it is based on deeper descriptions of service functionality and can be performed *ad hoc* and on demand. We call this *serendipitous interoperability* [Lassila 2002].

Eventually, the Semantic Web will 'break out' of the virtual realm and extend into the physical world. As we have observed, URIs can name anything, including physical entities, which means that we can use the RDF and DAML+OIL languages to describe people, their personal preferences, and various devices such as mobile devices and home appliances. Such devices can advertise their functionality – what they can do and how they are controlled – much like software agents. Being more flexible than low-level schemes such as Universal Plug and Play, the semantic approach opens up many exciting possibilities: Today, so-called *home automation* requires careful configuration for appliances to work together, but semantic descriptions of device capabilities and functionality will allow us to achieve such automation with minimal human involvement [Lassila 2002, Lassila & Adler 2002]. Concrete steps have already been taken here, with the work on developing a standard for describing functional capabilities of devices (e.g., screen sizes) and consumer preferences. Built on RDF, this standard called Composite Capability/Preference Profiles (CC/PP), will initially be used to allow mobile devices and other nonstandard Web clients to describe their characteristics so that Web content can be tailored for them on the fly [Klyne et al 2001]. With the full versatility of languages for handling ontologies and logic, devices could eventually automatically seek out and employ services and other devices for added information or functionality. Given all this, we can now imagine a Web-enabled microwave oven consulting the popcorn manufacturer's Web site for optimal popping parameters.

Conclusions

The power of the Semantic Web will be realized when we are able to automatically collect Web content from diverse sources, process the information, and exchange the results with other automated systems. The effectiveness of these systems (e.g., software agents) will dramatically increase as more machine-readable Web content and automated services (including other agents) become available. The Semantic Web promotes this synergy: even agents that were not expressly designed to work together can transfer data between each other when the data comes with semantics. This serendipity represents the ultimate form of system interoperability, and is a departure from the strict standardization approach towards mechanisms that are more flexible and allow operation in unanticipated situations.

References

[1] [Ankolenkar et al 2001] Anupriya Ankolenkar, Mark Burstein, Jerry R. Hobbs, Ora Lassila, David Martin, Sheila A. McIlraith, Srini Narayanan, Massimo Paolucci, Terry Payne, Tran Cao Son, Katia Sycara, and Honglei Zeng. "DAML-S: A Semantic Markup Language for Web Services." *Proceedings of the First Semantic Web Working Symposium* (SWWS'01), Stanford University, July 2001.

[2] [Berners-Lee et al 1998] Tim Berners-Lee, Roy Fielding, and Larry Masinter. "Uniform Resource Identifiers (URI): Generic Syntax." *Internet Draft Standard RFC 2396*, August 1998.

[3] [Berners-Lee et al 2001] Tim Berners-Lee, James Hendler, and Ora Lassila. "Semantic Web." *Scientific American* 284(5), (May 2001): 34-43.

[4] [Biron & Malhotra 2001] Paul V. Biron and Ashok Malhotra. "XML Schema Part 2: Datatypes." *W3C Recommendation 02 May 2001*, World Wide Web Consortium, Cambridge (MA) <http://www.w3.org/TR/xmlschema-2/>.

[5] [Brickley & Guha 2000] Dan Brickley and R.V.Guha. "Resource Description Framework (RDF) Schema Specification 1.0." *W3C Candidate Recommendation 27 March 2000*, World Wide Web Consortium, Cambridge (MA) <http://www.w3.org/TR/rdf-schema/>.

[6] [Burstein et al 2001] Mark Burstein, Jerry Hobbs, Ora Lassila, David Martin, Sheila McIllraith, Srini Narayanan, Massimo Paolucci, Terry Payne, Katia Sycara, and Honglei Zeng. "DAML-S 0.5 Draft Release." *DARPA Agent Markup Language Program draft document*, May 2001 <http://www.daml.org/services/daml-s/2001/05/>.

[7] [Dublin Core 1999] "Dublin Core Metadata Element Set, Version 1.1: Reference Description." *Dublin Core Metadata Initiative*, 1999 <http://purl.org/dc/documents/rec-dces-19990702.htm>.

[8] [Fikes & Kehler 1985] Richard Fikes and Tom Kehler. "The Role of Frame-Based Representation in Reasoning." *CACM* 28(9), (1985): 904-920.

[9] [Gruber 1993] Tom R. Gruber. "A translation approach to portable ontologies." *Knowledge Acquisition* 5(2), (1993): 199-220.

[10] [Heflin et al 2002] Jeff Heflin, Raphael Volz, and Jonathan Dale. "Requirements for a Web Ontology Language." *W3C Working Draft 07 March 2002*, World Wide Web Consortium, Cambridge (MA) <http://www.w3.org/TR/webont-req/>.

[11] [Hendler & McGuinness 2000] James Hendler and Deborah McGuinness. "The DARPA Agent Markup Language." *IEEE Intelligent Systems* 15(6), (November/December 2000).

[12] [Klyne et al 2001] Graham Klyne, Franklin Reynolds, Chris Woodrow, and Hidetaka Ohto. "Composite Capability/Preference Profiles (CC/PP): Structure and Vocabularies." *W3C Working Draft 15 March 2001*, World Wide Web Consortium, Cambridge (MA) <http://www.w3.org/TR/CCPP-struct-vocab/>.

[13] [Lassila 1998] Ora Lassila. "Web Metadata: A Matter of Semantics." *IEEE Internet Computing* 2(4), (1998): 30-37.

[14] [Lassila 2000] Ora Lassila. "The Resource Description Framework." *IEEE Intelligent Systems* 15(6), (November/December 2000): 67-69.

[15] [Lassila 2002] Ora Lassila. "Serendipitous Interoperability." In *The Semantic Web – Proceedings of the Kick-Off Seminar in Finland*, edited by Eero Hyvönen. University of Helsinki, 2002 (to be published).

[16] [Lassila & Adler 2002] Ora Lassila and Mark Adler. "Semantic Gadgets: Ubiquitous Computing Meets the Semantic Web." In "Spinning the Semantic Web," edited by Dieter Fensel et al. MIT Press, 2002 (to be published).

[17] [Lassila & McGuinness 2001]   Ora Lassila and Deborah L. McGuinness. "The Role of Frame-Based Representation on the Semantic Web." *Knowledge Systems Laboratory Report KSL-01-02*, Stanford University, 2001 <http://www.ksl.stanford.edu/KSL_Abstracts/KSL-01-02.html>.

[18] [Lassila & Swick 1999]   Ora Lassila and Ralph Swick. "Resource Description Framework (RDF) Model and Syntax Specification." *W3C Recommendation 22 February 1999*, World Wide Web Consortium, Cambridge (MA) <http://www.w3.org/TR/REC-rdf-syntax/>.

[19] [Merriam-Webster 1998]   *Merriam-Webster's Collegiate Dictionary*, Merriam-Webster, 1998.

[20] [Richard 2000]   Golden G. Richard III. "Service Advertisement and Discovery: Enabling Universal Device Cooperation." *IEEE Internet Computing* 4(5), (2000).

[21] [Sycara et al 1999]   Katia Sycara, Matthias Klusch, Seth Widoff, and Jianguo Lu. "Dynamic Service Matchmaking Among Agents in Open Information Environments." In *ACM SIGMOD Record* 28(1) (Special Issue on Semantic Interoperability in Global Information Systems), edited by A. Ouksel and A. Sheth, (March 1999): 47-53.

[22] [van Harmelen et al 2001]   Frank van Harmelen, Peter F. Patel-Schneider, and Ian Horrocks, eds. "Reference description of the DAML+OIL (March 2001) ontology markup language." *DARPA Agent Markup Language draft document*, March 2001 <http://www.daml.org/2001/03/reference.html>.

[23] [WebOnt] Available from: <http://www.w3.org/2001/sw/WebOnt/>.

IV MOBILE DOMAIN TECHNOLOGIES

- 4.1 Platform Layer
- 4.2 Mobile Internet Layer
- 4.3 Application Layer

Part 4.1

Platform Layer

- o The Symbian Platform
- o Carrier-Grade Linux
- o Wideband Code Division Multiple Access Technology
- o GSM/EDGE Radio Access Network
- o Wireless Local Area Networks
- o Bluetooth
- o Digital Video Broadcasting – Terrestrial Network

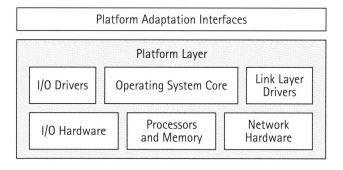

The Symbian Platform

In the past, most manufacturers of mobile devices and personal digital assistants implemented their own, proprietary operating system on their devices. There were few attempts to develop operating systems for devices with limited capabilities, and still today operating systems for mobile devices are not standardized. As the next generation of mobile devices is required to provide more and more features and capabilities, it is becoming increasingly difficult to extend these proprietary operating systems to support the additional features. Therefore, an increasing effort has been made by different mobile device manufactures to develop a standard operating system that would be suitable for a wide variety of mobile devices. The currently most successful non-proprietary operating systems are:

o   the Palm™ OS from 3COM®
o   the different versions of the Windows CE/Pocket PC/Pocket PC 2002 Operating System from Microsoft
o   the Symbian™ Operating System (Symbian OS™) from the Symbian Consortium

These operating systems are licensed to many Personal Digital Assistant (PDA) and mobile device manufacturers and new devices are being released almost every day. Recently, devices running a version of the Linux operating system have also appeared.

This article briefly introduces the Symbian OS, the Symbian Application Suite and the possibilities of software development on the Symbian platform.

Motivation

In recent years, the integration of computing and communication devices has increased enormously. Phones and computers became not only mobile, but are also more and more integrated into one (sometimes distributed) mobile device. Such mobile devices can be divided into two categories:

o   **primarily communication-oriented devices**, so-called *feature phones* or *smartphones* with limited means to input, process, store or transfer data or
o   **computing-oriented devices,** so-called *communicators* with enhanced data input capabilities, like a keyboard, more processing power and memory.

These two categories are also merging more and more, resulting in one mobile device with increasing capabilities for communication as well as computation. For such devices, the Symbian consortium developed the Symbian Operating System and the Symbian platform.

Mobile Domain Technologies – Platform Layer

Founded in June 1998, the Symbian consortium consists of the leading mobile device manufacturers Nokia, Motorola, Sony-Ericsson, Matsushita and the PDA manufacturer Psion. In addition, the Symbian platform is supported and licensed by an increasing number of manufacturers of mobile devices (as shown in the following figure).

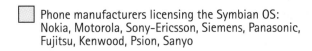

Phone manufacturers licensing the Symbian OS:
Nokia, Motorola, Sony-Ericsson, Siemens, Panasonic, Fujitsu, Kenwood, Psion, Sanyo

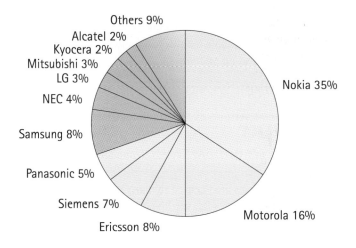

Worldwide Phone Sales in 2000 (Source: Gartner Dataquest)

Derived from the Psion PDA operating System EPOC, the Symbian OS is one of the longest coordinated efforts to develop an operating system for mobile devices. The current version of the operating system is version 6.1, which is used for example in the Nokia 7650, while an older version is used in the Nokia 9210 Communicator.

Overview of the Symbian Platform

The Symbian platform consists of

- o the operating system
- o a set of applications
- o support for different means of communications (e.g., voice, messaging and browsing)
- o support for personal computer (PC) connections (e.g., file browsing and conversion as well as synchronization and installation of additional software)
- o a framework for graphical user interfaces for different device families
- o an application architecture framework and application development support

To cope with the many different form factors of mobile devices, Symbian has defined different device families with different input/output capabilities (e.g., screens, pointing devices, data input devices). These so-called Device Family Reference Designs (DFRDs) enable the reuse of components within the same product family, the sharing of system resources and binary compatibility between different devices belonging to the same device family [EPOC5], [Task00].

In addition to the operating system, the Symbian platform provides an application suite for the most common PDA applications, like Personal Information Management (PIM), messaging, browsing, office applications and more.

The user interface framework eases the development and porting of applications and enables application to have the same look and feel as native applications [Symb01]. At the same time, the application architecture supports the development of document-centric and task-oriented applications. To support the development of further applications, the Symbian platform also contains Software Development Kits (SDKs) for applications written in C++ or Java™ as well as tools for the development of applications allowing a connection to a PC.

Symbian Operating System

At the beginning of November 2000, Symbian launched its sixth version of the operating system, optimized for pen- and keyboard based Communicators supporting color displays, Bluetooth, Wireless Access Protocol (WAP), HyperText Markup Language (HTML), Short Message Service (SMS), Java and e-mail [Symb01].

The Symbian OS is designed for devices with limited energy capacity, memory and processing power, especially Smartphones and Communicators. Current hardware requirements are a 32 bit CPU, like the StrongARM processor, and a processor frequency of 36 MHz or more [EPOC5].

The most important features of the Symbian OS are:

o   a component-based operating system that was designed from the *ground up* to be fully object-oriented

o   a robust, modern and well-architected 32-bit microkernel with pre-emptive multi-tasking, which allows processes to provide their own memory protection

o   support of client-server architecture, which ensures the stability of the system even in the case of a problem on the client side of an application

Due to these operating system features as well as a flexible object oriented application architecture, the Symbian OS can satisfy the consumer's need for an operating system which does not lose consumer data and provides a consistent and simple user interface.

Generic Technologies

Symbian's Generic Technology (GT) is the common core of Application Programming Interfaces (APIs) and technology, which is shared between all Symbian reference designs. The use of the same Generic Technology for different reference designs enables the simultaneous development of different reference designs for different mobile devices. Furthermore, it eases the porting of applications between reference designs, since those parts of an application which use Generic Technology APIs do not need to be changed [SyCr01], [SyQu01], [Sym71].

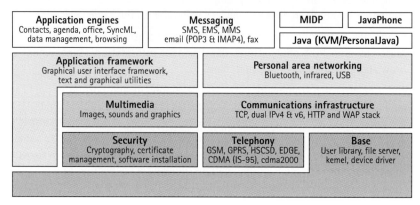

Architecture of Symbian OS, version 7

The major components of the Symbian's Generic Technology are illustrated in the previous figure [SDN01], [Task00].

o **Base:** The fundamental execution environment, the tools required to build it, and lowest-level security.

o **Multimedia server:** Audio recording and playback, and image-related functionality.

o **Communication:** Wide-area communications stacks including Transport Control Protocol/ Internet Protocol (TCP/IP), Global System for Mobile Communications (GSM), General Packet Radio Service (GPRS) and Wireless Application Protocol (WAP).

o **Application framework:** Middleware APIs for data management, text, clipboard, graphics, internationalization, and core Graphical User Interface (GUI) components

o **Personal Area Networking (PAN):** Personal-area communications stacks include infrared, Bluetooth and a serial port.

o **Application engines & services:** Engines for contact management, schedule and to-do list management, browsing engines and other applications

o **Messaging:** Internet e-mail, SMS text messages and fax.

o **Java execution:** Mobile Information Device Profile (MIDP™) as well as PersonalJava™ (Java Virtual Machine (JVM)-based execution environment with JavaPhone™ APIs.)

An additional component on both the PDA and PC side is responsible for Connectivity. It contains converters and viewers for foreign data formats (e.g., Microsoft Office documents) or e-mail attachments as well as a communications framework for connecting with a PC running Symbian Connect for synchronization purposes.

The key features of Symbian's Generic Technology are [SDN01]:

- support for two communicator reference designs - tablet- or keyboard-based
- support for four program and content development options (e.g., C++, Java, WAP and Web)
- close integration of functions required by mobile devices: contact information, messaging, browsing and mobile telephony
- communication protocols: TCP/IP, WAP, GSM, GPRS, Bluetooth, IrDA and serial
- security support: full-strength encryption and certificate management, secure communications protocols, certificate-based application installation
- a rich suite of application engines, including contacts, schedule, messaging, browsing, voice, office, utility and system control
- worldwide locale support through Unicode characters, flexible text input framework, and additional font and text formatting support
- support for data synchronization
- media server supporting audio recording and playback as well as imaging
- full support for the MIDP and PersonalJava application environment

Device Family Reference Designs

In order to deal with the wide variety of mobile devices, Symbian developed the concept of Device Family Reference Designs. The idea is the availability of similar resources within the devices that are running the same DFRD, thus allowing the binary compatibility and therefore the transfer of application between devices of the same family as well as the use of shared components within one device [SyCr01], [SyQu01].

Mobile Domain Technologies – Platform Layer

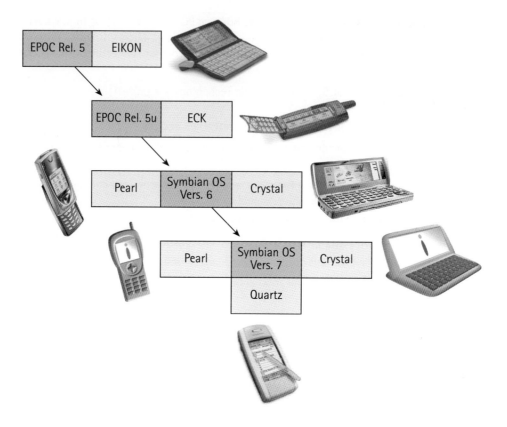

Symbian OS history and DFRDs

From version 6 onward, Symbian defined three different DFRDs for the Symbian OS (as shown in the previous figure):

- o the Crystal DFRD is Symbian's reference design for large-screen keyboard-oriented communicators. Recommended screen size is about half VGA (640x240 pixels) with options available for color or gray scale depending on the manufacturer. Crystal supports a full QWERTY keyboard and the options are built-in to have either a pen interface, which is similar to traditional Psion devices, or a button-oriented interface
- o the Quartz DFRD is designed for devices with a quarter VGA portrait color touchscreen screen, 240 pixels wide and 320 pixels high, that are pen operated and have no keyboard but instead hardware keys for cursor functionality. It also supports devices with a built-in phone, speaker, microphone, and perhaps additional keys
- o the Pearl DFRD, which is intended for devices with a rather limited user interface, such as a keypad and only a small screen, but that still have reasonable processing power. For such devices Symbian does not develop the user interface, which is done instead by the licensee

Nevertheless, the core of all these different DFRDs is General Technologies, including the kernel, middleware for communications, data management and graphics, the lower levels of the GUI framework and application engines.

Application Suite

The Symbian platform also includes an application suite optimized for the needs of mobile devices such as Smartphones and Communicators, and for handheld, battery-powered, computers [EPOC5]. The applications can be categorized into:

- **Communications applications:** e-mail, fax, WAP and Web browsing, SMS
- **Personal Information Management applications:** contacts, calendar, world map, alarm
- **Office applications:** word, spreadsheet, presentation viewer, database, spell checker, jotter
- **Utilities:** calculator, imaging, voice notes, video player, games
- **System:** control panel, file manager, print, installer
- **Connectivity:** PC connect, server synchronization, file conversion, application installation

Application Development

Together with the Symbian Application Framework, Symbian also offers Software Development Kits (SDKs) for C++ and Java. Furthermore, there is a separate development kit for the development of connectivity applications allowing data transfer between a mobile device and a PC [SDN01].

The C++ and the Java SDKs contain debug and release versions of emulators for one DFRD, development, debugging and deployment tools as well as tools for building installation images and localizing applications [SyCr01], [SyQu01]. All that is necessary to start developing your own applications is an Integrated Development Environment (IDE), as well as in the case of Java development, a Java compiler.

The Java execution environment implemented for the Symbian OS is a port of the Java execution environment from SUN. It offers a full enterprise and JavaPhone API set as well as full support of Java Native interfaces to offer C++ APIs to Java applications [Alli01].

To add additional devices and use their functionality in an application, it is necessary to write a device driver for that device for the Symbian OS. For this purpose it is necessary to obtain a Device Driver Development KIT from Symbian [SDN01].

Since the Symbian OS uses unique 32-bit identification numbers (UIDs), to identify file types as well as the structure of a file (e.g., executable, dynamically-loadable library or a type of storage or some other file format), it is necessary during the building process to provide such a UID for an application. For experimental purposes it is possible to use UIDs in the range from 0x01000000 to 0x0FFFFFFF, though to release an application it is necessary to be allocated a UID through the Symbian Developer Network [SDN01], [Uid01]. The release of applications with experimental UIDs is not allowed.

Conclusions

The Symbian platform depicts an operating system as well as an application suite, development tools and an application framework. Altogether the Symbian platform is extremely well-suited to mobile devices with limited computational communications and storage capabilities as well as limited energy resources. Supported by the major players in the mobile device market, it has very good potential to become the most important operating system for future mobile devices.

The next version of the Symbian platform, version 7, will extend the current version with additional technologies, such as advanced security features, GPRS-based packet data as well as full Bluetooth functionality [Symb01], [Sym71].

References

[Task00] Martin Tasker & al.: Professional Symbian Programming. Wrox Press, 2000

[FoNo01] http://www.forum.nokia.com

[Symb01] http://www.symbian.com

[SDN01] http://www.symbiandevnet.com

[EPOC5] EPOC R5 SDK documentation and sample applications

[SyQu01] Symbian Platform version 6 Quartz DFRD SDK

[SyCr01] Symbian Platform version 6 SDK for Nokia 9210 Communicator

[Uid01] uid_ew@symbian.com

[Sym71] Symbian OS Version 7; Functional description

[Alli01] Jonathan Allin: Wireless Java for Symbian Devices

Carrier-Grade Linux for the Mobile Domain

The telephony network was built to be reliable, and was designed so that you could always call for help when there was a fire or an accident. If you had a working connection to the network, it would connect your call. Such a network needed redundancy, so that a stray excavator would not bring it down. But also the nodes in such a network had to be reliable and function correctly, e.g., to connect a call and to be highly available to receive commands.

Systems satisfying such requirements are called carrier-grade. Initially, the requirement was that, excluding natural disasters, there should be no more than 30 minutes of unscheduled downtime per year [IT1993]. This means that the system should be available 99.994 % of the time. Today, this requirement is often rounded up to 99.999 %.

There are strict rules on how the unscheduled downtime is to be calculated [IT1993]:

o   Both hardware and software failures count for downtime. Since hardware failures are still quite common and faulty hardware is slow to replace - someone has to go to the site, analyze the fault, and replace the hardware - it should be replicated.

o   There are soft real-time deadlines for processing. These are in the millisecond range. In other words, the software is supposed to finish up whatever it starts doing within a few tens of milliseconds.

o   The strictest requirement is for availability. This means that under normal circumstances, the caller must be able to complete a call after a few tries, and failures count as unscheduled downtime.

o   Before the call attempt is completed, the system can drop the call with no penalty.

o   However, calls that are already ongoing should not be dropped. The requirement is such that a system should not drop all ongoing calls more than once a month.

o   Under overload, the system is expected to show graceful degradation in performance. This means that the system continues to connect calls until it is seriously overstressed. In practice, this is handled so that the system will do a minimum of processing before it decides whether it will continue to process the call attempt.

o   Emergency calls must always go through.

These requirements have some consequences for the architecture of such a system. First of all, the hardware redundancy means that there is some excess computing capacity. This can be used to protect against software failures, too. Software failures are difficult to protect against because they are by definition unexpected. This makes error recovery a challenging task: to recover from an error, one must understand what the error was in the first place and what caused it. Then it could be easier to prevent the error than to try to recover from it. In practice, error recovery code is so complex, is executed so seldom, and is so difficult to test that is has more bugs per line of code than the average.

Mobile Domain Technologies – Platform Layer

With the emergence of the Mobile Internet, the Web domain and the Mobile domain are coming closer together. As Linux has been used in Web domain servers with success, it is natural to ask how well it would now perform in the Mobile domain servers. This chapter introduces the major changes that are required.

Support for Replication

The customary way to build such reliable, highly available network elements is to use replication. This can be 2n (meaning that all hardware elements have a pair) or n+1 (meaning that there are some extra units). The replicated units are connected with a redundant bus, such as an Ethernet connection. For the software, this implies that it must support dual connections between units: if one of the buses stops working, the system must start using the other one without delay.

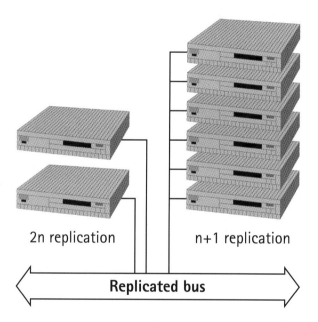

Replication schemes

The system must also support permanent names for the units. Then it will be possible to remove a defective unit and replace it with a good one bearing the same code, so that the other units will then recognize that this is the pair of a certain 2n replicated unit.

The kernel will also be affected by replication. A computer unit will have two network interface cards connected to two different local networks. Its clients are supposed to use one of these, depending on which one is working. However, normally Internet Protocol (IP) addresses are associated with network interfaces. There are now different possibilities:

  o   The client will use just one IP address. If one of the network interfaces fails, the other will take over its IP address. This is how the Linux Virtual Server [LV2002] works.

o   Each network interface will have a different IP address, and the client will use some query mechanism to find out which one it should be using.

o   The client will use the Stream Control Transmission Protocol (SCTP) [SC2002] to communicate with the server. It supports multi-homing, which means that a single SCTP endpoint can have several IP addresses.

All of these alternatives require some modifications to the standard kernel. Moving an IP address from one network interface to another requires some code to accomplish this, especially if it is to be done fast. Finding which IP address to use is also best done inside the kernel, as it could otherwise become a performance bottleneck. Finally, SCTP is a new protocol which must be added to the kernel.

Real-Time

As we discussed earlier, the carrier grade Linux kernel will need to support soft real-time properties. Ideally, the kernel should be preemptable and have predictable and low latency. However, the requirements for latency are not really challenging: Intel x86 hardware is capable of 20 microsecond response times [KD2002], whereas 1 millisecond latency is probably enough for carrier grade. Therefore, we are not willing to pay a performance penalty to get the best possible response time.

As to current Linux latency performance, the Linux 2.4.17 kernel response time can be over 100 milliseconds 0.01% of the time [RH2002], which is not good enough. A low-latency patch can improve this dramatically, so that the maximum in a short test is 1.3 milliseconds. Combining low-latency and preemption patches improves this a little bit more, and also reduces the worst measured latency in a long test to 1.5 milliseconds [RH2002].

Real-time programming also requires special operating system interfaces, many of which are in the POSIX™ standard [BG1994]. The requirements are that it must support real-time scheduling and it must have real-time primitives, such as semaphores which support priority queues.

The kernel should also have predictable performance characteristics when it interfaces with the hard disk. Thus, it should be possible to load programs completely into main memory when they are started and to keep them there. For access to data on the disk, the kernel should support a real-time file system.

The network driver Application Programming Interface (API) will need some changes to support real-time behavior. Currently, when a Linux system receives too many packets, it will spend too much time in interrupts. Then it cannot get any useful work done and does not even handle the optimal number of packets. A better solution was discussed earlier: the packets should be dropped as early as possible with little processing. In order to do this, the network driver will need to know when the system is overloaded. Then it should not generate an interrupt for every packet. These changes will probably be made to the 2.5 kernel [JC2001].

Symmetric Multi-Processing (SMP) support can also interfere with real-time performance. Process migration can make performance and scheduling analysis more difficult; thus it should be possible to lock processes to a named Central Processing Unit (CPU) or to a couple of CPUs. Even better would be to assign interrupt handlers to a single CPU: since an interrupt handler always has a high-priority part, the CPU is pre-empted from whatever it was doing. If a single CPU would take care of interrupt handling during an overload situation, the others can still get some useful work done.

Embedded

The carrier-grade Linux systems will be clusters of computer units. Most of these execute software with soft real-time requirements. This means that they have little use for virtual memory. In practice, most of them also do not need permanent storage. This means that the computer units would use a hard disk only for loading software when they are starting up. Since one large hard disk is cheaper than ten small ones, financial considerations imply that most computer units must be able to boot from the network. The software is also easier to manage and upgrade if it is stored centrally in the cluster.

The second requirement is that the computer units must start up fast. Embedded devices live in a fairly stable environment, unlike a desktop system where the consumer can plug in a new type of mouse at any moment. Thus, most options can be fixed when the system is created.

Monitored

A replicated system needs to know which hardware units are working and which are not. It should also notice if a software entity stops working correctly. Usually, watchdogs are used for this purpose: an application needs to send a message to a watchdog during every time interval. The watchdog can also poll the applications directly: the application then needs to respond to it within a given time. How often an application needs to assure its health to the watchdog and how quickly it needs to respond to watchdog queries must be configurable.

In a highly available system, the most important applications are running all the time. This makes it important that they do not leak any resources. To help build such systems, the kernel must provide visibility to all the finite resources it manages. If any of these runs out, at least one application must be restarted and its service is lost.

Kernel and device driver panics are also important to analyze, as they can point to complicated defects in the software. Therefore, there should be extended features to analyze system images after the unit has rebooted.

Fault Resistant

Of course, kernel and device driver panics should occur as seldom as possible. As many kernel panics as possible should be removed from the code. They should be replaced by code to analyze and remedy the failures that caused them. Only if this fails should a panic be used. In other words, the kernel and the drivers should be as hardened as possible.

To ensure that the system really is as fault-tolerant as is required, it is best to have some fault-injection mechanisms included. Only then is it also possible to test the error-correction software.

If the system fails despite all precautions, it is useful to find out precisely what happened. This is useful especially for hard-to-find bugs, which cannot be easily replicated. There needs to be some mechanism to allow the Linux kernel to write all important information to a persistent location after a kernel panic.

Conclusions

Replicating performance comparable to the fixed line telephone network with Linux servers may sound implausible at first. We have looked at the requirements that a carrier-grade Linux would have to satisfy. None of them are impossible. On the contrary, for most of them development effort is already ongoing. The success of Linux in carrier-grade business will now depend on how well and fast the carrier-grade features find their way to Linux kernels.

References

[RH2002] Clark Williams, "Linux Scheduler Latency." March 2002. Red Hat white paper, available from www.redhat.com.

[IT1993] "Digital Exchange Design Objectives - General." ITU-T Recommendation Q.541.

[JC2001] J. Corbet, "The Linux 2.5 kernel summit." April 2, 2001. LWN feature article. Available from LWN.net

[KD2002] Kevin Dankwardt, "Real-Time and Linux." ELJonline, January 2002. Available from LinuxDevices.com.

[BG2002] Bill O. Gallmeister, "POSIX.4: Programming for the Real World." O'Reilly 1994.

[SC2002] Lyndon Ong, "SCTP: An Overview." Available as www.sctp.org/sctpoverview.html.

[LT1999] Linus Torvalds, "The Linux Edge." Pages 101-111 in Chris DiBona, Sam Ockman, and Mark Stone (eds.), "Open Sources. Voices from the Open source Revolution." O'Reilly 1999.

[LV2002] www.linuxvirtualserver.org

Wideband Code Division Multiple Access Technology

This section briefly covers the main technical aspects of the Universal Mobile Telecommunication System (UMTS) Terrestrial Radio Access Network (UTRAN), specifically from the point of view of Wideband Code Division Multiple Access (WCDMA) technology. As the UMTS radio interface, WCDMA technology is going to be the most widely used technology for IMT-2000. WCDMA radio technology within the UTRAN is standardized by the 3rd Generation Partnership Project (3GPP).

Air Interfaces for IMT-2000

Within International Telecommunications Union (ITU), several different air interfaces are defined for third generation systems, based on either Code Division Multiple Access (CDMA) or Time Division Multiple Access (TDMA) technology. WCDMA is the main third generation air interface and will be deployed in Europe and Asia, including Japan and Korea, in the same frequency band, around 2 GHz. The wide deployment of WCDMA will create large markets for manufacturers and the providers of content and applications.

In North America, the spectrum around 2 GHz has already been auctioned to carriers using second generation systems, and no new spectrum is available for the new systems. Thus, third generation services there must be implemented within the existing bands by replacing part of the spectrum with third generation systems, an approach referred to as re-farming.

In addition to WCDMA, the other air interfaces which can be used to provide third generation services are Enhanced Data Rates for Global Evolution (EDGE) and multi-carrier Code Division Multiple Access (CDMA) (CDMA2000). These solutions mainly provide third generation services within existing frequency bands, for example, in North America.

The expected frequency bands and geographical areas where these different air interfaces are likely to be applied are shown in the next figure.

156 Mobile Domain Technologies – Platform Layer

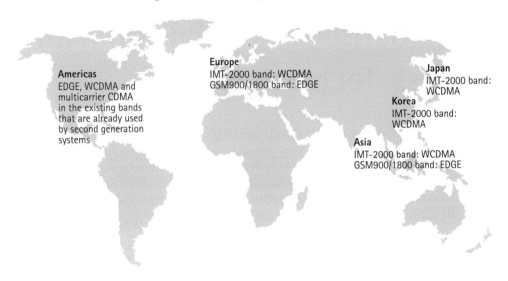

Expected air interfaces and spectrums for providing third generation services

Spectrum for Universal Mobile Telecommunications System

Work to develop third generation mobile systems started at the 1992 meeting of the World Administrative Radio Conference (WARC), where it identified the frequencies around 2 GHz which were available for use by future third generation mobile systems. Spectrum allocation in Europe, Japan, Korea and the USA is shown in the next figure. In Europe and in most of Asia, the IMT-2000 bands of 2 × 60 MHz (1920-1980 MHz plus 2110-2170 MHz) will be available for WCDMA Frequency Division Duplex (FDD). Availability of the Time Division Duplex (TDD) spectrum varies: in Europe it is expected that 25 MHz will be available for licensed TDD use in the 1900-1920 MHz and 2020-2025 MHz bands. The rest of the unpaired spectrum is expected to be used for unlicensed TDD applications (e.g., Self Provided Applications (SPA)) in the 2010-2020 MHz band. FDD systems use different frequency bands for uplink and for downlink, separated by the duplex distance, while TDD systems utilize the same frequency for both uplink and downlink.

WCDMA Technology

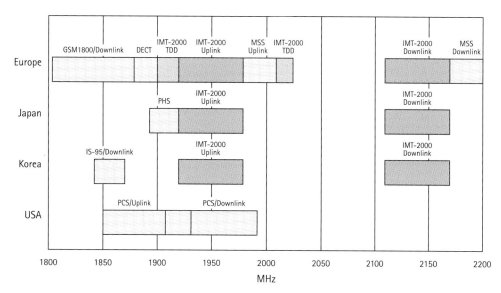

Spectrum allocation in Europe, Japan, Korea and USA

A few example UMTS licenses are shown in the table below for Japan and Europe. The number of UMTS operators per country is between 4 and 6.

Examples of the first phase UMTS licenses allocated:

Country	Number of operators	Number of FDD carriers (2 x 5 MHz) per operator	Number of TDD carriers (1 x 5 MHz) per operator
Finland	4	3	1
Japan	3	4	0
Spain	4	3	1
UK	5	2-3	0-1
Germany	6	2	0-1

Note: In Japan only 3 carriers per operator were released for use initially.

The UMTS licenses were given either based on the operator's capabilities to provide services (beauty contest) or were auctioned. Finland, Japan and Spain had a beauty contest while UK and Germany had an auction.

More frequencies have been identified for IMT-2000. At the WARC-2000 meeting of the ITU in May 2000, the following frequency bands were identified for IMT-2000 use:

- 1710-1885 MHz
- 2500-2690 MHz
- 806-960 MHz

It is worth noting that some of the bands listed, especially below 2 GHz, are partly used with systems like the Global System for Mobile Communications (GSM). What shall be the exact duplexing and other arrangements is under discussion at the moment. The feasibility studies for the different duplex arrangements with the new bands have been initiated as part of the 3GPP activities in connection with WCDMA evolution work.

Radio Access Network Architecture

WCDMA technology has shaped the WCDMA radio access network architecture due to the requirements of CDMA basic features, such as soft handover. In the next figure the basic UTRAN architecture is shown, indicating the connections to the core network for both circuit-switched traffic (Iu-CS) and packet-switched traffic (Iu-PS). The existence of an open, standardized Iur interface is essential for proper network operation, including soft handover support in a multi-vendor environment.

Another important trend on the architecture side is IP-based technology, which is making its way to the radio access side in addition to the core network. For example, IP-based transport is currently in the final stages of 3GPP standardization. Furthermore, UTRAN internal interfaces (Iub/Iur/Iu-CS) support IP as the transport in addition to the Asynchronous Transfer Mode (ATM) technology in 3GPP Release 99/Release 4. IP already exists on Iu-PS.

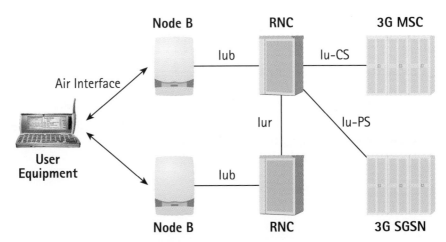

WCDMA radio access network architecture

WCDMA Basics

The WCDMA air interface is based on CDMA technology. All users share the same carrier, and also share this carrier's power. The characteristic feature is the wide 5 MHz carrier bandwidth over which the signal for each user is spread, as illustrated in the next figure. The transmission bandwidth is the same for all data rates, with the processing gain being larger for the smaller data rates than for the higher data rates. This processing gain protects against interference from other users active on the same carrier. In the receiver, despreading separates the transmitted and spread signal for data detection.

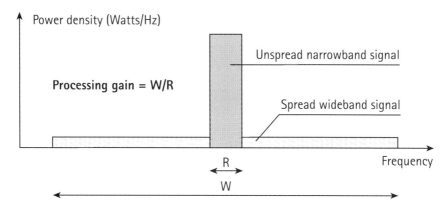

CDMA principle used in WCDMA

CDMA technology enables two key features, fast power control and soft handover. They both contribute to WCDMA system capacity but are also required for proper system operation. Fast power control, especially in the uplink, is required so that consumers do not generate extra interference and do not block the reception of the signals from other consumers. Without power control, a mobile device transmitting near the base station would block the reception of the other consumers further away if it exceeds the processing gain, also known as the near-far problem in CDMA. The power control command rate in WCDMA is set to 1500 Hz with typically a 1 dB step either up or down.

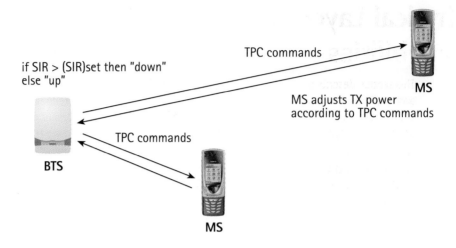

WCDMA fast power control principle

The soft handover is required for similar reasons. In a soft handover, a mobile device is connected simultaneously to two or more cells on the same frequency. Especially in the uplink, this is again vital since otherwise a mobile device between two cells could cause problems to the cell to which it is not connected. In a soft handover, all cells provide power control information to the mobile device and the near-far problem is avoided.

Soft handover

Physical Layer and Mobile Device Capabilities

This section describes the main differences between the third and second generation air interfaces and explains the key physical layer parameters of WCDMA. Also, it examines mobile device capabilities from the point of view of the air interface, and how these capabilities impact what services each type of mobile device can support.

Here we are considering GSM and IS-95 second generation air interfaces, which are based on TDMA and CDMA, respectively. Second generation systems were built mainly to provide speech services in macro cells. To understand the differences between second and third generation systems, we need to look at the new requirements of the third generation systems, which are listed below:

o   Bit rates up to 2 Mbps
o   Variable bit rate to offer bandwidth on demand
o   Multiplexing of services with different quality requirements on a single connection (e.g., speech, video and packet data)
o   Varying delay requirements
o   Quality requirements from a 10% frame error rate to a 10^6 bit error rate
o   Coexistence of second and third generation systems and inter-system handovers
o   Support of asymmetric uplink and downlink traffic
o   High spectrum efficiency
o   Coexistence of FDD and TDD modes

The WCDMA air interface has been defined to provide, in the first phase, data rates up to 2 Mbps in 3GPP Release 99 and Release 4. For Release 5, peak data rates up to 10 Mbps are possible with the High Speed Downlink Packet Access (HSDPA) feature.

The WCDMA way of sharing resources, with the main resource being the power shared among users, is key when dealing with variable bit rate services. When there is no fixed allocation of the physical resources that would be used by mobile devices, even when there is no data to transmit, capacity sharing on the carrier can be done very fast, even on a 10 ms basis, or, with HSDPA, even on a 2 ms basis. This power-based resource sharing is fully valid in the uplink, but for the downlink, more attention needs to be paid to code resource usage when the peak and average data rates differ. Methods have been defined for efficient use of downlink resources, for example, with a shared channel concept for downlink packet data. The fundamental principle of WCDMA resource sharing is illustrated in the next figure.

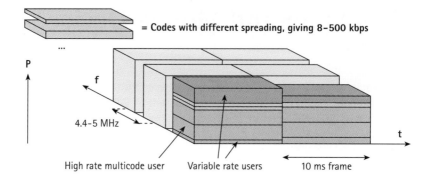

Dynamic resource sharing with WCDMA

For other service requirements, such as different delay requirements, WCDMA provides from the physical layer point of view different Transmission Time Intervals (TTI), the value being configurable to 10, 20, 40 or 80 ms. Multi-service capability is provided for fairly simple mobile devices by providing the means to multiplex different services with different QoS requirements on the same connection. The simplified WCDMA service multiplexing principle is illustrated in the next figure. In the actual multiplexing chain there are additional steps, such as rate matching and interleaving functions, which are necessary for proper system operation.

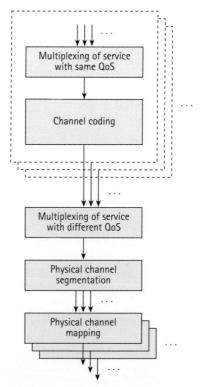

WCDMA principle of providing different QoS on a single physical channel

For co-existence with GSM, the handover as well as the necessary measurements have been specified to allow continued GSM service in WCDMA or vice-versa for the service set specified on the GSM side.

WCDMA can provide uplink and downlink data rates independently of each other, thus facilitating uplink and downlink asymmetry on a per connection basis. With variable duplex spacing it is also possible to configure less spectrum for the uplink if this is necessary, especially with some of the extension bands which will likely be available in the future. Since the current IMT-2000 frequency allocation also provides an unpaired spectrum, the TDD mode of operation has been specified where users share the same carrier in both the uplink and downlink transmission direction. There, the TDMA principle has been combined with CDMA with users sharing the time slot with spreading codes as illustrated in the next figure. Each time slot can be allocated either to the uplink or downlink direction with some limitations. Also, the (wideband) TDD mode of operation uses the 5 MHz bandwidth and the same solutions as WCDMA in FDD mode, and has been applied where applicable to the TDD mode as well for maximal commonality. Release 4 also includes a 1.28 Mbps TDD option. The schedule for commercial use of TDD options is currently open.

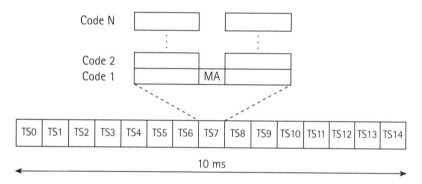

WCDMA TDD mode of operation with TDMA/CDMA principle applied

To achieve high spectrum efficiency, in addition to diversity methods due to the bandwidth, WCDMA also incorporates methods not used in second generation standards, such as transmit diversity, in the downlink direction. Also, advanced channel coding methods provide coding gain to achieve low Signal-to-Interference Ratio (SIR) requirements for the desired quality level.

WCDMA does not use the same principle as GSM with the Terminal class. Upon setup connection, WCDMA mobile devices tell the network a large set of parameters which indicate the capabilities of the particular mobile device. It is worth noting that mobile device capabilities can be given independently for the uplink and downlink direction. To provide guidance on which capabilities should be applied together, example capability classes have been specified:

o   **32 kbps class:** This is intended to provide basic speech service, including Adaptive Multi-Rate (AMR) speech service as well as some data limited data rate capabilities, but not together with the speech service.

- **64 kbps class:** This is intended to provide speech and data services, also with simultaneous data and speech capability, e.g., AMR speech and 32 kbps data simultaneously.
- **128 kbps class:** This class has the air interface capability to provide video telephony or various other data services.
- **384 kbps class** is being further enhanced from 144 kbps and has multicode capability, demonstrating WCDMA's support of advanced packet data methods. This class can provide, e.g., AMR speech simultaneously with packet data up to 384 kbps.
- **768 kbps class** has been defined as an intermediate step between the 384 kbps and 2 Mbps class.
- **2 Mbps class:** This is the state of the art class and the 2 Mbps capability has been defined for the downlink direction only.

In Release 5 there are also separate capabilities for HSDPA ranging from approximately 1.2 Mbps to 14 Mbps downlink peak data rates with the HSDPA feature.

The intention is also that each example class can provide the services provided by the classes with lower data rate capabilities. A mobile device can provide a particular capability level with FDD, TDD or both.

What is additionally worth noting is that in WCDMA there are several performance improvement features which are mandatory in the mobile device, allowing developers to take these features into account in the first phase network capacity/coverage planning as they are supported in all mobile devices from the beginning. An example of such a feature is downlink transmit diversity.

As part of the WCDMA evolution, mobile device capabilities are going to develop together with new features and take into account the possibilities of new features as well as provide information to the network on whether a particular enhancement is supported or not. As a part of Release 5 work, peak data rates beyond 10 Mbps are possible.

High Speed Downlink Packet Access

The major addition to the WCDMA radio in Release 5 is the HSDPA feature, with the impact to the network protocol coming from HSDPA shown in the following figure. Node B intelligence is increased for the handling of the retransmissions and scheduling functions, thus reducing the roundtrip delay between mobile device and the network entity handling retransmissions. Overall this makes retransmission combining feasible in the mobile device due to reduced memory requirements. The HSDPA feature and its performance are covered in more detail in [2].

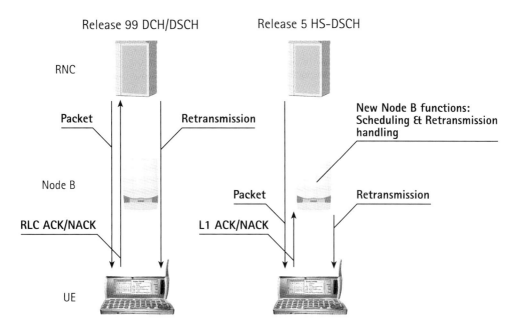

Release 5 HSDPA principle

Air Interface Performance

From the perspective of the consumer, the quality of mobile service may be limited by a weak signal, i.e., by the coverage, or by too many consumers trying the access the service at the same time, i.e., by the capacity of the system.

Coverage is important when the network is not limited by capacity, such as at the time of initial network deployment, and typically in rural areas. Macro cell coverage is determined by the uplink range from the mobile device to the base station, because the transmission power of the mobile device is much lower than that of the base station. The output power of the mobile device is typically 21 dBm (125 mW) and that of the macro cell base station 40-46 dBm (10-40 W) per sector.

To understand the typical cell areas of WCDMA cells, we can compare the WCDMA cell area to the existing GSM cell areas, a comparison which is also important since GSM operators will utilize their existing cell sites for their UMTS networks. With existing GSM1800 sites, WCDMA offers bit rates of 64-144 kbps in uplink with the same coverage probability as for GSM1800 speech. In downlink, the offered bit rates are higher, typically up to 384 kbps. With GSM900 sites, WCDMA offers bit rates of about 32 kbps in uplink with the same coverage probability as for GSM900 speech, assuming that simple smart antenna solutions are used in WCDMA. The cell size in GSM900 can be larger than in GSM1800 because the signal attenuation is lower at lower frequencies. On the other hand, the GSM site density in most urban areas is higher to provide more capacity, and therefore, reusing existing GSM sites will provide a high quality WCDMA network.

The typical site areas for GSM1800 and for WCDMA will be 2 km² in urban areas, 5-10 km² and 20 km² in rural areas, assuming a 3-sector site. The exact maximum site area depends on the environment and on the required coverage probability.

When traffic increases in the network, a higher base station density may be needed. The typical capacity of the WCDMA air interface is 1 Mbps per carrier per sector per 5 MHz carrier in 3-sector macro cells and 1.5 Mbps micro cell per 5 MHz carrier. This value represents a maximum total throughput per sector, which can be shared between users requesting service at the same time. Radio resource management algorithms in the radio network controller allocate the capacity between the users, and this allocation can be based on the requested services and their priorities.

Typical capacities of the WCDMA air interface (Release 99):

	Macro cell layer	Micro cell layer
Capacity per site per carrier	3 Mbps with 3 sectors	1.5 Mbps
Capacity per site per operator with 3 UMTS FDD carriers	9 Mbps	4.5 Mbps
Maximum site density in urban areas	5 sites / km²	30 sites / km²
Maximum capacity per operator	45 Mbps / km²	135 Mbps / km²

WCDMA will in most cases be deployed together with the GSM network, and handovers are supported from WCDMA to GSM and from GSM to WCDMA. In the initial deployment phase, WCDMA network coverage will not be as large as GSM network coverage and handovers from WCDMA to GSM can be used to provide continuous service even if the consumer leaves the WCDMA coverage area in the connected mode. Also, idle mode parameters are supported for inter-system cell reselection.

Inter-system handovers are triggered by the source radio access network, and the handover triggers are vendor specific. An inter-system handover can also be triggered due to high load to balance loading between WCDMA and GSM to fully utilize GSM and WCDMA networks together.

Conclusions

This chapter briefly covered air interfaces, spectrum, WCDMA basic physical layers and mobile device capabilities, as well as WCDMA performance. WCDMA technology offers a wide range of possibilities for service introduction, including means for providing different quality of service functions as well as means for providing multiple services simultaneously. The radio access network architecture has been shaped by WCDMA requirements and standardization in 3GPP has produced the specifications to enable proper system operation in a multi-vendor environment with open internal interfaces in the access network. The WCDMA networks can use advanced performance enhancement schemes from the start as they are mandatory for all mobile devices. The first step in the evolution of WCDMA technology during 2001/2002 has further boosted especially packet data capacity, making WCDMA more efficient to carry IP-based packet traffic.

The WCDMA radio access technology of UMTS is covered more in-depth in [1] and the references therein. The latest WCDMA radio specifications developed by the 3GPP TSG RAN can be found on the 3GPP Web site [2].

References

[1] H.Holma & A.Toskala, "WCDMA for UMTS," Second Edition, John Wiley & Sons, 2002

[2] www.3gpp.org

GSM/EDGE Radio Access Network

The Global System for Mobile Communications (GSM)/ Enhanced Data Rates for Global Evolution (EDGE) standard is taking steps to define a fully capable third generation (3G) mobile radio access network. Third Generation Partnership Project (3GPP) Release 5 specifies a GSM/EDGE Radio Access Network (GERAN) which can connect to a 3G mobile core network through the Iu interface and is capable of offering the same set of services as the UMTS Terrestrial Radio Access Network (UTRAN).

GSM Evolution towards 3G/UMTS

The GSM standard has evolved from a basic voice service to a wide variety of speech and data services over the course of a decade. The GSM/EDGE Release 1999 introduced 8-Phase Shift Keying (8-PSK) modulation providing a threefold increase in the peak data rates. The addition of 8-PSK modulation was made without modifications to the established GSM/General Packet Radio Service (GPRS) architecture. Enhanced GPRS (EGPRS) is an enhancement to GPRS and uses the Gb interface (which connects GSM Base Station Controller (BSC) to Serving GPRS Support Node (SGSN)) in communication with the packet switched core network.

The next major step in the GSM/EDGE evolution path is Release 5 from 3GPP. In this step, the functional split between the GSM/EDGE Radio Access Network (GERAN) and the Core Network (CN) will be harmonized with UTRAN, enabling GERAN to connect to the same 3G core network with the Iu interface and provide the same set of services as UTRAN. The new architecture implies significant modifications to the GERAN radio protocols.

The largest new functionality in 3GPP Release 5 is the IP Multimedia Subsystem (IMS) domain of the core network which incorporates completely new services. In later 3GPP releases (Release 6 and onwards), IMS services with real-time requirements are defined. Connecting GERAN to the IMS domain and being able to provide IP based services efficiently requires:

- o Use of a real-time capable interface towards the packet switched domain of the core network, such as Iu-ps.
- o Definition of a header adaptation mechanism for the Real-Time Protocol (RTP)/User Datagram Protocol (UDP)/Internet Protocol (IP) traffic.

Mobile Domain Technologies – Platform Layer

Other major enhancements provided by Release 5 are:

o   Wideband Adaptive Mult-Rate - Wideband (AMR-WB) speech for enhanced speech quality.
o   Half Rate 8-PSK speech for improved speech capacity.
o   Enhanced Power Control for speech.
o   Location Services enhancement for Gb & Iu interfaces.
o   Inter BSC & BSC/Radio Network Controller (RNC) Network Assisted Cell Change (NACC).

Architecture

The basic principles used in the design of the Release 5 GERAN architecture are:

o   Separating radio-related and not-radio-related functionalities between the core network and the radio access network, thus enabling development of the platform to allow the provision of services independent of the access type.
o   Ensuring support services for pre-Release 5 terminals.
o   Maximizing commonalities between GERAN and UTRAN yet maintaining backward compatibility for GERAN. The key alignment point between GERAN and UTRAN is a common interface to the 3G core network (Iu interface).
o   Standardizing the GERAN Release 5 should support a true multivendor environment.

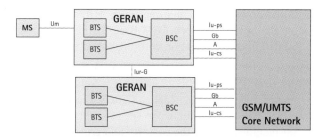

The proposed new GSM/EDGE radio access network architecture

Legacy Interfaces

The protocol model for legacy interfaces is based on Release 4 for GSM/EDGE and the functionality split between the radio access and the core network remains unchanged.

The Gb interface is used in GSM/GPRS to connect SGSN and the Base Station System (BSS). This interface is needed in GERAN for supporting pre-Release 5 devices.

The A interface, a traditional interface in GSM used to connect BSS and the Mobile Services Switching Center (MSC), will be supported by GERAN.

New Interfaces

The general protocol model for new interfaces is based on the protocol model used in UMTS and, as shown in the next figure. The layers and planes are logically independent of each other, offering additional flexibility. For example, if needed, transport network protocol layers may be changed in the future with virtually no impact on the radio network layer.

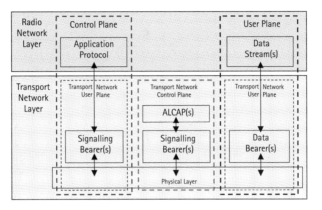

General UMTS protocol model

The Iu interface is common for UTRAN and GERAN, and is specified at the boundary between CN and the Radio Access Network (RAN). The protocol model for the Iu-ps interface is shown in the next figure.

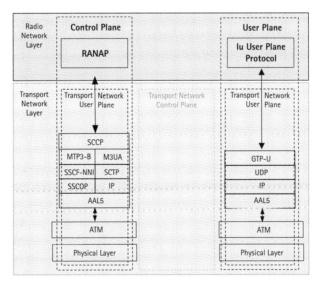

Protocol model for the Iu-ps interface

The Iur-g interface provides a logical connection between any two GERANs and supports the exchange of signaling information between them. The Iur-g interface is based on the UTRAN Iur interface [utranIur99]. However, only a subset of the functionalities is needed for the GERAN of Release 5, mainly related to cell-level mobility management. Therefore, it is proposed that only Radio Network Subsystem Application Part (RNSAP) Basic Mobility Procedures and RNSAP Global Procedures be adopted from UTRAN and modified as necessary. These procedures do not require the user plane, as illustrated in the protocol model in the following figure.

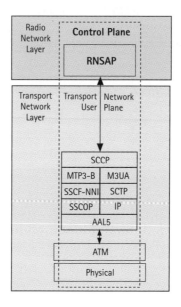

Protocol model for Iur-g interface (subset of UTRAN Iur)

The functionality split across these interfaces is aligned with the UTRAN-CN split. For GERAN, this means that functionalities like ciphering and header adaptation are moved from CN and are part of the radio access network. In addition, the mobility management tasks are solely part of the radio access network.

The main benefit of this harmonization is that a carrier deploying both radio access technologies may operate one single core network, and both access technologies can provide the same set of services. This includes provision for all the UMTS traffic classes through GERAN. For network and device manufacturers, the harmonization provides benefits in the form of synergies in protocol and interface implementation.

The harmonization of the GERAN and UTRAN architectures simplifies future development of these radio access technologies. The IP transport is expected to benefit carriers by bringing down transmission costs.

Quality of Service

In Release 5, GERAN and UTRAN Quality of Service (QoS) management frameworks are fully harmonized, since GERAN adopts the UMTS architecture with the 3G-SGSN being connected to both radio access networks through the Iu interface. The Iu interface towards the Packet Switched (PS) domain of the core network is called Iu-ps, and the Iu interface towards the Circuit Switched (CS) domain is called Iu-cs.

The harmonization of GERAN and UTRAN QoS management frameworks implies that the same QoS parameters can be deployed in lower level protocols. The harmonization of the QoS management for both 3GPP radio access technologies is very convenient, as services will be seamlessly provided to consumers regardless of the radio access technology used. It will be convenient for carrier network management as well, as an integrated management tool can be used for both technologies. There are four different classes of traffic:

- o Conversational
- o Streaming
- o Interactive
- o Background

The main difference between these classes is how delay sensitive the traffic is: the Conversational class is meant for traffic which is very delay sensitive while the Background class is the most delay insensitive.

Conversational and Streaming classes are mainly intended to carry real-time traffic flows. Conversational real-time services (e.g., video telephony) are the most delay sensitive applications, so these data streams should be carried in the Conversational class.

The Interactive and Background classes are mainly meant to be used by traditional Internet applications, such as Web, e-mail, Telnet, File Transfer Protocol (FTP) and News. Due to looser delay requirements, compared to conversational and streaming classes, both provide a better error rate by means of retransmission schemes. The main difference between the Interactive and Background classes is that the Interactive class is mainly used by interactive applications, e.g. interactive e-mail or interactive Web browsing, while the Background class is meant for background traffic, e.g. background e-mail or file downloading. Responsiveness of the interactive applications is ensured by separating interactive and background applications. Traffic in the Interactive class has higher priority in scheduling than Background class traffic. Background applications use transmission resources only when interactive applications do not need them. This is very important in the mobile networks where the radio interface bandwidth is small compared to fixed networks.

A similar functional split between the Control plane and the User plane as in UTRAN has been adopted for GERAN. Control plane functions relate to bearer-level service management (e.g. bearer activation, resource activation, and deactivation), whereas user plane functions relate to packet level service management (e.g. packet classifying, dropping, scheduling). Every level has a different set of management functions.

User Plane QoS mechanisms in GERAN:
- o Link adaptation, providing a means to avoid fast changes in radio conditions, such as dynamic selection of channel coding, data rate, interleaving depth and modulation depending on the service requirements.
- o Traffic conditioning, allows conformance between the negotiated QoS for a service and the data unit traffic (packet conditioner). Depending on the type of service it should work in a different way.
- o Packet Scheduler, a core function of the QoS provisioning since the radio interface is the most likely bottleneck in the whole network. Thus, efficient handling of radio blocks based on priorities leads to better performance. The scheduling of the radio resources for different Radio Access Bearers (RABs) is performed both in uplink and downlink directions based on radio interface interference information and the different QoS profiles.
- o Power Control function tries to minimize the interference in the network while the QoS requirements of the connections are maintained, which effectively increases the capacity of the system. In the uplink direction, the power control is used to decrease the power consumption of the mobile station.

Control Plane QoS mechanisms in GERAN:
- o Connection Admission Control, which is related to the RAB/Packet Flow Context (PFC) and the Radio Bearer (RB) establishment procedure.
- o QoS preserving, which takes place once the bearers are established. These mechanisms should try to maintain the negotiated QoS profile, so the load and quality functions have to be available to perform such control.

Radio Protocols

The following figures show the user plane and the control plane protocol stacks in a packet-switched domain, when connected through Iu-ps to the 3G CN. The Protocol stacks connecting to the circuit switched CN domain with Iu-cs are not shown, nor are the protocol stacks, when connecting to the CN through legacy interfaces, A and Gb.

Packet Data Control Protocol

The Packet Data Convergence Protocol (PDCP) allows for the transfer of user data using services provided by the Radio Link Control (RLC), and header adaptation of the redundant network Protocol Data Unit (PDU) control information in order to make transport over the radio interface spectrum efficient. Header Compression consists of compressing transport and network level headers so that the decompressed headers are semantically identical to the original uncompressed headers. Header compression is suited to generic Internet applications, especially multimedia applications.

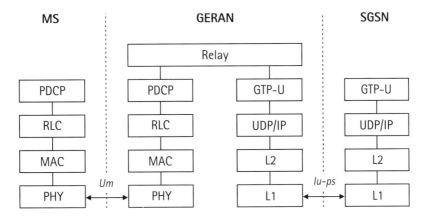

User plane protocol stack towards the PS CN domain through Iu-ps

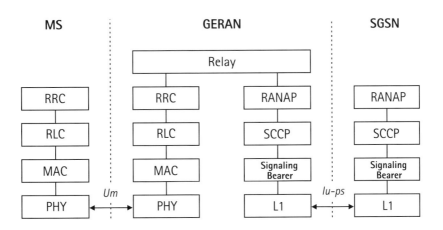

Control plane protocol stack towards the PS CN domain through Iu-ps

Radio Resource Control

The Radio Resource Control (RRC) protocol is based on both GSM Radio Resource (RR) [rrcp99] and UTRAN RRC [rrcps99] specifications. Many new RR concepts introduced to UTRAN (compared to GSM Release 4) are adopted from UTRAN RRC to GERAN RRC, such as the Radio Bearer, the RRC connection and the RRC connection mobility management concepts. Other RRC functions include the broadcast of information related to the non-access stratum (core network) and the access stratum, paging/notification, routing of higher layer PDUs, control of requested QoS, initial cell selection and re-selection in idle mode, support for location services, control of security functions and Mobile Station (MS) measurement reporting and control of such reporting.

RRC implements the QoS requirements of the RAB by establishing a Radio Bearer (RB) between MS and GERAN. For each RB, RRC allocates either a dedicated or a shared channel. In the case of dedicated channels, RRC also takes care of radio resource allocations, though in the case of shared channels, Medium Access Control (MAC) takes care of radio resource allocations.

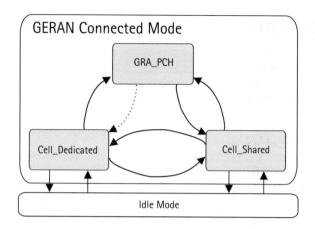

RRC states and state transitions

RRC connection mobility means that in the RRC connected mode, the GERAN tracks the location of MS without CN involvement. When there is no RRC connection, i.e., in idle mode, MS is identified by non-access stratum identities and the GERAN does not have its own information about the individual MS. The RRC connected mode has two levels of accuracy for RRC connection mobility: Cell and GERAN Registration Area (GRA). On the cell level, MS makes cell updates after each GERAN cell reselection and the network tracks the MS's location on the cell level. For inactive users, MS may fall back to the GRA level of accuracy. In this state the MS location is known only with GRA accuracy.

Radio Link Control

The RLC protocol allows for data transfer in transparent, acknowledged or unacknowledged modes, and notifies unrecoverable errors to the upper layer.

When in the transparent mode, the RLC has no functionality and does not alter the data units of the upper layer.

In the non-transparent mode, the RLC is responsible for ciphering RLC PDUs in order to prevent any unauthorized acquisition of data. The RLC non-transparent mode is built based on (E) GPRS if the Radio Access Bearer is mapped onto a Packet Data Traffic Channel (PDTCH). Alternatively, the RAB may be mapped onto an Enhanced Traffic Channel (E-TCH). Signaling is performed on a Packet Associated Control Channel (PACCH) when PDTCH is used, and on a Fast Associated Control Channel (FACCH) and Slow Associated Control Channel (SACCH) when (E- TCH) is used. The mapping of a RAB onto the proper traffic channel is dependent on the MAC mode and the targeted traffic class. In the acknowledged mode, Backward Error Correction (BEC) procedures are provided which allow error-free transmission of RLC PDUs through two possible selective Automatic Repeat Request (ARQ) mechanisms: Type I ARQ and Hybrid Type II ARQ (Incremental Redundancy). In the unacknowledged mode, no BEC procedure is available.

Medium Access Control

The MAC sublayer allows transmission over the physical layer of the upper layer PDUs in dedicated or shared mode. A MAC mode is associated with a physical subchannel for use by one or more mobile stations (dedicated or shared mode respectively). The MAC layer handles the access to and multiplexing onto the physical subchannels of mobile stations and traffic flows.

Physical Layer

Physical layer in GERAN interfaces with the MAC and the RRC protocols, offering logical channels and associated transmission services. Logical channels are divided in two categories:

o   Traffic channels, and
o   Control channels.

Traffic channels are intended to carry either encoded speech or user data, while control channels carry signaling or synchronization data. Logical channels are multiplexed either in a fixed predefined manner, or dynamically by the MAC on physical subchannels. Physical SubChannels (PSCH) are the units of the radio medium. Some are reserved by the network for common use, while others are assigned to dedicated connections with Mobile Stations (i.e., Dedicated Physical SubChannel (DPSCH)), or are assigned to a shared usage between several Mobile Stations (i.e., Shared Physical SubChannel (SPSCH)). The types of logical channels allowed on a PSCH depend on its mode.

Traffic channels of type TCH are intended to carry either encoded speech or user data on a physical subchannel in dedicated MAC mode only, i.e., for one MS only. The TCH use a circuit-like connection over the radio interface, and can be either Gaussian Minimum Shift Keying (GMSK) or 8-PSK modulated. One main characteristic of the channel coding for the TCH is the use of diagonal interleaving. The two categories of TCH are Speech TCH and Data TCH.

The channel coding of Speech TCH is optimized to maximize the perceived speech quality. Output bits from the speech codec are divided into different classes based on their subjective importance and the bits of the different classes are protected unequally in channel coding. Earlier releases of the GSM standard already introduced Full Rate (FR) and Half Rate (HR) speech on GMSK channels, an Enhanced Full Rate (EFR) speech on GMSK channels, and finally the AMR codec on both full rate and half rate GMSK channels.

The next GERAN releases will introduce the wideband AMR codec for high quality speech services, and will also introduce narrowband AMR on half rate 8-PSK channels to increase speech capacity [qramr00].

Unlike Speech TCH, the channel coding of Data TCH includes equal error protection for the payload bits. Furthermore, due to the more relaxed delay requirements, channel coding is enhanced with longer interleaving.

A wide range of data TCHs already exist in the GSM/EDGE standard, using both GMSK and 8-PSK modulation, and offering data rates from 9.6 kbps to 43.2 kbps in a single slot configuration. In E-TCH/F32.0 multislot configuration, e.g. the 64 kbps data rate can be provided with two time slots. The GERAN physical layer will reuse these Data TCHs. Typically GERAN will use Data TCHs for services with relaxed delay requirements (e.g. streaming services) in dedicated MAC mode

TCH	Modulation	Code rate	Data rate (kbps)
TCH/F9.6	GMSK	0.53	9.6
TCH/F14.4		0.64	14.4
E-TCH/F28.8	8PSK	0.42	28.8
E-TCH/F32.0		0.47	32.0
E-TCH/F43.2		0.64	43.2

Data TCH modulation and coding schemes

MCS	Modulation	Code rate	Data rate (kbps)
MCS9	8PSK	1.0	59.2
MCS8		0.92	54.4
MCS7		0.76	44.8
MCS6		0.49	29.6
MCS5		0.37	22.4
MCS4	GMSK	1.0	17.6
MCS3		0.80	14.8
MCS2		0.66	11.2
MCS1		0.53	8.8

EGPRS modulation and coding schemes

PDTCHs are intended to carry user data on physical subchannels in either dedicated or shared MAC mode. PDTCH is temporarily dedicated to one MS but one MS may use multiple PDTCHs in parallel for individual packet transfer (multislot configuration). PDTCH allows several MSs to be multiplexed on the same SPSCH, and also allows several traffic classes from the same MS to be multiplexed on the same PSCH (shared or dedicated).

The channel coding of PDTCH uses short rectangular interleaving in order to ensure efficient multiplexing. In GPRS, four different coding schemes based on GMSK modulation were standardized. In EGPRS an additional set of 9 coding schemes, using both GMSK and 8-PSK modulation was introduced. The available data rates run from 8.8 kbps to 59.2 kbps in a single slot configuration. In multislot configuration the maximum data rate can be brought up to 473.6 kbps. PDTCH can be used by GERAN for any traffic class.

The fast block level power control technique was introduced in Release 99 for the Enhanced Circuit Switched Data (ECSD) to increase spectral efficiency, especially for the transparent circuit switched service. In Release 5, the fast power control technique, i.e., Enhanced Power Control (EPC), is defined for the channels where SACCH is allocated.

GERAN Performance

Some performance aspects of EGPRS performance (mainly Release 99 specifications) are presented in this section and evaluated via network level simulations where only downlink user traffic is simulated. The objective is to give an idea about the performance that GERAN networks can offer to carriers.

Incremental Redundancy

Incremental Redundancy (IR) automatically adjusts the code rate to actual channel conditions by incrementally transmitting redundancy information until decoding is successful. IR does not require any channel quality measurements and can operate based solely on acknowledgement messages sent by the receiver. Typically the same Modulation Coding Scheme (MCS) is used for the initial transmission and also for all retransmissions, but with different puncturing schemes. Combining a retransmitted block with the previous transmissions of the same block significantly reduces the block error probability of the retransmitted block. This combining reduces the effective coding rate and therefore high MCSs can be used even in relatively poor radio conditions.

Link level simulations for MCS-8, MCS-6 and MCS-2 with and without IR combining can be seen in the figure below. Appreciable performance increases are envisaged from the link level results.

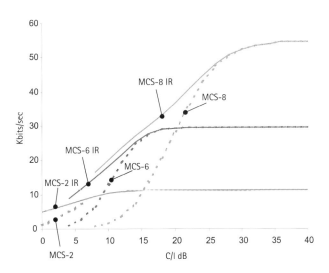

Incremental Redundancy performance in EGPRS. TU3 channel, ideal FH.

For boosting EGPRS performance further, network vendor specific enhancements like the use of smart antennas in the base station (including downlink delay diversity) can be used.

EGPRS Link Adaptation

The Link Adaptation (LA) functionality makes the decision to change or not to change the coding scheme by comparing the estimated channel quality with certain threshold values in order to obtain maximum throughput. Different algorithms are possible based on the available channel quality estimates specified for EGPRS (Carrier-to-Interference Ratio (CIR), Block Error Ratio (BLER) and Bit Error Probability (BEP)).

The user data rate for single TimeSlot (TS) with ideal LA can be seen in the next figure.

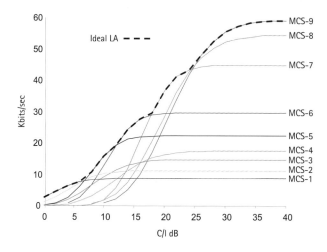

Throughput experienced for the different MCS (TU3 channel and ideal FH conditions).

The use of LA is more efficient in the case of EGPRS compared to GPRS due to the IR. LA and IR are similar in the sense that they adapt the coding rate to the radio environment, so both features must be coordinated for optimum performance.

In the following picture the proportional throughput gains of both IR and LA over the individual MCS-7 are depicted [smart01]. These results show how much bigger throughput can be achieved with the combination of IR and LA compared to the best-performing EGPRS individual MCS.

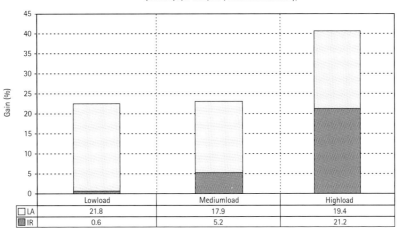

Relative gains provided by IR and LA over the individual MCS-7

From the previous sections we have been able to conclude that the best configuration for a GERAN network from a performance point of view includes the coordinated use of LA and IR mechanisms. System level evaluation from capacity (spectrum efficiency) and quality (throughput) is presented here for two different deployment scenarios: non-hopping and Radio Frequency (RF) hopping. Only best effort traffic has been simulated.

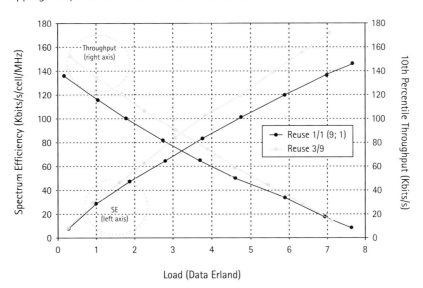

Evolutions of spectrum efficiency and 10th percentile throughput in both non-hopping and RF-hopping macro cell EGPRS networks

In order to be fair, the analysis was performed with one Transceiver (TRX) configuration and with the same bandwidth [geran02].

Conclusions

The alignment of the GSM/EDGE Radio Access Network (GERAN) architecture with the UTRAN architecture allows GERAN to:

- o connect to the 3G core network through the Iu interface, and
- o provide the same set of 3G services as UTRAN as well as similar QoS provisioning mechanisms for them.

Further, the alignment will make smoother future developments in both radio access technologies (GSM/EDGE and WCDMA) towards IP-based multiradio networks.

The other significant developments in GERAN Release 5 include support for IP-based services and spectral efficiency and quality enhancements for voice services - in terms of half rate 8-PSK speech channels, Enhanced Power control and Wideband AMR speech codecs.

References

[tdmaconv99] R. Pirhonen, T. Rautava, J. Penttinen, "TDMA Convergence for Packet Data Services," IEEE Personal Communications, 1999.

[edge99] A. Furuskär et al, "EDGE: Enhanced Data Rates for GSM and TDMA/136 Evolution," IEEE Pers. Commun., Vol. 6, No. 3, June 1999, pp. 56-66.

[advmeas00] Y. Zhao, G. Sébire, E. Nikula, "Advanced Measurements for EGPRS Link Quality Control," in the proceedings of the 6th Asia-Pacific Conference on Communications APCC 2000, October 30th - November 2nd, 2000, Seoul, Korea.

[linkadp99] K. Balachandran et al, "A Proposal for EGPRS Link Control Using Link Adaptation and Incremental Redundancy," Bell Labs Technical Journal, July-September 1999.

[servint00] X. Qiu, K. Chawla, L.F. Chang, J. Chuang, N. Sollenberger, J Whitehead, "RLC/MAC Design Alternatives for Supporting Integrated Services over EGPRS," IEEE Pers. Commun., Vol. 7, No. 2, April 2000, pp. 20-33.

[utranIur99] 3GPP TS 25.420, UTRAN Iur Interface General Aspects and Principles, Release 1999.

[rrcp99] ETSI GSM 04.18, Radio Resource Control Protocol, Release 1999.

[rrcps99] 3GPP TS 25.331, RRC Protocol Specification, Release 1999.

[qramr00] "T. Bellier, M. Moisio, E. Nikula, R. Pirhonen and H. Rantala , "Quarter rate AMR channels - A speech capacity booster in GSM/EDGE," WPMC '00; November 12-15, 2000; Bangkok, Thailand.

[ecsd99]   ETSI SMG2 "ECSD-concept evaluation," Tdoc SMG2, 670/99, 31 May-4 June, 1999, Tucson.

[smart01]   Harri Luukkonen, Sami Nikkarinen. "EGPRS Network Level Simulation Results. Phase I - Part A," January 2001.

[geran02]   Timo Halonen, Javier Romero, Juan Melero. "GPRS/EDGE performance - GSM Evolution Towards 3G/UMTS." 2002, Wiley.

Wireless Local Area Networks

Wireless Local Area Networks (WLAN) can be used as an extension or as an alternative to fixed LANs. The first products were limited to the former applications and enabled special portable terminals to access fixed network resources, for example, databases in warehousing and health care applications. Currently, complete local area networking can be set up very conveniently by WLAN. WLANs are rapidly gaining significance as mobile devices (e.g., laptop computers) are replacing desktop personal computers. At the same time, major technological advances, e.g., in terms of increased data rates, enhanced security and improved support for Quality of Service (QoS), are taking place. The overall trend is towards higher bit rates, interoperable networking with other systems, international standards, and interoperable products between different WLAN product manufacturers. WLAN is also seen to have a lot of potential as a complementary local high speed access method for traditional mobile services.

The purpose of a wireless LAN is the same as that of a wired LAN: to convey packets of information among the devices attached to the LAN. Unlike wired LANs, wireless LANs may connect mobile devices, which have the freedom to roam in the radio coverage area. Wireless LANs are targeted for slow mobility indoor communication, and achieve data rates similar to the existing wired LANs, ranging from 1 to 54 Mbit/s with small cell sizes.

Wireless LAN systems are basically planned for local, in-house and on-premises networking, providing short distance radio links between computer systems. The demand for wireless high speed data communication is a logical consequence of the rapid growth in personal computing and mobile communications in recent years. Consumers demand the same freedom for data communication as they have for telephone services. Mobile devices can offer the speech quality of fixed networks, plus the added mobility. The requirement for a good wireless LAN is to provide throughput and security equal to wired LANs.

Frequencies

The traditional mobile systems operate on licensed frequency bands. Carriers are given a piece of the spectrum for their sole use from the total allocation of a particular system. This scheme works nicely with systems covering wide areas, such as Global System for Mobile Communications (GSM). However, given the limited amount of spectrum this means that only a few carriers can have licenses in a certain geographical area.

For WLAN networks, licensed frequencies are clearly not a feasible solution. By their nature, these systems are operated by companies, other organizations, or individuals. Furthermore, in order to be able to provide bandwidths over 50 Mbit/s these systems will demand a relatively large frequency band.

Typically, WLAN networks operate on an unlicensed spectrum. This means that a certain piece of spectrum is available for a set of systems, provided that they fulfill the agreed requirements. In terms of operating environment and co-existence, this arrangement creates a very different set of problems compared to traditional mobile networks. For example, traditional network planning is not possible because anyone has the right to install base stations and therefore co-coordinating their physical locations is very difficult.

We will here discuss the 2.4 GHz and 5 GHz bands, as they are the most essential bands because of their global availability and because complexity and the cost of technology goes up with the frequency.

2.4 GHz Industrial, Scientific and Medical Frequency Band

The 2.4 GHz Industrial, Scientific and Medical (ISM) band is currently the most popular band for wireless short-range devices. It is globally available, although there are some exceptions (e.g., fewer channels available) in some countries. However, recent developments in Japan and France indicate that strong market pressure and demand for these systems is leading to the harmonization of regulations in these countries as well.

The application category utilizing the 2.4 GHz band is quite wide, including, e.g., high-frequency heaters, telemetry and paging systems, wireless LANs, cordless phones, and microwave ovens. From this point of view, it is likely that the band will eventually get congested. The introduction of Bluetooth devices, and increased interest in WLANs in general is likely to foster this development. Therefore, very high performance WLAN devices will gradually move to the 5 GHz band.

5 GHz Frequency Band

In Europe, the European Radiocommunications Committee (ERC) Decision (ERC/DEC/(99)23) designates the frequency bands 5150-5350 MHz and 5470-5725 MHz (455 MHz in total) to High Performance Radio Local Area Networks (HIPERLANs). For the first band between 5150-5 250 MHz the maximum mean Effective Isotropic Radiated Power (EIRP) is 200 mW, and only indoor usage will be allowed. For the band 5470-5725 MHz the maximum mean EIRP is 1W, and both indoor and outdoor usage will be possible. Furthermore, dynamic frequency selection is mandatory, thus providing uniform loading of HIPERLANs across the minimum of 330 MHz or 255 MHz in the case of equipment used only in the second band of 5470-5725 MHz. In addition, transmitter power control is required to ensure a mitigation factor of at least 3 dB for uplink and downlink.

Currently the 5 GHz band is available for unlicensed devices in the United States, where the Federal Communication Commission (FCC) has allocated the Unlicensed National Information Infrastructure (U-NII) band. U-NII contains 300 MHz in three bands: 5150-5250 MHz, 5250-5350 MHz, and 5725-5825 MHz. The transmission power limits are 200 mW, 1 W and 4 W EIRP, respectively.

In Japan, the 5150-5250 MHz band is currently allocated for radio LAN use. The band is restricted to indoor use in order to protect the Mobile Satellite Services (MSS) feeder links. Furthermore, the transmission rate should be equal to or greater than 20 Mbit/s, while the transmission power should be less than 10 mW/MHz. In the future, the 5150-5250 MHz band will be allocated for WLAN services. To facilitate frequency sharing, Japan requires that the spectrum not be monopolized by a single system, so consequently a carrier sense method is required in medium access technologies.

To summarize, the frequencies of the 5 GHz band allocated in each region overlap, which allows interoperability of devices. Harmonization of the physical layers in 5 GHz WLAN technologies also promotes this development. Unfortunately, it seems that the upper layer protocols of the 5 GHz systems will lack functional compatibility.

IEEE 802.11

The Institute of Electrical and Electronics Engineers (IEEE) initiated the 802.11 working group in 1990 to develop a WLAN standard for the 2.4 GHz unlicensed ISM frequency band. This band was selected because of its global availability and maturing radio technology. The IEEE 802.11 standard was approved in 1997, but the first products had already appeared before that, based on a draft standard.

The IEEE 802.11 standard specifies the physical layer and the MAC protocol layer. The network architecture supports both an infrastructure based topology and a distributed topology with independent peer-to-peer connections. In the former, a master station that can also operate as an access point to wired network controls the network. The standard includes support for authentication of peer devices, data encryption, and power saving functionality. Both asynchronous and time-bounded data transfer services are specified.

Today, the IEEE 802.11 family of standards includes specifications for five types of physical layers. Four of these are based on radio technologies and one on diffuse infrared technology. Still, all physical layer alternatives utilize the same MAC protocol. Three of the radio standards utilize the 2.4 GHz frequency band and one is designed for the higher 5 GHz frequency band.

Reference Model

Like IEEE 802.3 (Ethernet) or 802.5 (Token Ring), IEEE 802.11 also covers the two lowest layers of the Open Systems Interconnection (OSI) reference model. The 802.11 reference model consists of two main parts, Medium Access Control (MAC) and Physical (PHY) layers. MAC is mapped onto the Data Link layer and PHY onto the Physical layer of the OSI model. The figure below presents the 802.11 reference model.

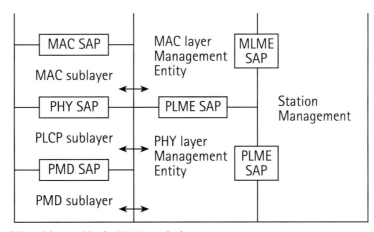

Portion of the OSI model covered by the 802.11 standard

802.11 is required to appear to the Logical Link Control (LLC) layer as an 802.X LAN. This requires that 802.11 handles station mobility within the MAC layer. To meet the reliability assumptions that LLC makes about the lower layers, however, it is necessary for 802.11 to incorporate functionality which is not traditionally included in the MAC layer.

Architecture

The standard defines the logical architecture of an 802.11 network. The architectural basic components are the Station (STA), wireless medium, Access Point (AP), distribution system and portal. These components are used to form the 802.11 network. There are two fundamental alternatives for the 802.11 architecture, namely, independent and infrastructure networks. These are depicted in the two figures below, respectively.

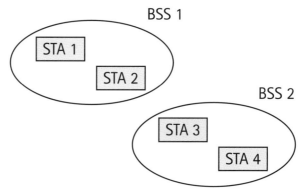

Independent networks BSS1 and BSS2

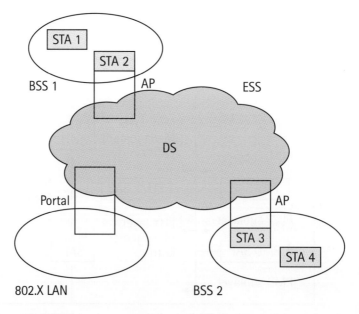

Infrastructure network (ESS)

The independent network, or Basic Service Set (BSS), consists of two or more stations that communicate directly with each other through the wireless medium. This kind of network can also be referred to as an ad hoc network. Physical limitations determine the direct station-to-station distance which can be supported. For some networks this distance is adequate, though other networks require increased coverage.

An infrastructure network, or Extended Service Set (ESS), is constructed from multiple BSSs, access points and a Distribution System (DS). Since 802.11 defines only the logic of the network architectures, the standard does not discuss the actual implementation.

DS enables mobile device support by providing the logical services necessary to handle address to destination mapping and seamless integration of multiple BSSs. DS extends the network and enables communication between stations which are too far from each other to communicate directly.

AP has station functionality and provides access to the distribution service. Data moves between BSS and DS via AP.

The key concept of ESS is that it appears to the LLC layer as an independent network. Stations within an ESS can communicate with the other stations and mobile stations may move from one BSS to other BSSs within the same ESS, transparently to LLC. The IEEE 802.11 architecture does not assume anything about the physical locations of the architectural components.

Integration with an existing wired LANs is possible through the portal, the logical point at which a MAC Service Data Unit (MSDUs) from a non-802.11-wired LAN enters the 802.11 DS. The portal can be implemented in the same physical device as the AP but not necessarily. Furthermore, the portal is not equivalent to a traditional bridge in the strict sense, since a bridge is used to connect similar type MAC layers.

Services

The 802.11 architecture allows for the possibility that the DS may not be identical to an existing wired LAN. DS can be implemented with many different technologies, including current 802.X-wired LANs. 802.11 does not constrain the DS to be either Data Link or Network layer based, nor does it constrain DS to be either centralized or distributed in nature. Instead of specific implementations, 802.11 defines services. There are two categories of 802.11 services: Station Services (SS) and Distribution System Services (DSS). Both categories are used by the 802.11 MAC layer.

Station services support transport of MSDUs between stations within the same BSS. They are present in every 802.11 station. The station service subset is:

a) Authentication
b) Deauthentication
c) Privacy
d) MSDU delivery

Distribution system services are provided by the DS and enable the MAC to transport MSDUs between BSSs within an ESS. The DSS service subset is:

a) **Association**
b) **Disassociation**
c) **Distribution**
d) **Integration**
e) **Reassociation**

Distribution delivers MSDUs within the DS. It is the job of the DS to deliver the message in a such way that it arrives at the appropriate DS destination for the intended recipient. How the message is distributed within DS is not specified by 802.11. The necessary information for the message distribution is provided by the three association related services.

Integration enables delivery of MSDUs between DS and an existing wired network. The message is delivered via the portal and the integration service is responsible for accomplishing whatever is needed to deliver a message from the DS to the wired LAN media. Details of the integration are DS-specific and not specified by 802.11.

The different association services support different categories of mobility. The basic association service establishes association between a station and an AP, as such information is needed for message delivery via DS. At any given moment, a station may be associated with no more than one AP while an AP may be associated with many stations at any time. Association is always initiated by the mobile station.

Reassociation enables the transfer of an existing association from one AP to the other within an ESS. Reassociation also enables changing the attributes of an established association while the station remains associated with the same AP. The existing connections must be maintained during the reassociation.

Disassociation voids an existing association. It can be invoked by either party of the association and cannot be refused by the other party.

Authentication is used when stations identify each other. Since wireless media is not bound as wired media are, 802.11 supports several authentication processes although it does not mandate the use of any particular authentication service. 802.11 provides link level authentication between stations. It does not provide end-to-end or user-to-user authentication. 802.11 authentication is simply used to bring the wireless link up to the assumed physical standards of a wired link. 802.11 also supports shared key authentication with the Wired Equivalent Privacy (WEP) option. Authentication is void by the Deauthentication.

Medium Access Control Layer

One of the fundamental ideas of 802.11 is to have one MAC over multiple radio physical layers. This MAC is implemented with a Carrier Sense Multiple Access/Collision Avoidance (CSMA/CA)-based mechanism, which is closely derived from Ethernet (CSMA/Collision Detection (CD)). The Distributed Coordination Function (DCF) sub-layer is the core of the MAC, and its goal is to provide each node with equal opportunity to an access medium via CSMA/CA. Both connectionless service and time-bounded services are built on top of DCF with contention free Point Coordination Function (PCF) and contention based services (see the figure below).

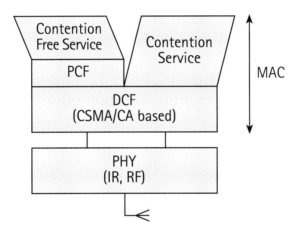

802.11 MAC architecture

Basic Channel Access

In 802.11, protocol performance is affected by two main factors, the basic channel access mechanism and the MAC frame format.

The 802.11 MAC data path is based on the concept of carrier sensing. Before transmission, a station senses the medium to determine if it is free. The medium must be detected to be free for an interval which exceeds the time of the Distributed InterFrame Space (DIFS). Otherwise, transmission is deferred until the end of the on-going transmission. After successful DIFS, a station will enter a contention phase by setting its backoff timer. The backoff timer value is decremented until the medium is sensed to be busy or the timer reaches zero and data transmission may begin. If the data transmission attempt is unsuccessful (i.e., a collision takes place), the backoff timer value is increased exponentially to reduce the probability of collision.

The transmitter cannot determine if the receiver successfully received the data, so explicit acknowledgements are required to provide feedback on the status of a data transmission. A short ACK-packet ends the medium access cycle. The ACK must be sent immediately after the Short InterFrame Space (SIFS). If the transmitter does not receive an ACK-packet or receives a negative acknowledgement, it will perform a MAC level recovery retransmission. One successful channel access cycle is presented in the following figure.

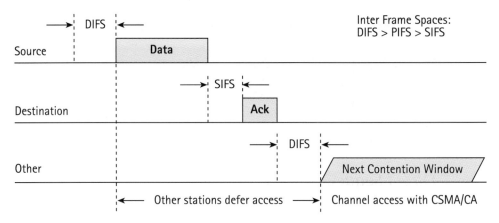

Basic channel access cycle

The basic channel access cycle does not provide any QoS related support. CSMA/CA is purely aimed at providing equal and fair access to the channel for all contending nodes. All QoS-related traffic handling must be done above the basic channel access layer in an 802.11 implementation.

802.11 is a distributed system, which can be configured into a centralized mode. Therefore, the design of 802.11 WLANs is further complicated by two phenomena called hidden node and capture effect. A pair of nodes in the same WLAN are referred to as being hidden from each other if they cannot hear each other's transmissions while trying to send data to a third node simultaneously. Capture refers to the capability of a receiver to successfully receive a transmission from one station when multiple stations are transmitting simultaneously. DCF contains an alternative way of transmitting data involving short Request To Send (RTS) and Clear To Send (CTS) frames prior to data transmission. The RTS-CTS procedure is a handshake to agree on the duration of time that the medium is reserved for data transmission. Stations, which are able to hear either the transmitter or the receiver, will defer from transmission, thereby increasing the probability of successful data transmission. This increase in the probability of successful data transmission is achieved at the expense of increased protocol overhead, however. This overhead is particularly significant for short data frames.

Physical Layer

The IEEE 802.11 PHY layer consists of two sublayers. The *upper* sublayer is the Physical Layer Convergence Protocol (PLCP) layer and the *lower* sublayer is the Physical Medium Dependent (PMD) layer. PLC adapts the capabilities of PMD into PHY service and offers them through the PHY-Service Access Point (SAP). PMD transmits and receives data via the wireless medium between two or more stations.

For implementation of the PMD sublayer, 802.11 defines three basic alternatives: Frequency Hopping Spread Spectrum (FHSS), Direct Sequence Spread Spectrum (DSSS) and Infrared (IR). These implementations are orthogonal and cannot communicate with each other.

Frequency Hopping Spread Spectrum

FHSS uses the 2.4 GHz ISM-band for transmission. The number of channels is 79 in the USA and Europe and 23 in Japan. The basic data rate is 1 Mbit/s and channel spacing is 1 MHz. The number of different hopping sequences is 78 (USA and Europe) or 23 (Japan). The modulation scheme is 2-level Gaussian Frequency Shift Keying (GFSK) with BT=0.5. The channel dwell time is 400 ms, i.e., the channel is changed 2.5 times per second.

Optionally, a higher data rate of 2 Mbit/s can also be used. This higher data rate is implemented by the use of 4-level GFSK modulation.

Direct Sequence Spread Spectrum

DSSS uses the same 2.4 Ghz frequency band as FHSS. The data spreading is accomplished with the 11 bit Barker code. The number of channels is 11 in the USA, 9 in Europe and 1 in Japan. Channel spacing is 5 MHz, which results in some overlapping of channels. The data rates are the same as FHSS: for 1 Mbit/s the modulation method is Binary Phase Shift Keying (BPSK), while 2 Mbit/s is achieved with Quadrature Phase Shift Keying (QPSK).

Infrared

The infrared implementation of PMD uses wavelengths between 850 and 950 nm which are similar to the ones used in remote controllers and Infrared Data Association (IrDA). The system is not directed, i.e., it does not require line-of-sight from the transmitter to the receiver. This can be implemented with diffuse IR technology. The range of the communication path is limited to 10 meters, or with a more sensitive receiver to 20 meters. Naturally, IR light cannot penetrate walls and therefore communication is limited to inside one room. The data rates are the same as with FHSS and DSSS.

Conclusions

Currently, most WLAN terminals are laptops or Personal Digital Assistants (PDAs) with separate WLAN network adapters. The WLAN device business is gradually moving towards integrated devices, as device categories are expanding with, for example WLAN PDA devices and PDA devices with integrated WLAN.

WLAN can also be used to complement carriers' traditional wide-area network service portfolios by offering cost-efficient wireless broadband data solution indoors. Target places for the WLAN access services are airports, railway stations, hotels, business parks and office buildings where most laptop users typically work.

The 802.11 technology and standards are constantly evolving to meet new requirements.

Bluetooth

Bluetooth technology is an open de facto industry standard for short-range wireless voice and data communications. As the technology has acquired broad industry acceptance, the specifications have been also adopted by Institute of Electrical and Electronics Engineers (IEEE) in the 802.15.1 work group. Bluetooth technology allows devices to communicate over a single air interface up to a distance of 10-20 meters. The Bluetooth Special Interest Group (SIG) is an industry consortium with more than 2000 members.

Connectivity Scenarios

Bluetooth technology will be used as a short-range wireless technology mainly in 3 different scenarios:

- o Personal domain
- o Ad-hoc community
- o Networked services

In the personal domain, personal devices, e.g., mobile devices, digital cameras, and music players, are interconnected by Bluetooth technology. Personal devices are frequently connected to other devices, and therefore some static security information is used to prevent misuse.

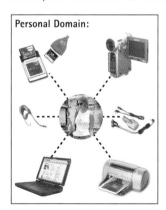

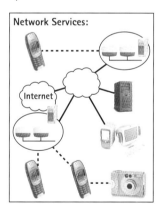

Usage scenarios of Bluetooth technology

In the ad-hoc domain, devices of different consumers are typically interconnected to run applications (e.g., games or chat). The devices are frequently connected to other devices, though access rights to the device's internal data are limited to the actual use case.

Bluetooth technology can also be used to access local services, or services which are offered by a network behind the access point. Due to the limited range of Bluetooth, local services are quite attractive as the consumer's location is close to the access point. The service provided can be free of charge, or some authorization is needed for royalty bearing services.

Technical Overview

Lower Layers

Bluetooth operates in the 2.4 GHz Industrial, Scientific and Medical (ISM) band. It uses a fast frequency hopping scheme with 79 frequency channels, each being 1 MHz wide. The following table summarizes the main parameters of the Bluetooth radio and baseband specification:

Frequency	2401.5 – 2480.5 MHz
Modulation	GFSK, BT = 0.5, Index: 0.28 – 0.35
Bit Rate	1 Mbps
Hopping Rate	1600 hops/s (typical) 3200 hops/s (during inquiry, paging)
Output power	Class 3: 0 dBm (typical) Class 2: 4 dBm (optional) Class 1: 20dBm (optional)
Receiver sensitivity (@ 0.1% BER)	–70 dBm
Topology	Star (Master-Slave)
Duplexing	Time Division Duplexing (TDD)
Connection Types	Synchronous, Asynchronous
Encryption	1 bit stream cipher with key up to 128bit

The star topology of the system requires one master device, which can handle active connections with 7 slave devices simultaneously. However, it is possible to temporarily suspend active communication with a slave device (park mode). Parked devices are also part of the piconet, so the number of devices in a piconet can exceed seven.

The BT Radio module is specified to be simple, resulting in low cost implementations, as well as having relaxed requirements for sensitivity and implementation accuracies, which typically allow low current consumptions in implementations. A typical current consumption for a state-of-the-art implementation is in the area of 10-30 mA for a 64 kbps full-duplex connection. Furthermore, sophisticated low-power modes also allow the current consumption to be reduced during idle times.

Baseband defines different packet types for both synchronous and asynchronous transmissions, as well as packet types supporting different error handling techniques (e.g., error correction or detection, and encryption). Also, data whitening is used to reduce any DC offsets in the receiver due to special payload characteristics.

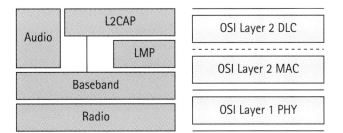

Bluetooth lower layers

The previous picture shows the lower layers of the Bluetooth protocol stack in comparison with the International Organization for Standardization/Open Systems Interconnection (ISO/OSI) reference model. Bluetooth specifications do not follow the ISO/OSI model, but the Bluetooth functions can be related to the functions in that model.

The Logical Link Control and Adaptation Protocol (L2CAP) layer shields the middleware and applications protocols from the complex lower layer Bluetooth functions. It offers a transparent transmission of higher layer data over the Bluetooth air interface. L2CAP supports functions like protocol multiplexing, assembly/reassembly of large data packets received from higher layers, and the control of the quality of service.

The Link Manager Protocol (LMP) is responsible for controlling the connections of a device. This includes functions like connection establishment and link detachment, security management (e.g., authentication, encryption) and power management of different low power modes. There might be multiple instances of Link Managers if a device maintains multiple connections.

Middleware Protocols

The middleware protocols of the Bluetooth stack comprise references to existing middleware protocols, modified existing protocols as well as protocols explicitly developed for the Bluetooth technology. One of the important protocols is the Service Discovery Protocol (SDP), as Bluetooth focuses on ad-hoc networks of different devices and different capabilities. This dynamic environment is managed by the SDP, which defines a standard way for devices to describe supported services offered to other devices.

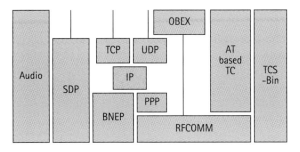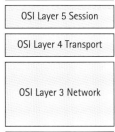

Bluetooth middleware protocols

The RFCOMM protocol defines a virtual serial port to applications, which makes it relatively simple to reuse existing applications. Especially for cable replacement, serial port is an important functionality of the Bluetooth specifications. The definition of RFCOMM incorporates major parts of the European Telecommunications Standards Institute (ETSI) TS 07.10 standard, which defines multiplexed serial communications over a single serial link. As many of the other transport protocols in Bluetooth, RFCOMM uses an L2CAP connection to establish a link between two devices.

Another key protocol in the specifications is the OBject EXchange (OBEX) protocol, which provides an easy and spontaneous mechanism to transfer objects over the channel provided by RFCOMM. This session protocol is based on a specification defined in the Infrared Data Association (IrDA), called IrOBEX, used for object exchanges over infrared channels. In principle, the OBEX protocol can also run over Transmission Control Protocol/User Datagram Protocol (TCP/UDP) but this is not clearly specified in the Bluetooth specifications. In addition to the session protocol, the OBEX specifications also define four object formats to be transmitted over Bluetooth. These are:

- **vCard** — format to describe electronic business cards, managed by the Internet Mail Consortium
- **vCalendar** — format to define electronic calendar and schedule entries
- **vNote** — format defined by IrDA organization, representing short electronic notes
- **vMessage** — also defined by IrDA, this format describes electronic messages and electronic mail

The OBEX protocol and the OBEX applications are supported by the Service Discovery Protocol, which offers information about the supported OBEX services in the SDP service records.

While RFCOMM and OBEX based applications are important for ad-hoc networking and non-reprogrammable small devices, the industry trend is moving towards IP transport based applications. In the Bluetooth 1.1 core specifications, the Bluetooth IP specification was solely based on IP on top of PPP, but in the summer of 2001 a new profile for Bluetooth was released, which allows IP packets to be encapsulated into a protocol running directly over L2CAP. This new protocol, Bluetooth Network Encapsulation Protocol (BNEP) allows the application to exploit the full benefits of IP, as the restrictions of an underlying Point-to-Point Protocol (PPP) no longer exist. Especially if IP version 6 (IPv6) is used, IP network formation becomes quite simple.

Bluetooth technology also supports audio & telephony applications. Telephony Control System (TCS) specifications define how audio is transferred as well as how telephony functions are controlled over a Bluetooth link, i.e., how call control functions like dialing or call acceptance can be controlled by a remote device. While the control functions are defined in the TCS specifications, the audio is routed directly to the baseband layer, without being routed through L2CAP or other layers. In addition to TCS, the specifications also define AT commands between two devices for AT based telephony control.

Profiles

While the Middleware and the Lower Layer functionality are defined in the core specifications of Bluetooth, profiles are defined in a separate document in the profile specifications. Profiles determine the mandatory and the optional parts of the specification for the defined use cases. Defining profiles fosters interoperability between devices from different manufacturers.

The Bluetooth profiles can be classified into 4 categories:

- o Generic Bluetooth Profiles
- o Serial Port Profiles
- o Telephony Profiles
- o Networking Profiles

There are two generic profiles: since they cannot be directly related to a special use case, they are mandatory profiles for all Bluetooth devices. The Bluetooth Generic Access Profile (GAP) is the root of all other profiles and defines which functions are needed for basic Bluetooth connectivity. The other generic profile Bluetooth Service Discovery Application Profile (SDAP) maps to the SDP defined in the core specifications and must be supported by all Bluetooth devices.

The serial port profiles are used for applications which can be described as cable replacement use case, therefore mapping directly to the RFCOMM protocol. The profiles are Bluetooth Generic Object Exchange Profile (GOEP), Bluetooth Synchronization Profile (SP), Bluetooth Object Push Profile (OPP), Bluetooth File Transfer Profile (FP) and Bluetooth Serial Port Profile (SPP).

Telephony profiles include profiles defining typical telephone use cases, such as cordless telephony or headset usage. The profiles are Bluetooth Cordless Telephony Profile (CTP), Bluetooth Headset Profile (HSP) and Bluetooth Intercom Profile (IntP).

For data applications, the networking profiles are of interest. They include profiles for Bluetooth Dial-Up Networking Profile (DUNP), Bluetooth Fax Profile (FaxP) and Bluetooth LAN Access Profile (LAP). Lately a new profile was approved, which defines how IP is used directly on top of L2CAP. This Personal Area Networking profile is going to be a new, interesting access alternative for Mobile Internet applications.

Future Developments

Bluetooth technology is a mature technology today, both from the point of view of standardization as well as in terms of technology and products. The Bluetooth SIG has numerous industry members and many companies are working on products and technical solutions. The key for Bluetooth to become a real volume business are products, which are easy to use, without any interoperability problems between devices of different manufacturers and for a low price. The material cost for Bluetooth functionality is expected to go down dramatically, when the volumes of products increase to the tens of millions.

While today a majority of Bluetooth products support functionality in the area of cable replacement, we will see a transition to the area of Personal Area Networking types of applications. The low current consumption and the low cost structure of Bluetooth allows integration to many personal devices, so it allows consumers to create their own short range networks.

Conclusions

Bluetooth technology is an integral part of the Mobile Internet. Emerging from widespread use in personal ad-hoc networking and cable replacement scenarios, the technology will migrate into usage models of Internet access. Local services, especially, will be provided by the use of Bluetooth technology. The role of IP over Bluetooth will gradually increase and even ad-hoc networking use cases and cable replacement scenarios will see IP as a unifying transport mechanism for several applications. IP networking with Bluetooth will support multiple access and implementation of different applications, which will be independent of the underlying access technology.

Digital Video Broadcasting – Terrestrial Network

IP Datacasting (IPDC) is a technology to provide access to popular content for large audiences simultaneously. Examples of such content include the Internet, news services, updates to commonly used software packages, as well as audio and video transmissions. IP Datacasting is based on an IP multicasting paradigm, with some conceptual additions to make it usable in one-directional networks and/or service concepts. One possible wireless transport for delivering IPDC services are digital broadcast networks, such as the Digital Video Broadcasting Terrestrial (DVB-T) network.

It is envisioned that for some applications the best way to provide IP Datacasting services is to utilize the existing DVB-T standards and frequency space. In addition, IP Datacast services in the DVB-T access network may interoperate with mobile devices or networks.

The potential of using a broadcasting channel for mass media (mostly audio-visual content) delivery has been around for a long time in the form of television and radio. Television broadcast networks are currently being converted from analog to digital in Europe and elsewhere in the world in order to increase network efficiency. In addition to the traditional distribution of TV programs, digital television networks enable high-bitrate data transfer to one or more receivers simultaneously. Now that there is a new technology entering the broadcast domain, namely DVB-T enabling wireless IP multicast delivery, hence there is a new potential access technology for IP Datacast services to be created.

Technical Overview

Radio Characteristics

The DVB-T standard is based on Coded Orthogonal Frequency Division Multiplex (COFDM) technology. The system is designed to operate within the UltraHigh Frequency (UHF) spectrum and can be used with 6, 7 or 8 MHz channel bandwidths depending on the regional demands. Operations on the Very High Frequency (VHF) band are also possible.

The modes in the DVB-T standard provide built-in flexibility for different operating conditions. There are two FFT sizes (8k and 2k) available as well as several inner modulation schemes, 5 different code rates and 4 different guard intervals. Together these parameters can be used to select the right mode for a specific need. On the other hand, if the need is to provide data services for slower mobility, i.e., in a hot spot area, the parameters can be selected to provide up to 30 Mbps.

Service type	Modulation	Data Speed [Mbps]
Hot Spot	8k 64QAM CR=5/6 GI=1/32	30.16
Fixed	8k 64QAM CR=2/3 GI=1/8	22.12
Mobile	2k 16QAM CR=1/2 GI=1/8	11.06

The DVB-T system was designed to cope with severe multipath propagation. Therefore, it is capable of coping not only with a Gaussian channel, but also with Ricean (fixed) and Rayleigh (portable) channels. It can withstand high-level (up to 0 dB), long delay echoes which makes it possible to use the spectrum-efficient Single Frequency Network (SFN) for broadcast networks. It can also cope with dynamic multipath distortion, which makes it possible to set up mobile services. It therefore works well in dynamic ghost channels, associated with in-door reception.

DVB-T also includes several hierarchical modes, which can transfer two separate bit streams simultaneously, with different robustness. Therefore, the DVB-T offers an efficient way of combining fixed and mobile services or otherwise classifying the services according to the required quality of service or service area.

Mobility

The DVB-T system can offer a wide range of possibilities for mobile use. Tolerance for Doppler shift is very good in general and excellent in some modes. The actual achievable speeds depend on the mode selected and the frequency used. Even the 8k 64QAM high capacity mode can tolerate speeds of up to about 40 km/h (25 mph). If a suitable mobile mode like 8k or 2k 16QAM is selected, the maximum speeds are in the range of 100 km/h (60 mph) to more than 300 km/h (190 mph). With the most robust 2k QPSK mode, speeds even higher than 600 km/h (370 mph) are possible. It should also be noted that these results were achieved with more or less standard open market receivers, without any special features developed for mobile use. With advanced receivers, performance can be even better.

DVB-T Mode	Bit Rate [Mbps]	C/N for Mobile [dB]	Max Speed [km/h]
2k QPSK ½ ¼	4.98	12	>600
2k 16QAM ½ 1/8	11.06	17	300
8k 16QAM ½ 1/8	11.06	17	100
8k 64QAM 2/3 1/8	22.12	26	40

Radio Network Design Issues

Since the DVB-T system is robust to interference from delayed signals, which can be either echoes of terrain or building reflection, or signals from distant transmitters operating at the same frequency, it is possible to build Single Frequency Networks. The concept of SFN is efficient in that it saves the spectrum required to offer a service to a certain geographical area. By having both 2k and 8k modes as well as several guard intervals, the DVB-T system can offer efficient tools for planning SFNs for various purposes including Mobile Internet.

One benefit of the SFN principle of DVB-T in the Multi Frequency Network (MFN) type of network is the possibility to enhance the receiving conditions locally with cheap gap fillers and repeaters using the same frequency. These are basically simple selective amplifiers which reradiate the original signal and offer better field strength to difficult locations. When using gap fillers, the possibilities for portable and mobile reception are even better. These can also replace local in-house distribution cables.

Although the DVB-T has originally been designed for traditional high power, large cell broadcasting purposes, it is fully scalable. It is possible to build networks with low power, small cell base stations. DVB-T offers some unique tools for network planning in these cases, such as the possibility to use the SFN-principle to save spectrum and enhance the coverage areas by using cheap gap fillers.

One useful cost-efficient topology for low power networks is to use directional sectorized antennas from one physical base station site. This will downscale the cell size, but will not increase the number of required base station sites as every site serves several cells from one antenna mast.

Datacasting Profiles

Five protocol profiles are defined in the DVB specification for data broadcasting, with different application areas and requirements. The profiles which can be used to provide DVB data services are Data piping, Data streaming, Data/Object carousels and Multi-Protocol Encapsulation (MPE). The following figure illustrates the protocol stacks of the different DVB data broadcasting profiles. Of the data broadcasting profiles specified by DVB, the DVB multiprotocol encapsulation, is best suited for generic Internet access, as it provides standard encapsulation for IP-based protocols.

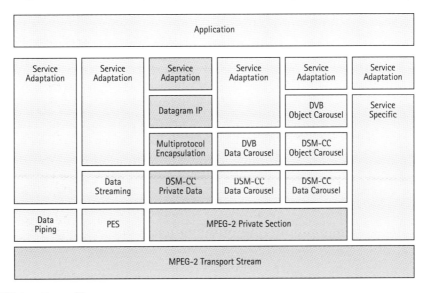

DVB-T Datacasting profiles

IP Multiprotocol Encapsulation

The DVB MPE profile is intended for sending datagrams of non-DVB protocols over DVB networks. The encapsulation provided by the DVB MPE profile is closely tailored after the International Organization for Standardization (ISO) Local Area Network/Metropolitan Area Network (LAN/MAN) standards. Thus, the DVB network can be considered an Open Systems Interconnect (OSI) Layer 2 data link between an MPE broadcast service provider and DVB data receivers. However, there are differences between DVB MPE and more traditional OSI data link layer technologies, such as Ethernet: data links over DVB MPE are unidirectional, provide virtual broadcast channels with the use of different Program Identifier (PID) values, and often include a much larger number of receiving hosts than a normal LAN/MAN segment.

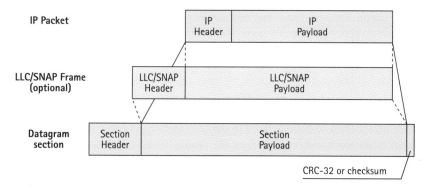

Encapsulation of IP packets in DVB Datagram Sections

While datagrams of other protocols can be fragmented and sent over multiple sections, no fragmentation is done for IP packets in MPE. Thus, each IP packet must fit into a single datagram section which can be up to 4097 bytes in size. This sets an upper limit on the size Maximum Transmission Unit (MTU) of IP packets which can be transmitted using multiprotocol encapsulation: 4074 bytes if Logical Link Control/SubNetwork Access Protocol (LLC/SNAP) framing is used, or 4080 bytes without LLC/SNAP framing.

The DVB Data Broadcasting specification [ref DVB Technology] defines the datagram section format used with MPE. This section format is based on the DSM-CC section format which in turn is based on the MPEG-2 private section format.

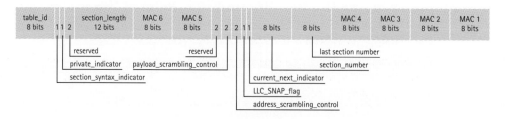

Header of a DVB Datagram Section showing the structure of the datagram section header

The table_id and MAC address fields shown in the figure above are most useful when filtering multiprotocol encapsulated data from a DVB transmission. The table_id field enables the receiver to filter out sections not containing multiprotocol encapsulated data. The destination MAC address is then used to filter out datagrams not addressed to a particular receiver or group of receivers. In addition, higher protocol layers (e.g. in the IP stack) can further filter out datagrams received over MPE before they are passed to application software.

The format of the 48-bit MAC address is not defined by DVB. However, the DVB Data Broadcasting specification states that the widely-used scheme of mapping IPv4 multicast addresses into 48-bit Ethernet MAC addresses can optionally be used to convert an IPv4 multicast address into a DVB MAC address. The scheme allows receivers to filter multicast IP packets already on the section layer.

IP Datacasting

Enabling multicast

Broadcast and multicast capabilities are inherently supported in the Internet Protocol stack, making it perfectly suitable for delivery over a broadcast medium. In fact, with multicast the channel switching can be implemented by the consumer simply by subscribing to a multicast group. In essence, this would be the same feature which is being implemented in traditional television broadcasting by tuning into an appropriate physical frequency. In the unidirectional (broadcast only) datacast solution, the client does not have to have any kind of return channel. Still, it is possible to provide a limited set of IP multicast-based services in this case.

The other possible scenario is to use normal multicast functionality, in order to enable rich, scalable IP Datacast services. In this case, there is a need for a return link via some other network as well. Thus, there is a need for hybrid devices and networks enabling true convergence.

The existence of multiple different access methods creates the possibility for generating new innovative services and solutions. By making the cell size smaller, we can enable consumers to take advantage of localized services and even mobile services.

Hybrid solution

DVB-T access can be combined with the mobile system in several different ways, ranging from device-based integration to network-level integration. The different solutions for hybrid devices and networks are discussed, for example, in the UMTS/DVB forum [rcf UMTS Forum]. In this chapter we discuss how to combine DVB-T access with mobile networks at the network level, considering what we call here a common core scenario.

The common core refers to a case where DVB-T is used as another radio access technology in UMTS or in other future mobile systems. This could be achieved with different integration levels, depending on the mobile system in question.

From the application's point of view, the situation is fully transparent because the hybrid nature of the system can be taken care of by the mobile network, below the IP layer. Applications can simply rely on the availability of IP multicast functionality. From the consumer's point of view, the only difference between the services might be the quality or the number of services available, depending of the availability of the DVB-T access in certain areas.

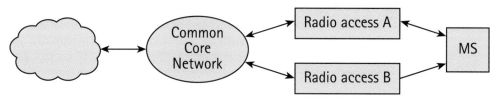

Common core network. Radio access B could be IP over DVB-T and access A UMTS [walsh].

Conclusions

In the near future, there will be an opportunity to ramp up the market for mass media content delivery in the Mobile Internet by using IP datacast technology. Datacast is a service platform for wireless IP multicast. Flexibility, scalability, and mobility of DVB-T combined with a large bandwidth ensures that the technology cannot be omitted as Mobile Internet technologies are selected.

References

DVB Techology   http://www.dvb.org/dvb_technology/index.html

White Papers, e.g. Convergence between DVB-T & UMTS

Documentation and presentations, e.g. The DVB Cookbook

UMTS Forum   http://www.umts-forum.org

Report #14, "Support of Third Generation Services using UMTS in a Converging Network Environment," February 2002

M.Grundström, H.Hakulinen, M.Kalervo, "Providing High-speed Internet Access using DVB broadcast Channels," Telecom'99 Forum, International Telecommunications Union, Geneva, Switzerland, October 10-17, 1999.

J.P.Luoma, "Internet Access over DVB Networks," Master of Science Thesis, Tampere University of Technology, January 2001.

R.Walsh, L.Xu, T.Paila, "Hybrid Networks - A Step beyond the 3G, EU/Drive", The Third International Symposium on Wireless Personal Multimedia Communications, Bangkok, Thailand, November 12-15, 2000.

Part 4.2

Mobile Internet Layer

- o Mobility Support for IPv6
- o Multimedia Sessions in the Mobile Domain
- o Java in the Mobile Domain
- o Data Synchronization in the Mobile Domain
- o Middleware in the Mobile Domain
- o Authentication Methods and Technologies

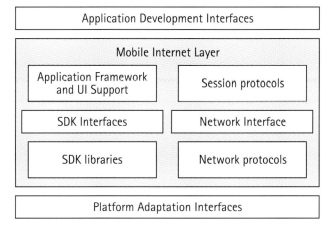

Part 4.2

Mobile Internet Layer

Mobility Support for IPv6

Mobile IP for IPv6 (Mobile IPv6) has been designed to handle the mobility management needs for mobile nodes (typically mobile devices) in the IPv6 Internet. The design of Mobile IPv6 has been motivated by a desire to remain as close as possible to the guidelines motivating the design of the underlying IPv6 protocol, and building on the experience gained from the design and implementation of Mobile IP for IP version 4 (RFC 2002). This chapter describes the relevant aspects of the design of IPv6, followed by the design of Mobile IPv6, and some of the most recent areas that are under discussion for further development and standardization.

The main motivation for producing a new version of IP has been to enable global addressability for a much larger population of Internet devices than is possible using IPv4, which has been the mainstay of the Internet for almost two decades now. When IPv4 was first designed in the early 1980s, few people imagined that the Internet would grow beyond a hundred thousand nodes, much less the hundreds of millions of network devices that make up the Internet today. One result was that the total size of the address space was set at 4 billion individual addresses. Worse, though, was the chaotic manner in which networks were assigned. There were only three sizes of networks specified - class C networks with 256 nodes, class B networks with 65,536, and class A networks with slightly over a million addressable nodes. Thus, mid-size organizations with a thousand computers would routinely consume 60 times the necessary address space.

By the mid 1990s, projections were made that this practice would consume the entire IPv4 address space within 10 years. This set into motion several complimentary activities which still today are having dominant effects on the evolution of the global Internet.

Most immediately, the use of subnetting, originally offered as an optional feature of Berkeley Unix systems, was made mandatory. This feature, called Classless Inter-Domain Routing (CIDR), allows networks to have sizes of any power of two, and thus allows much more efficient allocation of IPv4 addresses. As the use of CIDR spread across the Internet, the rate of allocation of the original classful network prefixes slowed very substantially, effectively delaying the exhaustion of network prefixes for quite a few years. We might have had yet another 10 years at current rates of consumption, except for the anticipated growth of mobile networking.

The consumption of network prefixes has slowed, but not the rate of actual address consumption. Thus, there is still great concern that not enough IPv4 addresses will be available to fulfill the needs of the Internet at its current rate of growth. The situation becomes drastically worse if we include the expected number of mobile devices among the population of devices needing Internet addresses. While current mobile devices do not use the Internet to any great extent, it is assumed that this will change, and third generation standards have been written with this expectation. Unfortunately, we cannot expect to achieve anywhere near 100% utilization of the IPv4 address space (nor the IPv6 address space, but for different reasons). Theories have been advanced suggesting that a natural limit to address utilization exists, and that the limit (the so-called "H-factor") appears to be about 25-30%. The H-factor arises because network sizes are constrained to be a factor of two, because of the natural effects of network aggregation, and because of the needs for network administrators to plan for growth. Consequently, unless something unexpected happens, the total number of computers assigned to have global IPv4 addresses is unlikely to ever exceed 1 billion. As the Internet continues its exponential rate of growth, that hard limit looks frightfully close.

IPv6, with its vast address space, offers a straightforward solution to this problem. Furthermore, care has been taken so that the IPv6 network prefixes can be assigned in a much more orderly fashion (initially, as well as ongoing). It seems realistic to assume that at some point in the future the Internet will be populated mainly by IPv6-addressable devices, with a smaller subset of devices that are addressable with both IPv6 and IPv4 addresses, and a smaller yet collection of legacy devices that for one reason or another can not be easily upgraded. However, just because we can design a future Internet with sufficient addressability does not at all mean that we will have a clear idea of how to solve the new problems of scale that are almost guaranteed to arise. Examples abound to illustrate this sort of unpredictability. For instance, the rise of the Web, the economic upheaval and transformation worked by the dot.coms, and the commercial exploitation of the Internet by Internet Service Providers (ISPs) are all having long-term effects that could not have been clearly foreseen ten years ago. Although we probably cannot predict the transformations that might be attributable to an Internet with 4 billion nodes, we can pretty confidently state that IPv6 offers the best model for global addressability and end-to-end communications. This is especially important for voice applications, which need both of these important features.

Mobile IP

Mobility introduces a problem with today's Internet addressing model that can be stated very simply. An IP address is used for two different purposes: routing and host identification. The routing prefix forms the initial bits of the IP address. The remaining bits indicate the specific identity of the host among the collection of network nodes collected together as part of the network, which has been identified by the routing prefix. The routing infrastructure of the Internet operates to deliver packets to the specific network indicated by the routing prefix, and the router for that network delivers the packet to the host identified by the rest of the bits.

This model works fine for nodes that do not change their location. For such computers, the routing does not change. Since the identity of the host computer is also unlikely to change very often, if at all, the same IP address can serve both functions for an indefinite amount of time. When the network node can move, however, this simple relationship is no longer valid. The IP address should stay the same so that the host can keep its identity, but the IP address would have to change in order for packets to be routed to the host computer's new location.

For network nodes that move around, Mobile IP solves this problem by associating a care-of address with the network node at each of the node's new points of attachment. The node can keep its permanent IP address (called a "home address"), but vary the care-of address to conform to the routing needs of the Internet. Thus, Mobile IP transforms (part of) the mobility management problem into a routing problem. Circumscribing the exact design space for a network-layer mobility support has been a fascinating development over the last 13 years, and details are still emerging; some of these will be described at the end of this chapter.

A specialized router on the home network, called the *home agent*, is given the responsibility of forwarding packets to the mobile node when it is away from its home network (i.e., when it has a care-of address at some new point of attachment). The home agent, then, has to receive information about every new care-of address acquired by the mobile node so that it can carry out this forwarding operation. The association between the mobile node's home address and its care-of address is called a *binding*; the association also includes a lifetime value, after which the care-of address information is to be considered expired and must be deleted. The message, which contains the new care-of address (or the refreshed lifetime, in case the care-of address has not changed), and which the mobile node delivers to the home agent, is called a *Binding Update*.

Using this terminology, we can say that the essential part of Mobile IP is the protocol by which a Binding Update is delivered. This turns out to be a very powerful tool. When a correspondent node (i.e., an arbitrary node within the Internet) needs to send a packet to the mobile node, it can just send it to the mobile node, which is identified to the rest of the Internet just by its home address. The mobile node's care-of address is not typically used to identify the mobile node, but only to forward packets to its current point of attachment to the Internet. In this way, communications to the mobile node can continue without disruption even while the mobile node changes from one care-of address to the next. As far as the application on the correspondent node is concerned, the mobile node is still addressable at its home address, and that does not change, so communications can proceed transparently. In fact, the transport protocol (e.g., Transmission Control Protocol (TCP)) does not detect the change of care-of address.

From the above description, it is possible to formulate a pretty good plan for designing Mobile IP.

1. The mobile node must detect that it has moved.
2. The mobile node must discover or configure its new care-of address.
3. The mobile node must inform the home agent about its new care-of address.
4. The home agent must forward packets from the home network to the mobile node's care-of address.

IPv6 offers native features that enable a mobile node to accomplish some of these needs directly, and the other requirements, although not already provided, are still greatly simplified by IPv6 design points.

This powerful method of associating a topologically correct, but changeable, care-of address to the mobile node's care-of address goes a very long way towards solving the mobility management problems for Mobile Internet nodes. However, there are new classes of problems which are uncovered after the routing problems are solved, including resource discovery, profile management, context and environmental controls, billing, and adaptive programming. None of these problems are realistically solvable using Mobile IP, and they all can be of great importance for improving the consumer experience. Even given that they are important problems, they must be considered beyond the scope of the discussion in this chapter.

IPv6 Design Points

The general discussion of Mobile IP in the last section applies equally well to IPv4 or IPv6. In order to understand Mobile IPv6 better, it is important to first understand some additional details about IPv6.

As has been described, the main reason for using IPv6 addresses for the future Internet is that the available address space becomes huge. The IPv6 address space (128 bit) is 2^{96} times as large as the IPv4 address space. Given that the IP header required modifications anyway to accommodate address fields of this size, a certain amount of effort was expended to repair problems which had been identified over the 10 years of experience with IPv4. Some of these efforts were made to correct what were considered to be design flaws, and some of the others to enable improved performance.

Router Advertisement

Although Router Advertisement messages are specified for both IPv4 and IPv6, the former advertisements are not extensively used in today's Internet. Furthermore, the advertisements play a much more important role in the address autoconfiguration features for IPv6, as will be described shortly. Since home agents are typically considered to be routers, and since mobile nodes need to find out the identity of the home agents on the home network, the Router Advertisements are specified to be used for this purpose, and extended with additional information which is needed for Mobile IPv6. For instance, there is a new bit which indicates that the router is also willing to serve as a home agent for (mobile) nodes on the network. There is also another extension that supplies the global IP address of the home agent, for situations where that information is not otherwise available.

Router advertisements also supply information about how long the IPv6 addresses may be used. Every IPv6 address has a lifetime; if the lifetime expires, use of the address must be discontinued. Typically, each new Route Advertisement is considered to renew the lifetime, so that a node which is attached to a network with typical lifetime parameters in the advertisements may continue to use the same IPv6 address indefinitely. This use of address lifetime does help with renumbering.

Address Autoconfiguration

One of the important improvements in IPv6 is the development of Stateless Address Autoconfiguration. This relies on first acquiring a *link-local* address. Link-local addresses have the special routing prefix "FE80::/64," which means a bit-string 64 bits long, with 0xFE80 in the first two bytes, followed by 48 bits of zero. A packet with link-local destination IP address cannot be forwarded by any router. Once the node has exclusive use of a link-local address, it can form a globally valid IPv6 address by replacing the FE80:: prefix with the routing prefix; the routing prefix is obtained from the advertisements issued by the local (default) router(s).

This method for getting IPv6 addresses works as follows:

1. The node first attempts to acquire a *link-local* address. It forms a 64-bit interface identifier, and probes on the link to find out if anyone is using the IPv6 address formed by prefixing FE80:: to that interface identifier. This process is called *Duplicate Address Detection (DAD)*.

   It is very unlikely that the link-local address selected by the mobile node would already be in use, since there are 16 quintillion such link-local addresses available. However, if DAD fails for this reason, the mobile node can easily retry the process with another choice for its interface identifier on the new link.

   Typically, using the node's MAC address on every new link will assure a unique link-local address.

2. The node monitors the link for a Router Advertisement. When the advertisement is detected, the mobile node extracts the routing prefix, along with the lifetime information for the addresses on the link, the MTU for the link, the layer-2 address for the default router, and any other pertinent information. The advertisement also specifies whether Stateless Address Autoconfiguration can be used. If so, the mobile node just replaces the upper 64-bits of its routing prefix with the advertised routing prefix.

3. If the advertisement prohibits the use of Stateless Address Autoconfiguration, then the mobile node must use some other means of acquiring an address which is topologically correct for the current point of attachment. The exact means is beyond the scope of the discussion in this chapter, but one possibility would be to use Dynamic Host Configuration Protocol v6 (DHCP for IPv6).

The mobile node can use this method for obtaining addresses each time it establishes a new point of attachment to the IPv6 Internet. Note that, using these rules, the mobile node has to wait for the completion of Duplicate Address Detection at every new point of attachment. For some applications, this delay is unacceptable. A recent effort has been undertaken to eliminate this delay, as part of an overall specification for Fast Handovers. That work will be described in a later part of this chapter.

Security

One advantage of using IPv6 is that every network node is guaranteed to be able to compute authentication data as well as perform encryption and decryption. These are powerful features for enabling e-commerce and keeping the Internet safe from the threats of many kinds of malicious attacks. This is in line with recent efforts by the IETF leadership to ensure that all newly standardized protocols are able to be operated securely.

The mobile node must be able to deliver new care-of addresses to all intended recipients securely. In addition, for the particular case of delivering care-of addresses to the home agent, the protocol must also be reliable in addition to being secure, because the home agent must always have up-to-date information about the mobile node's care-of address.

In order for the recipient to be assured that the care-of address it has received is truly the same as was intended to be reported by the mobile node, Mobile IPv6 requires that all such messages be accompanied by some data that can only have been produced by the mobile node - and, thus, is unforgeable. For this reason, Mobile IPv6 in its original design mandated the inclusion of an IPv6 Authentication Header (AH) in every packet reporting a new care-of address. Recent activities surrounding the exact formulation for inclusion of authentication data will be described later in this chapter. Since IPv6 network-layer information is organized as a series of extension headers, it is quite feasible to include additional header information in those packets. These additional headers are organized independently and located after the main IPv6 header. The Mobile IPv6 care-of address is inserted into one of these optional headers, and then AH follows the previous headers with authentication data, which assures that all of the header information (not just the Mobile IPv6 data) is authentic.

The other part of security which is required in any operation setting is key exchange and distribution. This is a much harder problem; in contrast to the mandated features for authentication and encryption, no specific protocol has been mandated for key distribution for IPv6 network nodes. Probably, the most widely implemented such protocol for IPv6 network nodes has been Internet Key Exchange (IKE), but that protocol is considered too heavyweight for the general needs of Mobile IPv6. For most of the last five years, the mobile-ip working group was not considered to have the responsibility for developing such key distribution protocols, and we basically assumed that they would be developed elsewhere. Now, as part of the ongoing re-evaluation of what is acceptable for standardizing major new protocols, it has been required to identify a key distribution protocol.

Destination Options

As mentioned previously, some of the features of IPv4 were redesigned during the development of IPv6. One such redesign involved the layout of the network-layer options. In IPv4, the options were directly included as part of the main header. Parsing the options is quite time-consuming, compared to parsing the other header fields. This caused most IPv4 routers to be processed along the *slow path* of the intermediate routing points. Consequently, very few IPv4 packets carry any optional fields at all, and in the end one could almost say that IPv4 effectively does not offer any useful optional protocol features.

In contrast, IPv6 options are more easily parsed, since they are not located inside the main IPv6 header. Moreover, specific measures have been taken to eliminate all processing requirements on intermediate nodes, in many cases. This has been done by introducing a new class of network-layer options, called *destination options*. Other options, called hop-by-hop options, retain the IPv4-style option semantics for features that need to supply information to all intermediate routers. Actually, other IPv6 options (e.g., IPSec options and the IPv6 fragment header) could just as well been classified as *destination options*, but they have their header number and so do not have to be organized as sub-parts of the general Destination Option header structure.

ICMPv6 Improvements to Encapsulation

Once the home agent has the care-of address information for the mobile node, it is able to reroute packets as they arrive on the home network so that they arrive instead at the care-of address. This is done by use of IPv6 encapsulation, which is very similar to IPv4 encapsulation for most purposes. Tunneling works very well in the usual case, as long as some care is taken to avoid exceeding the Maximum Transmission Unit (MTU) for all links along the path from the home network to the mobile node's current point of attachment (i.e., from the home address to the care-of address). Sometimes, however, things go wrong. In this case, the home agent will likely receive an Internet Control Message Protocol (ICMP) error (perhaps, Host Unreachable) indicating that the care-of address may no longer be a suitable locator for the mobile node. Such events should be reported to the node that originally sent the packet to the mobile node at its home address, i.e., the correspondent node.

This tunneling operation is done, using encapsulation, in practically the same way for both IPv4 and IPv6 mobility support by the home agent. However, for IPv4 there is an important difference, because of the way ICMP works for IPv4. For either version of IP, ICMP is specified to return the entire IP header of the undeliverable packet, along with some part of the payload. For IPv4, this is specified to be at least 8 bytes of payload. For IPv6, it is at least 512. Unfortunately, 8 bytes of payload for an encapsulated packet is not enough for the home agent to determine the IP address of the sender of the encapsulated packet (i.e., the IP address of the correspondent node). This happens because the encapsulated IP header appears as payload at the intermediate node (i.e., router) that generates the ICMP message back to the home agent. If the home agent discards the encapsulating (outer) header, the subsequent 8 bytes of payload is not enough to contain the source IP address of the encapsulating (inner) header.

In contrast, IPv6 processing by the home agent for such ICMP messages is practically guaranteed to supply the address of the correspondent node. Thus, the home agent can, with very little overhead, relay the ICMP message to the correspondent node, and hope that the correspondent node will cease producing traffic for the now unreachable mobile node.

In order to have the same effect for IPv4, the home agent has to store quite a bit of state information for each tunneled packet (called *soft-state*), for at least the duration indicated by the Round Trip Time (RTT) between home agent and care-of address. The stored state information would typically include most of the header information for each tunneled packet. Additional processing overhead would also be required to match up the incoming ICMP messages with the collection of stored state for the recently tunneled packets to the care-of address indicated in the ICMP message. If, on the other hand, the IPv4 home agent does not implement the collection of soft-state as has just been described, correspondent nodes (especially with User Datagram Protocol (UDP) applications like streaming audio) are likely to continue to pump out volumes of data that can not be delivered, thus adding unnecessarily to congestion at the home network.

Mobile IPv6 Protocol Overview

With the previously described IPv6 features in mind, it is now straightforward to describe the Mobile IPv6 protocol. Recall that the basic operation of Mobile IP requires that the mobile node acquire a care-of address at each new point of attachment, and that this care-of address has to be supplied to a router (called the *home agent*) on the mobile node's home network. With this general description, one can outline the general requirements for a Mobile IPv6 protocol as four general subrequirements:

1. (Detection) The mobile node has to find out that it has moved,
2. (Address Configuration) The mobile node has to configure a care-of address,
3. (Binding Update) The mobile node has to inform its home agent about the new care-of address, and
4. (Tunneling) The home agent has to reroute packets from the home network for delivery at the care-of address.

In this section, we will describe the protocol operations designed to provide a solution for the above four sub-requirements.

Detection

Mobile IPv6 allows a mobile node to make use of the basic IPv6 Router Solicitation and Advertisement messages defined in RFC 2461. When a mobile node can no longer detect any periodic broadcast Router Advertisements at a particular point of attachment to the Internet, it can either configure a new care-of address from another Advertisement that it has already received, or it can issue a Router Solicitation in order to trigger the transmission of an advertisement for its own use.

Mobile nodes should typically make use of mechanisms from RFC 2461 for movement detection only when they do not have better mechanisms from lower-layer protocols. With many well-designed physical media, the mobile node can obtain information about the noise level and signal strength for packets received from various access points. When one access point begins to drift out of range, the mobile node can attempt the process of establishing a link with some other access point that has better signal. The link establishment could trigger the transmission from the mobile node of a Router Solicitation. Then the Router Advertisement from the new router would have all the information needed by the mobile node to discover the necessary details about its new point of attachment.

Address Configuration

Once the mobile node establishes its link, it can configure a care-of address as already described. No changes for Mobile IPv6 are necessary.

Binding Update

Once the mobile node determines its new care-of address, it has to report this information to its home agent. The message in which the mobile node sends this information is called a Binding Update. A *binding* is, naturally, the association between a mobile node's home address, and the mobile node's care-of address, which locates the mobile node at its current point of attachment. The home agent maintains a Binding Cache with one entry as needed for each of the mobile nodes that it serves but which are currently attached to some other network other than the home network. As mentioned earlier, the Binding Update has to be verified (using cryptographic techniques) as authentically transmitted by the mobile node. Once the home agent receives a Binding Update from one of its mobile nodes, the home agent takes steps to ensure that it can redirect packets destined to the mobile node's home address, as described in the following subsection about Tunneling.

The mechanism by which mobile nodes can inform their home agents about the care-of address is a very powerful tool, forming the basis for all mobile networking support at the network layer. This tool can be also used to enable correspondent nodes to deliver packets directly to the mobile node without requiring the services of the home agent.

Tunneling

In order to redirect packets from the mobile node's home address to its care-of address, the home agent has a bit of work to do. Its work items include the following:

1. Capturing packets that are addressed to the mobile node's home address.
2. Participating in DAD to protect the mobile node's addresses on the home network against reallocation to another network node.
3. Redirecting the packets that it captures when they are forwarded to the home address, by using encapsulation.
4. Other tasks to inform the mobile node about renumbering events on the home network.

The home agent performs task (1) by following the protocols specified in RFC 2461. In particular, the home agent has to listen for Neighbor Solicitations and respond on behalf of the mobile node with a Neighbor Advertisement, with the *O* bit set. That bit setting requires the mobile node's erstwhile neighbors to update their Neighbor Cache with the information provided by the home agent. Naturally, the result is that the mobile node's neighbors on its home network are instructed to use the home agent's layer-2 address as the appropriate resolution of the mobile node's home address.

In order to preserve the functionality of Duplicate Address Detection, the home agent must also respond to address probes for the mobile node's link-local address. Otherwise, another new node on the home network could conceivably autoconfigure the same link-local address as the mobile node had already done when it was attached to its home network. This would violate a basic presupposition of the IPv6 protocol, with undoubtedly horrible effects. Experience dictates that infrequent, unpredictable address-level conflicts must be prevented by taking all due measures during the protocol design stages. Mobile IPv6 has done so by a natural use of the already existing Duplicate Address Detection and Neighbor Advertisement features.

The last piece of the tunneling problem, after all the preconditions have been met on the home network, is determining the encapsulation protocol. Here the natural choice is IPv6-within-IPv6 encapsulation, RFC 2473. This choice enables the home agent to avoid making any changes whatsoever to the network-layer packet as it was transmitted by the correspondent node. This is important, because changes to the original packet would likely make it impossible to maintain the correctness of authentication data that may have been supplied by the correspondent node.

Route Optimization

If we wish to design for the future mobile Internet, we have to make scalability a major goal. There are many dimensions in which scalability is important, but the first major dimension is to ensure that all network entities involved with the transaction will be able to handle the processing load. Otherwise, long delays will be unavoidable at the overloaded network element. Another more subtle way to improve scalability is to involve as few network entities as possible in the protocol operation, at least in all of the typical cases of interest. Rarely performed operations may reasonably involve additional processing or more network entities.

Applying this general design principle to Mobile IPv6, we can see that the home agent represents a single point of failure, unless additional steps are taken, which are currently outside the scope of the Working Group document. The home agent could also very well become a point of congestion, if large numbers of correspondent nodes wish to communicate with many of the mobile nodes on the associated home network at the same time. For the applications envisioned in the future (e.g., streaming video and multicast radio to hundreds of thousands of consumers per home network), this would amount to snatching defeat from the jaws of victory. Fortunately, we can utilize the Binding Update to avoid this major scalability problem. If the mobile node supplies a Binding Update to its communication partners (i.e., correspondent nodes), then all subsequent traffic from the correspondent node can travel directly to the mobile node's care-of address without the intervention of the home agent.

The procedure is conceptually almost trivial and requires practically no additional protocol, because the mobile node can use the same Binding Update message format with any node on the Internet as it uses with its home agent. Just as the home agent does, the correspondent node maintains a Binding Cache with entries for mobile nodes that have supplied care-of addresses. The actions required from the correspondent node are much like that of a home agent, except that the correspondent node does not have to tunnel packets to the mobile node from other correspondent nodes, whereas the home agent does have to do this. The other difference is that usually the correspondent node does not have to acknowledge the Binding Update (i.e., send a *Binding Acknowledgement*), whereas the home agent always does.

Even though the protocol for receiving Binding Updates is substantially the same, the operational details are quite a bit different because of the needs for key distribution. Since the correspondent node has to have a security association with the mobile node before it can verify the authenticity of the Binding Update, it has a prior need to exchange key information with the mobile node; the key exchange is the basis upon which the security association is built. The original supposition within the *mobile-ip* working group was, as suggested earlier, that such key exchange algorithms would be designed as needed and probably in other working groups. Recent events have precipitated a pioneering effort to create a lightweight key exchange protocol, which creates a minimal security association between mobile node and correspondent node - a security association, which is only intended for the simple task of verifying authentication data sent along with Binding Updates.

Note that, as a side-effect, these route optimization techniques help to avoid any need for the home agent to relay ICMP messages back to the correspondent node. This results from the fact that when correspondent nodes send packets to the mobile node's care-of address without the assistance of the home agent, ICMP messages are directly sent to the correspondent node instead of the home network.

This use of the Binding Update has the following additional advantages:

- o Reduces network utilization, since fewer links are traversed. This could mean much less congestion in all the routing fabric of the Internet.
- o Reduces the load on the home agent, and the potential for complete isolation of the mobile node due to loss of the home agent.
- o Reduced round-trip times for application endpoints
- o Better predictability of routing paths, which could be crucial for QoS management.
- o More symmetric delay characteristics for incoming and outgoing traffic between mobile node and correspondent node.

A lightweight key establishment protocol has been developed for inclusion as part of the base Mobile IPv6 specification.

Ingress Filtering

In today's Internet, there are many security concerns. Packets received from arbitrary correspondent nodes should not be trusted without some further assurance about the origin and integrity of the data within the packets. This applies equally strongly to the fields in the IP headers of packets; IP source addresses can often be easily forged by mischievous or malicious users to give misleading information to applications running on the computer receiving the forged IP header. Since IP addresses typically identify end nodes, as well as containing routing information about the end nodes, a forged header could allow masquerading nodes to cause a wide range of damaging effects. Ingress filtering is one of the techniques, which have been developed to reduce the number of opportunities for mischief and malicious damage. While engineers debate its effectiveness, no one debates the fact that ingress filtering is being deployed at many sites on today's Internet.

Stated simply, an ingress filtering border router prevents packets from entering the global Internet if the source IP address does not match any of the expected routing prefixes for the network domain originating the packet. This allows the network administrator for the domain to inhibit malicious nodes from utilizing that domain as a source of masquerading attacks against network nodes in other domains. Note that ingress filtering does not, offer any additional protection to the nodes in the administrator's domain; it may offer some legal protection, if it can be proved that masquerading attacks could not have emanated from the administrator's domain.

Ingress filtering is problematic for Mobile IP, because the whole point of the protocol is to enable communications with correspondent nodes to proceed undisturbed even as the mobile node acquires new care-of addresses. Since applications typically use the Source IP address to identify the mobile node, and the mobile node uses its home address to identify itself for persistent connections, the easiest procedure would be for the mobile node to insert its home address as the Source IP address of its transmitted packets. Unfortunately, Ingress Filtering causes this not to work.

In order to avoid this problem without introducing a need for further negotiation, Mobile IPv6 specifies the Home Address destination option. This new option contains the home address, to be supplied to the application at the correspondent node, while at the same time allowing the care-of address to be used as the topologically correct Source IP address for all transmitted packets. This is perfect, because the care-of address is, by definition, an address which matches the admissibility criterion at the ingress filtering border router.

The Home Address destination option introduces additional complexity at the network layer for several reasons. The recipient of a packet containing this option has to supply the home address to the higher-level protocol layers (e.g., the transport layer) instead of the address in the Source IP field of the IPv6 header. This necessitates a change in the traditional way of programming the IP-level processing. More importantly, it also necessitates changes in the way that security data is verified. For instance, IPSec headers must be processed *as if* the home address were the address originally contained in the Source IP address field of the IPv6 header, instead of the care-of address which was typically in that header field.

Since it is important for all nodes to supply the correct IP address to higher-level protocols, the Mobile IPv6 specification mandates that the Home Address destination has to be implemented by all IPv6 nodes. This unequivocal mandate will have to be seriously studied by all IPv6-related IETF working groups when the Mobile IPv6 specification is promoted to Proposed Standard.

Home Agent Discovery

If a mobile node is away from home, and its home agent crashes or discontinues operation, the movement of the mobile node would cause it to become unavailable to all those Internet nodes except the ones which receive explicit signals from the mobile node. Thus, in order to remain available for answering incoming application connectivity, the mobile node should have a way to discover another home agent on its home network, which could then take over the duties for the mobile node's previous home agent.

There is a reserved range of well-known anycast addresses on every IPv6 network. This enables more useful implementation for IPv6 features compared to IPv4, and is also another example of how IPv6 makes it easier to design the features needed for use with mobile networking. Members of an *anycast group* agree to carry out (by whatever means) the following semantics:

- o Each packet sent to an anycast address is to be processed as if it were received by exactly one of the members of the anycast group.

o Packets (currently) may not be transmitted with an anycast address as the Source IP address of the IPv6 header.

To enable Home Agent Discovery, a well-known anycast address has been allocated for use by all the home agents on any network. The availability of the well-known *All Home Agent's* anycast address makes the procedure very straightforward.

Mobile IPv6 defines two new ICMP messages for the purpose. The Home Agent Address Discovery Request message is sent by the mobile node to the All Home Agent's anycast address. One of the home agents on the home network receives this message, and responds to it with an ICMP Home Agent Address Discovery Reply message. The remaining details about the operation of this discovery mechanism are involved with how the home agents maintain lists of all the available home agents on the home network. In this way, any home agent in the anycast group can supply the greatest amount of useful information to the mobile node. Constructing the list is simple and does not require additional protocols, because home agents are required to advertise their presence on the home network.

Renumbering

IPv6 renumbering is intended to simplify the process by which a network domain can change one or more of its network prefixes. This is a huge problem on today's Internet; renumbering can take many days to accomplish, and the results are usually quite chaotic as unpredictable bugs turn up long after the scheduled end of the renumbering process. Consequently, renumbering rarely occurs. This has the further consequence that the routing tables for the Internet become more and more difficult to aggregate properly. It would be much better if the addressing and aggregation properties of the local administrative domain could follow the business processes of acquisition and merging, as well as retargeting for new Internet Access Providers (IAPs), without the fear of disruption. The processes of address autoconfiguration in IPv6 (e.g., RFC 2461) are designed to expedite this renumbering task. When a router does not renew the lifetime of a previous routing prefix, and it begins to advertise a new routing prefix, the nodes on the network take that as a signal that they should reconfigure their IPv6 addresses so that they will properly match the newly advertised IPv6 routing prefix.

If the home network is renumbered while the mobile node is attached to some other network, all mobile node home addresses on that network may become invalid. This is important information that has to be disseminated in some efficient manner. The exact algorithms are fairly complicated, because of the need to avoid sudden bursts of control information at those times when renumbering events are initiated. The general idea is to allow the home agent to supply to the mobile node all the information which, if it were on the home network, it would receive in the local Router Advertisements. The home agent does this by sending an ICMP Mobile Prefix Advertisement message to the mobile node. If the mobile node detects that the lifetime of its home address is almost due to expire, it can trigger the transmission of the advertisement by sending an ICMP Mobile Prefix Solicitation message to the home agent.

Recent Directions

The basic Mobile IPv6 specification, while incomplete, has features relevant and needed in a very broad realm of applicability. But it is known that there are quite a few situations where additional features are needed that cannot be easily included as part of the base specification. These new areas include:

- Interactions with QoS
- Interactions with Authentication, Authorization and Accounting (AAA), especially third-part authorization
- Firewall management
- Localized Mobility Management
- Seamless Mobility
- Context Transfer
- Paging
- Mobile Routers

In this section, brief sketches of these activities will be presented. Many aspects of these new features are geared in some way to improve system performance for handovers. There are several general ways to classify handovers, and it is important to consider specific solutions in light of each of these different classification schemes:

Intra-domain vs. Inter-domain

Handover signaling can be restricted to occur only between network entities in the same administrative domain, or alternatively it might be allowed between entities in independent administrative domains. In the former case, security needs can much more easily be met, and manual configuration of security associations between network elements is feasible. Indeed, it might even be possible to rely on aspects of the physical security of the domain to simplify or eliminate the need for security altogether. Between domains, no such simplifications are likely, and manual configuration is decidedly unscalable.

Mobile-controlled vs. Network-controlled

Deciding when to change points of attachment from one access router to another, and which new access router to use for connectivity, are important parts of the handover process. These decisions can either be taken by the mobile node itself (*mobile-controlled*), or by network elements within the access network on behalf of the mobile node (*network controlled*). Usually, the decision occurs based on information provided by either the mobile node or the network, so we might describe a particular handover protocol as being "mobile-controlled, network-assisted" if the mobile node makes the decision based in information provided to the mobile node for that purpose by a network element within the access network.

Predictive vs. Reactive

The handover can be attempted in anticipation of certain network events (e.g., loss of signal strength or excessive load). This is called a *predictive handover*, and it always involves an attempt to identify, with a high degree of certainty, the identity of the new access router. Alternatively, any necessary handover operations can be attempted after the mobile node establishes a physical link with a new access router.

In any discussion about the various handover schemes, it is usually very instructive to compare the features of each scheme according to the above-mentioned classifications. If it is possible to have a single scheme work equally well according to all eight potential sub-classifications defined by the above three major classifications, system complexity in support of handovers could be reduced in very important ways. Otherwise, the mobile node and network elements are likely to be deployed with similar but different solutions to nearly alike problems, adding to development time, expense, maintenance problems, and increasing the likelihood for errors.

Seamless Mobility

Of all the recent technology areas listed above, Seamless Mobility is perhaps the most crucial. There are two fundamental aspects of importance. A handover should be fast, meaning that a mobile node should experience very little delay between the time that a link becomes unavailable at one access router, and the time at which a link becomes established at a new access router. The first objective for fast handovers is to meet the delay criterion which will enable handovers for voice connections, i.e., Voice over IP (VoIP). The second objective for seamless mobility is to make the handovers smooth. This has been interpreted to mean that the number of dropped packets should be zero, or if that is not possible, absolutely minimized. Seamless mobility is defined to require handovers that are both fast and smooth. Note that the two ideas are not independent, because when a handover is very, very fast there is less opportunity for packets to be lost or damaged. Nevertheless, it has proved beneficial to consider them separately.

The current discussions about fast handovers for Mobile IPv6 center around the Internet Draft produced by a design team early in 2001. The design team document focuses on predictive handovers only since reactive fast handovers are expected to be handled reasonably well by features already existing in the Mobile IPv6 base specification. The fast handover document specifies the following protocol features:

- Parallelized inspection of router tables to validate prospective new care-of addresses, replacing time-consuming Duplicate Address Detection operations.

    Since the mobile node's new care-of address can often be taken to be the same as its existing care-of address but with a new routing prefix, detecting address collisions can handily be done before the mobile node arrives at the New Access Router (NAR), assuming that NAR is known before the handoff begins operation.

- **Proxy Router Advertisement and Solicitation.**

    The Previous Access Router (PAR) can give the mobile node all the information it would otherwise have to acquire after arrival at the new access network. This includes the MAC address and the IP address of the new access router, so that the mobile node does not have to perform any Neighbor Discovery operations at all. For the mobile-controlled case, a Proxy Router Solicitation sent to PAR is expected to elicit a Proxy Router Advertisement.

- **Handover Initiate and Handover Acknowledgement messages between access routers.**

    These newly specified ICMP messages enable handover information between routers to be passed in parallel during the time that the mobile node is making the physical and MAC-layer handovers to obtain connectivity at the new point of attachment.

PAR sends all relevant information about the mobile node as part of the Handover Initiate (HI) message, and gets an handover acknowledgement (HAck) with positive or negative status indications.

With these features, it is expected that Mobile IPv6 can support fast handovers in the case of predictive handovers.

Context Features for Transfer

The seamless handover functions just described have mostly to do with fixing up the IP-layer (i.e., packet forwarding) operations, as is appropriately within the purview of the Mobile IP working group. Once we can get packets to go to the new access router instead of the previous access router, we have solved the connectionless handover problem. This is, however, not enough, because the expensive radio spectrum requires very special handling.

In order to make the most economical use of the available radio bandwidth, other information related to the mobile node is typically established and associated with the link between an access router and mobile node. When the mobile node moves to a new access router, this other information should also logically follow the mobile node to the new access router. Eventually, this ends up requiring an additional state associated with the link between the mobile node and the access router. The existence of this state information motivates development of ways to transfer context information between access routers.

For instance, as has been mentioned before, header compression is required for VoIP packets sent over the air to a mobile node. It is very important to avoid overhead from IPv6 headers, which are typically 40 bytes, and associated higher level protocol headers (e.g., UDP and Real-time Transport Protocol (RTP), or TCP). Otherwise, over 60% of the available bandwidth could be wasted just on transmitting protocol headers, instead of what is really wanted, the payload.

Header compression works by allowing the receiver to *predict* the contents of the entire header, and thus eliminate the need to actually transmit them. This is quite feasible, since many of the header fields are the same each time, and in typical cases all of the remaining fields can be predicted very accurately if the previous packet's header fields are known. Thus, after one or a few packets have been transmitted with full headers, subsequent packets only require a small header that says, essentially, that the prediction will work, or sometimes that it will work as long as certain small pieces of additional information are taken into account.

The result of this is that a link with efficiently operating header compression engines has states built up at both the compressor and the decompressor. Note that most traffic has at least some traffic in both directions, so that both the mobile node and the access router will have both compressor and decompressor. When the mobile node undergoes handover, it is profitable to transfer this built-up state from the previous access router to the new access router, because the alternative would be to rebuild the state from scratch; this takes time and additional bandwidth, however, both of which may be in short supply.

Examples of context features that should be transferred between access routers include, but are not necessarily limited to, the following:

- o   Header Compression, as just explained
- o   Security Association between mobile node and access router
- o   Buffers, stored to assist with smooth handovers
- o   QoS reservation and accounting information
- o   Multicast group membership

Some context features may not be appropriately handled as part of the smooth handover - for instance, Web caches, and disk shadows, but more generally any bulky or non time-critical data, especially that will not fit as suboptions to the HI and HAck messages. Those handover messages, since they are already being used as signaling for IP, seem very appropriate as containers for the needed additional context feature data.

Context Transfer Framework

We have developed a framework for carrying out the context transfer operations. It was originally conceived as an adjunct to Mobile IPv6 and the Fast Handover mechanism, but the framework itself is more general and can be instantiated in many ways. For instance, there is no dependence on specific Mobile IPv6 messages, so that the framework could work with another mobility management protocol. We have built and tested framework-derived messages with the Mobile IPv6 Fast Handover ICMP messages (Handover Initiate and Handover Acknowledge), but the information could as well be carried as UDP, TCP, or Stream Control Transmission Protocol (SCTP) payloads, with the corresponding changes in retransmission handling and so on. When transported along with HI and HAck, to facilitate seamless handover, our context framework data is carried as suboptions to those ICMP message types.

This all works because the context transfer framework focuses on data definitions, instead of protocol and message flow. The data for various context features is organized according to a generic profile type, which can then be specialized for each of the various kinds of feature context. For example, header compression profile types would represent a subcollection of the generic profile type; RTP over UDP over IPv4 would be an example of a specific header compression profile type. For easier bookkeeping, and perhaps even for programming convenience, we expect that all header compression profile types will share the same leading prefix bits for the profile type. Similar considerations hold for other profile types, for instance, those listed in the previous section. Each individual profile type would be subject to standardized definitions, and the type numbers assigned by Internet Assigned Numbers Authority (IANA) just like other protocol numbers are today. Note that if the data is organized according to standardized formats defined by a profile type, then we do indeed get protocol independence, transport independence, and even can have the same data definitions for reactive and predictive protocol operations.

One of the benefits of the profile type approach is that, for many profile types, the relevant context state is implicit from the profile type, and thus does not need to be explicitly carried as part of the data structure. Other profile types have typical values for some relevant data, but need the flexibility to supply atypical values in some circumstances. We describe these data fields by default values, and allow for any data which conforms to its default value specification to be omitted. For this purpose, we have a *presence* vector; a *1* in the vector means that the data is present, and by inference it has a non-default value. Of course, for many fields, there is no default value: for instance, the SPI field of a security context would have no default value. For programming convenience, we have specified that the presence vector bits are defined the same way for all profile types of a give classification of feature contexts. So, for instance, the *required bandwidth* field of all Quality of Service (QoS) profile types is to be located in the same field relative to the start of the feature context data, regardless whether the profile type specifies defaults for video data, or audio data, and no matter what the encoding is for the data stream that needs the particular QoS context.

Localized Mobility Management

When a mobile node is traveling in a well-defined foreign network, it would be nice to be able to manage all the necessary signaling for mobility management locally, without causing control traffic to traverse administrative network boundaries. This turns out to be feasible, if steps are taken to create additional state information at certain regional routers when the mobile node first establishes connectivity and authorization within the newly visited foreign domain. If signaling can indeed be restricted to having only local effects, better response times and reduced network load within the visited domain can be expected.

Binding Security Association Establishment

When the Security Area Directors (ADs) mandated that the mobile-ip working group reconsider its security model for Binding Update, the problem they observed was that we did not have a realistic method by which general IPv6 nodes might be able to carry out the necessary authentication for the mobility cache management messages (e.g., Binding Update). Authentication is critical, as has been noted previously. In the working group, the general sense had been that someone in the IPSec working group would produce the needed key establishment protocol for the needs of verifying such cache management IP options. That expectation was never realized, and thus Mobile IPv6 was put on hold while a security design investigation was carried out. It has taken quite a while, but the results are very promising. For the first time, there is hope that we can in fact establish the necessary security association (i.e., a Binding Security Association (BSA)) between any mobile node and any correspondent node in the IPv6 Internet. Furthermore, this BSA establishment can be carried out without reliance on a PKI, and without the need for preconfiguring pairwise keys between the nodes which are involved.

The basic idea for establishing the key is that we can rely to a certain extent on the routing integrity of the Internet. After all, if that goes wrong, we are going to have a lot of other problems that are suddenly a lot more important than the integrity of mobile cache management messaging. Since the message flow depends for its validity on the ability of the mobile node to get packets delivered to the various addresses that it claims, the new key establishment method has often been described in terms of a *return routability* test. Most importantly, a correspondent node has to be sure that its protocol partner (i.e., the alleged mobile node) can receive packets that are addressed to the claimed Home Address. Furthermore, for added safety, the correspondent node takes steps to ensure that the supposed mobile node is also addressable at the claimed care-of address. Only when these reachability properties have been ascertained, is the correspondent node expected to carry out the necessary calculations to establish the key for the BSA.

Furthermore, the key establishment protocol has to be designed in such a way that the prospective mobile node does most of the work. Otherwise, it would be all too simple for a malicious network node to trigger complicated processing at the correspondent node, at little cost to itself. This is a variation of a so-called *Denial of Service* attack; if the correspondent node is made to attend to useless time-wasting activities, it will be also unable to carry out its intended functions.

The usual considerations of protocol scalability assume added importance for a future Internet containing a billion IP-addressable mobile nodes. We need to have a protocol by which one server can manage to create security associations with many thousands of mobile nodes; this will be important for mobile commerce, and further emphasizes the need to avoid denial-of-service attacks (for instance as just described) against the mobile commerce servers. The opposite extreme of scalability also deserves equal attention. If various household gadgets, and personal wireless toys, can carry on application communications with some mobile node (e.g., a mobile device) on the Internet, then processing a Binding Update from that mobile node should be cheap. A lot of IPv6 devices are likely to be tiny and computationally limited, and yet must be prepared to be application endpoints for communications with a mobile node. We expect that Binding Cache management messages will be made mandatory for all IPv6 devices, regardless of capacity.

The last design guideline we had to work with, as enunciated by the Security ADs, was 'First, do no harm.' We interpreted this to mean that a mobile node and correspondent node, after running the protocol for establishing a Binding Security Association, would not be any more vulnerable to attack than if the two nodes were stationary and had not run the protocol. Since we are only required to establish security enough to authenticate Binding Updates, which may be viewed as routing control messages, the Return Routability messages are sufficient. But they do not necessarily supply enough security assurance against other kinds of attacks. We believe that by allowing the mobile node to prove that it is reachable at the home address, and separately reachable at the prospective care-of address, the routability concerns are satisfied.

The messages by which these proofs are supplied are as follows:

- o   Home Address Test Initiate (HoTI)
- o   Care-of Address Test Initiate (CoTI)
- o   Home Address Test (HoT)
- o   Care-of Address Test (CoT)

The mobile node sends the CoTI message from its care-of address (i.e., used as the Source IP address). At about the same time, it reverse tunnels the HoTI message back to the home agent. The HoTI message is sent from the mobile node's home address, but the encapsulating header for the reverse tunnel operation contains the mobile node's care-of address so that ingress filtering can work.

When the correspondent node gets the CoTI message, it sends back a CoT message to the care-of address (directly) containing a random number which supplies part of the key material which will form the basis of the BSA. Similarly, when the correspondent node receives the HoTI messages, it sends back another random number to the mobile node's claimed home address. This second number is the other part of the key material needed by the mobile node to establish the BSA. In order to avoid keeping per-mobile state, the correspondent node just keeps track of two *current* nonces, and consumes them whenever it receives a verifiable Binding Update from the mobile node which utilizes those nonces. Using nonces at the correspondent node in this way eliminates most of the threat of a Denial of Service attack. We also need to make sure that the computation that produces the keys from the nonce is simple enough, which is done by using Secure Hash Algorithm 1 (SHA-1), a relatively lightweight one-way hash function. Since the return routability protocol also avoids other bottleneck functions, we believe that it is scalable, and can be shown to *do no harm* by straightforward case analysis.

Mobile IPv6 Status

Mobile IPv6 has undergone a great deal of testing for interoperability and protocol correctness. One of the first events was at INRIA on Sept 15-17, 1999, and there have been many more. Significantly, European Telecommunications Standards Institute (ETSI) has hosted a bake-off, on October 2-6, 2000. This is one indication of how important Mobile IPv6 may become for future communication standards. There have been three years of interoperability testing at Connectathon, with success growing each year. All the interoperability test events have provided valuable insight into the viability and correctness of the specification.

Conclusions

We believe that the future Internet will be largely wireless and mobile. IPv6 addressability will be needed for billions of mobile devices. Mobile IPv6 builds on the successful design of IPv6 to offer better and more efficient service than Mobile IPv4. Autoconfiguration is suitable for the Mobile Internet, and has been preserved in the design for Mobile IPv6. Security is a key component for success, and security design improvements have occupied the recent attention of the mobile-ip working group with some fair amount of success. Nevertheless, Mobile IPv6 is neither a complete solution for Voice over IP, nor for the needs of third generation mobile networks. In the former case, additional work is needed for smooth handovers and context transfer. In the latter case, a lot of development is occurring to establish an authorization and authentication infrastructure for mobile communications carried over very expensive spectrum.

Multimedia Sessions in the Mobile Domain

The section "Multimedia Sessions in the Web Domain" covered the basics of using Session Initiation Protocol (SIP) and other Internet Engineering Task Force (IETF) protocols to control voice calls and other multimedia services in the Web Domain. The purpose of this section is to describe how these protocols are suited for the same task in the Mobile Domain.

Characteristics of the Mobile Domain

There are certain issues when applying SIP and other Internet protocols to the Mobile domain, mainly stemming from the fact that in mobile networks the quality of the last hop link is not as good as in wireline networks in terms of capacity and packet loss rate. Also, the quality may have big variations, and even total connectivity disruptions are regular. In order to make real-time services over IP efficient in mobile networks, various compression and robustness mechanisms need to be applied. Engineering a VoIP device is much harder than a regular handset with circuit-switched voice.

On the other hand, the Mobile domain has several advantages over the general Internet as well. Mobile devices have a slot for the Subscriber Identity Module (SIM), which can be used to identify and authenticate the consumer. Also, carriers have clearly-defined business relations with each other allowing roaming and charging agreements. Currently these only cover Circuit Switched networks, but General Packet Radio Service (GPRS) roaming is also starting to emerge. The same mechanisms can be applied to SIP and even to other application level protocols.

3GPP IP Multimedia Subsystem

Currently the Third Generation Partnership Project (3GPP) is specifying an IP Multimedia Subsystem (IMS) [1] on top of 3G GPRS access networks to offer voice and multimedia services for 3G mobile devices. IMS is basically completely access independent, so it can also be used with 2.5G mobile networks or Wireless Local Area Network (WLAN). Even the GPRS core could be replaced with, e.g., a Mobile IP-based approach without major impact on the IMS. Naturally, the access network must provide Quality of Service support if voice or other real-time applications are to be utilized. IMS first appears in 3GPP Release 5 specifications. In this phase it will provide basic session setup and media delivery services, as well as a platform for presence and instant messaging. In Release 6, all the features required for complete Circuit Switched network replacement, such as emergency calls or legal interception, will also be supported.

IMS uses several of the protocols described previously, such as SIP, Session Description Protocol (SDP) and Real-time Transport Protocol (RTP). It can be assumed that high-end mobile devices will also include protocols like HyperText Transfer Protocol (HTTP) and Real-Time Streaming Protocol (RTSP). The IMS architecture is shown in the following figure. The principle is that only IPv6 addresses are used in the IMS, as otherwise there would not be enough public addresses for all devices. In practice, IPv4 is also supported.

232 Mobile Domain Technologies – Mobile Internet Layer

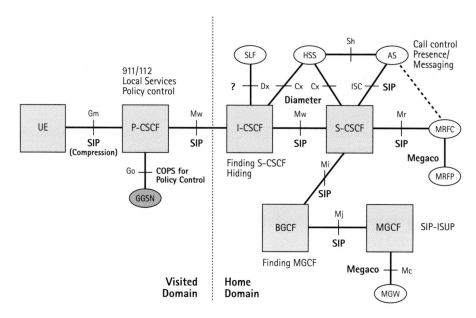

3GPP Release 5 IP Multimedia Subsystem Architecture

The figure only shows the IMS components; the radio network or GPRS core is not present except for the Gateway GPRS Support Node (GGSN), which has a direct interface to IMS. One would be situated between User Equipment (UE) and Proxy-Call State Control Function (P-CSCF) providing basic IP connectivity and mobility management below it. UE is basically the mobile device. It talks with SIP to P-CSCF which is very similar to a SIP proxy server. If the consumer is roaming in a visited domain, P-CSCF is located there. The role of P-CSCF is to provide emergency call and other such services requiring specific knowledge of the visited domain.

The serving CSCF (S-CSCF) is always located in the consumer's home domain, and it takes the role of the SIP registrar and proxy servers, so a consumer registers there using SIP via P-CSCF. Interrogating CSCF (I-CSCF) is another SIP proxy responsible mainly for finding the correct S-CSCF for a given consumer. S-CSCFs can be dynamically allocated per registration in order to achieve efficient load balancing and error resiliency. The Breakout Gateway Control Function (BGCF) and Media Gateway Control Function (MGCF) are used for telephony network interconnection. Media Resource Function Control and Processing (MRFC/MRFP) are related to multi-party conferencing, network announcements and media transcoding. The Application Server (AS) is a SIP element dealing with services, such as advanced call control, presence or instant messaging. There can be separate ASs for different purposes. AS can also have, e.g., an HTTP interface in addition to SIP. Home Subscriber Server (HSS) and Subscriber Locator Function (SLF) are related to consumer profile management and authentication.

Basically, IMS is just an instance of a SIP network. It has a number of proxies and a registrar, although they have strange acronyms. The User Agent is situated in the UE. When two devices establish a session they talk to each other via the CSCF elements in the same way as presented in the previous figure. RTP media flows do not go via CSCFs but go directly between the devices.

There are, however, some differences. As mentioned, one of the most pressing issues with the mobile network is the scarcity of bandwidth. This means that each protocol should be as compact as possible when sent over the air. In practice, Internet protocols do not meet this requirement as such. For example, when voice is carried in Real-time Transport Protocol/User Datagram Protocol/Internet Protocol (RTP/UDP/IP) packets, over half the size of the packets may be header information. In order to cope with this, the Internet Engineering Task Force (IETF) has defined Robust Header Compression (ROHC) [2] for RTP/UDP/IP, reducing the size of the headers to a couple of bytes in most scenarios. In this way, VoIP transport becomes fairly efficient compared to traditional cellular voice. 3GPP has adopted ROHC for this purpose.

Further methods, such as unequal error protection for speech frames over the air interface, can be applied if even more efficiency is needed. The Adaptive Multi-rate (AMR) codec is suitable for the mobile networks, as it has several modes with different bitrates, allowing it to be used in a dynamically changing radio channel.

The efficiency requirement applies to signaling as well. All application layer signaling protocols related to the multimedia sessions are text based and rather large compared to traditional mobile signaling protocols. For example, an SIP INVITE can be almost 1000 bytes long and it takes a while to send that amount of data over a narrow radio link. Also, a number of round trips are needed to setup the session with guaranteed QoS. A typical call setup sequence is shown in the following figure.

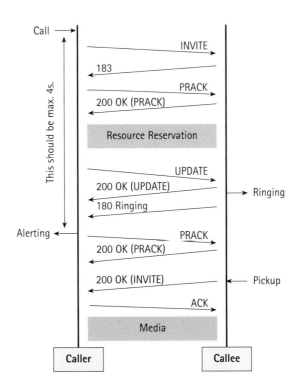

SIP session establishment with resource reservation

In order to make the call setup time comparable to current Global System for Mobile Communications (GSM), signaling compression is needed. In this type of compression, dynamic dictionaries can be used to boost the compression efficiency. Thus, a relatively good compression ratio can be expected. The signaling compression will be placed on the application layer in between transport and the actual signaling protocol. The IETF ROHC working group has defined a generic signaling compression framework, supporting a wide variety of different compression algorithms through the revolutionary Universal Decompressor Virtual Machine (UDVM) approach. The actual compression algorithm is run in UDVM using a standardized instruction set. It is expected that the most efficient algorithms might achieve compression ratios up to 7:1.

In IMS, the consumer is authenticated using an SIM module during SIP registration. This is achieved by running 3GPP's Authentication and Key Agreement (AKA) algorithm on top of the SIP REGISTER transaction. S-CSCF accesses HSS for authentication information using the Diameter protocol and SIM is used to perform the calculations in the device. In addition to authentication, AKA also provides keys for the integrity and confidentiality protection of the further SIP messages.

As geographical location is readily available in mobile networks, it is possible to include it in SIP based services too. It is possible that the terminal always includes Cell Identity information in every SIP request, and the request can be then handled based on this information. In this way it is possible, for example, to connect calls to the geographically closest service centers.

Conclusions

In principle, Internet protocols work regardless of the link layer. However, in the mobile networks, specific enhancements, such as compression for SIP and RTP, are needed to make the protocols efficient enough for the air interface. 3GPP is the standards body working on the IP-based architecture for multimedia communication. The SIM module in the handset can be utilized to authenticate the consumer on the SIP-level as well, making the system easy to use for the subscriber.

References

[1] 3GPP Release 5 IP Multimedia Subsystem, 3rd Generation Partnership Project; Technical Specification Group Services and System Aspects; IP Multimedia (IM) Subsystem - Stage 2; (3G TS 23.228 version 5.0.0).

[2] Robust Header Compression (ROHC), IETF RFC 3095.

Java™ in the Mobile Domain

The Java™ platform has developed into three main types of platform editions. The Java 2 Platform, Enterprise Edition (J2EE™), is used for developing Java-based, reliable and scalable Web and enterprise applications. The Java 2 Platform, Standard Edition (J2SE™), is commonly found in desktop workstations and personal computers. The Java 2 Platform, Micro Edition (J2ME™) is designed for use in resource-constrained mass-market mobile devices. The figure below gives an overview of these platforms. This chapter concentrates mainly on the role of the J2ME in the Mobile domain.

Because a wide range of mobile devices exists for a variety of different types of uses, and because such devices can be very resource constrained, further refinement of the J2ME platform is needed. For this reason, the concepts of *Configurations* and *Profiles* are used in the J2ME platform to define specific families of resource-limited devices and their capabilities.

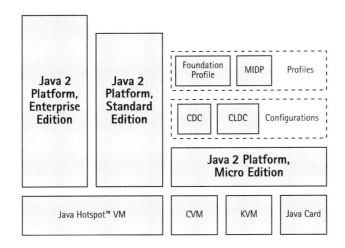

Java platform editions

Mobile Devices and Resource Constraints

Mass-market consumer devices are typically more constrained than fixed desktop and enterprise devices, in terms of the following resources:

o   Memory and persistent storage may be limited depending on the price-point, intended use and physical dimensions of a device. In multi-purpose devices, memory may be shared between the different features supported on a device, which may further limit its availability per feature.

- o The CPU used in such devices can be chosen based on the intended price-point and use of a device, and also for power consumption.
- o The physical dimensions of a mobile device generally limit its display size, hence smaller devices have smaller display sizes. On some small devices, a display might not even exist. In addition, a device's screen may be black/white, grey scale or color. Color screen capabilities may also vary (e.g., the number of colors).
- o The physical dimensions, the locale and intended use of a device can place limits on a keyboard's size and layout. For example, many mobile devices have a characteristic *one handed* keypad, which is very different from the keyboards found on desktop workstations.
- o Networking is also rather different than in fixed workstations. Mobile devices tend to be used in a way that emphasizes mobility, leading to mobile networking. Mobile networking typically differs from fixed broadband networking in terms of lower bandwidth, greater latency (and jitter), and lost connections when network coverage is disrupted. Use of wireless networking may also affect battery consumption, which is typically not a concern for a fixed device that is plugged into an appropriate fixed power supply.
- o Power consumption is often an issue for *unplugged* small mobile devices, since they generally operate on batteries. Power consumption affects how long consumers may use a device for specific purposes before they have to recharge its batteries.

Java 2 Platform, Micro Edition

The previous section gave an overview of some important properties of mobile devices. The Java 2 Platform, Micro Edition is a highly optimized runtime environment targeting such products. It covers a wide range of potential devices, (e.g., mobile devices, set-top boxes, and car navigation systems). It is an adaptation of the existing Java platform to smaller and more constrained environments.

An important goal of Java is *write once, run anywhere* portability within an edition (and also, as much as possible, between different editions of the Java platform). This goal of portability within an edition, across different manufacturer implementations, is more easily achieved in the more resource rich J2SE and J2EE than in the J2ME. This is because the enormous diversity of device types and constraints which may be encountered are much greater than are found in the larger footprint versions of Java.

To cope with this reality and to still achieve the goal of portability, the J2ME is further sub-divided into smaller entities called *Configurations* and *Profiles*. A configuration defines a basic set of application programming interfaces (APIs) and features of a virtual machine that are expected to be the same for devices with similar amounts of resources. Profiles enable portability by defining domain-specific sets of APIs.

Currently, there are two configurations defined. Both of them target networked devices. Networking means a device uses either a fixed or mobile communication link to a network.

o   **The Connected Limited Device Configuration (CLDC)** is for devices with very tight resource constraints. For the CLDC 1.0 virtual machine, several features of the virtual machine such as floating point math (the forthcoming version 1.1 of CLDC will include floating point support), object serialization and reflection have been excluded. In the product portfolio of Sun Microsystems, the virtual machine supporting CLDC is called KVM.

o   **The Connected Device Configuration (CDC)** is for devices with more resources than CLDC. The underlying virtual machine is assumed to have full virtual machine support. In the product portfolio of Sun Microsystems, the virtual machine supporting CDC is called CVM.

While configurations are targeted at devices with particular amounts of available resources, the domain-specific sets of APIs are defined in profiles. Some examples of profiles include:

o   **The Mobile Information Device Profile (MIDP™)** is an API set designed for resource constrained devices (e.g., mobile phones, two-way pagers, PDAs). It uses the Connected Limited Device Configuration as its basis.

o   **The Personal Profile** provides a J2ME environment for devices with a high degree of Internet connectivity and Web fidelity. It is a successor to Sun's PersonalJava™ environment, and so has the explicit backwards compatibility requirements for that.

o   **The Foundation Profile** is a set of APIs meant for applications running on small devices which have some type of network connection, but are not required to have a graphical display. It is uses the Connected Device Configuration as its basis.

Because of their obvious relation to the Mobile domain, the remainder of this chapter concentrates on the Connected Limited Device Configuration and the Mobile Information Device Profile. Many good starting points [JAVAMOB1] [JAVAMOB3] [JAVAMOB4] exist that provide more detailed information on the Connected Device Configuration and the other profiles listed above.

Mobile Information Device Profile Version 1.0

The first version of the Mobile Information Device Profile was finalized in 2000. The main features of MIDP 1.0 are:

o   The core java.lang and java.util libraries are defined by CLDC.

o   Timer APIs for delayed and periodic execution of events are supported.

- A Networking API based on the Generic Connection Framework of CLDC is included. The idea of the Generic Connection Framework is to provide an abstraction layer over several communication mechanisms, such as Transport Control Protocol/Internet Protocol (TCP/IP) sockets.

An application using the Generic Connection Framework need only provide the destination Uniform Resource Locator (URL) and the system returns a connection object for the application. For most communication, the application does not need to know the detailed connection class and thus the application can encapsulate the connection-dependent issues in the URL string.

- MIDP 1.0 requires that the *HttpConnection* class must be implemented. *HttpConnection* is defined in such a way that it can be implemented either by using the Wireless Application Protocol (WAP) protocol stack or by using a TCP/IP protocol stack.
- The user interface library (LCDUI) was specifically designed with the constraints (screen and keypad) of typical mobile devices in mind. The API design is based on two layers. The high-level API is designed for business applications whose client parts run on mobile devices. For these applications portability across devices is important. To achieve this portability, the high-level API employs a high level of abstraction and provides very little control over look and feel. Also, by using the high-level API, the application can utilize the native look and feel of the system so the consumer is immediately familiar with the look and feel.
- The low-level API, on the other hand, provides very little abstraction. This API is designed for applications that need precise placement and control of graphic elements, as well as access to low-level input events. Some applications also need to access special, device-specific features. A typical example of such an application would be a game.
- The Record Management Store (RMS) API was defined for storing small data items in persistent storage.

In order to maintain the integrity of the system, all MIDP applications are run under the control of the Application Management Software. It is responsible for installation of applications, starting and stopping of applications and ensuring that the basic functionality of the mobile device is not compromised by the Java applications.

An application that conforms to the MIDP and CLDC specifications is called a MIDlet.

Mobile Information Device Profile Version 2.0

The second version of MIDP adds a number of improvements and new features. Some of the feature candidates are:

- User Interface: Several improvements and enhancements to the lcdui package were added. A Game API was defined with a rich set of features for developing two- dimensional

(2D) games. A minimum set of audio capabilities was added, and optional support for other types of sounds was defined.

o Security and Networking: In MIDP 1.0, all MIDlets are untrusted from the device's point of view. A standard "sandbox" security model is used. In MIDP 2.0, support for signed applications and a security model was defined, as well as support for secure HTTP and socket connections.

It is important to note the MIDP 2.0 has many important new features, but still maintains backward compatibility with MIDP 1.0.

MIDlet Provisioning

There are several possible ways for a consumer to discover and install MIDP applications onto their mobile device. Applications can be pre-installed at the factory, downloaded over a serial cable from a PC, or downloaded over-the-air by an application on the mobile device, such as a browser or other provisioning application. Over the Air (OTA) downloading is expected to be the most important means for downloading new MIDlet applications to mobile devices.

OTA downloading is covered by a *recommended practices* document for MIDP 1.0. The MIDP 2.0 specification formally includes an *Over the Air User Initiated Provisioning* specification.

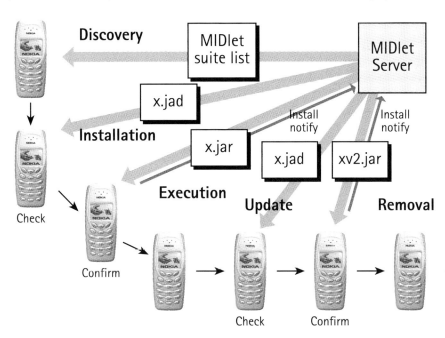

MIDlet deployment and lifecycle

The previous figure illustrates the main interactions involved in over-the-air downloading of MIDlet content to a device. The process involves the following phases:

Discovery

The consumer uses a browser on his mobile device to browse Web pages on a remote server. The consumer finds a browser page containing a link to an interesting MIDP application. Actually, the link points to a Java Application Descriptor (JAD) which contains the core attributes of the application.

Installation

When the consumer clicks on the link, the browser downloads the descriptor and provides it to the Application Management Software (AMS).

The AMS uses the attributes in the descriptor to check if the device is able to run the application, and also checks if there are other reasons that would prevent downloading. If application downloading can be done, the AMS uses the URL given in the descriptor to locate and download the Java Archive (JAR) file from a MIDlet Server.

Next, the AMS installs the MIDlet to the device, perhaps after asking for permission from the consumer. At this point, the AMS may also send a confirmation notification to the server. The server uses this confirmation for billing and consumer care purposes.

Execution

The application is then started and used as a separate consumer action. A MIDlet may be standalone or networked.

Update

An application may be updated in a similar fashion as above, with a newer version of the MIDlet.

Removal

Eventually, the consumer may wish to delete the application. The AMS will provide services to allow the consumer to do this. The supplier of the MIDlet may also wish to be notified of this event.

MIDlet Networking

Client to server communication in MIDP is supported on top of HyperText Transfer Protocol (HTTP) communication, as a mandatory feature of the specification. HTTP itself can be supported on top of a variety of underlying transport protocols.

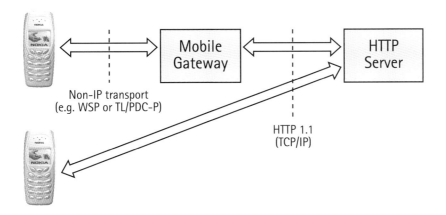

MIDlet networking

A networked MIDlet client uses services contained in a backend HTTP Server. The device establishes a client-role HTTP connection for such connections. The details of how the network connection is setup depend on the type of network and mobile device used. While a connection is open, the MIDlet converses with the server by means of performing simple HTTP GET and PUT operations.

A common technology choice for the HTTP Server will be the J2EE. The communication protocol between a MIDlet and Web tier or Business tier component can be an ad-hoc protocol, or an Extensible Markup Language (XML) messaging based protocol, or an XML-based remote procedure call (RPC).

However, when technologies such as XML or Simple Object Access Protocol (SOAP) are used for messaging or RPCs, a MIDlet must include the associated parsers and other support libraries. This may affect the amount of time needed to download a MIDlet over-the-air, or significantly impact the amount of space consumed by the MIDlet when it is stored on the device.

A Java Specification Request (JSR) has been initiated to define an optional package that would provide standard support from J2ME™ to Web Services [JAVAMOB10]. If included in a J2ME device, it might help to provide standardized support for XML parsing and XML-based-RPCs. This could reduce the size of downloaded J2ME applications, such as MIDlets.

Mobile Media

In addition to the new user interface capabilities defined in MIDP 2.0, some MIDP devices may eventually support the Mobile Media API [JAVAMOB9]. It defines a multimedia API for J2ME™. This small-footprint API allows for easy and simple access and control of basic audio and multimedia resources. It also addresses scalability and support for more sophisticated features.

Conclusions

This chapter gave a broad overview of the role of Java technology in the Mobile Domain. It concentrated on the benefits of using the J2ME.

As a programming platform, Java brings many benefits to mobile device programming. But more importantly, it enables application development for a large new market of mobile device owners. From the consumer's point of view, Java brings new types of services with rich graphics and interaction, such as games and other types of applications. Java also enables increased personalization of the mobile device.

The number of Java-enabled mobile devices has increased rapidly in the past few years. Combined with the popularity of Java-based Web tier and Business tier servers, Java technology forms an important foundation for development of end-to-end applications in the Mobile domain.

References

[JAVAMOB1]	The Java™ 2 Platform, Micro Edition, http://java.sun.com/j2me
[JAVAMOB2]	Connected Limited Device Configuration (CLDC) and the KVM Virtual Machine, Java™ 2 Platform, Micro Edition, http://java.sun.com/products/cldc
[JAVAMOB3]	Connected Device Configuration (CDC) and the CVM Virtual Machine, Java™ 2 Platform, Micro Edition, http://java.sun.com/products/cdc
[JAVAMOB4]	Foundation Profile, Java™ 2 Platform, Micro Edition, http://java.sun.com products/foundation
[JAVAMOB5]	Mobile Information Device Profile (MIDP), http://java.sun.com/products/midp/
[JAVAMOB6]	The Java™ 2 Platform, Enterprise Edition, http://java.sun.com/j2ee
[JAVAMOB7]	Nokia and Java™ Technology, http://www.nokia.com/java
[JAVAMOB8]	JSR118: Mobile Information Device Profile 2.0, http://www.jcp.org/jsr/detail/118.jsp
[JAVAMOB9]	JSR135: Mobile Media API, http://www.jcp.org/jsr/detail/135.jsp
[JAVAMOB10]	JSR172: J2ME™ Web Services Specification, http://www.jcp.org/jsr/detail/172.jsp

Data Synchronization in the Mobile Domain

An important factor in the increasing popularity of mobile computing is the possibility to use several copies of the same information. Personal information (e.g., Calendar, Contacts, and Tasks) is a prime example of this kind of important shared data. Several consumers access the same information, make updates, and act based on the information. The majority of modifications to the data are made on personal copies of the information, and these are then updated to the general copy available to everybody. The process of making the databases identical is called data synchronization.

Unfortunately, there are still major challenges to be resolved, beginning with data synchronization protocols, which are definitions of the workflow for communication during the synchronization. There have been a myriad of proprietary synchronization protocols, each of which have had only a limited number of supported applications, operating systems and transport protocols. As a result, consumers have not been able to share their information with others in a practical and efficient way.

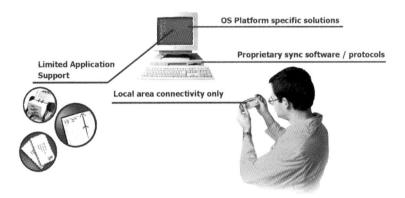

The synchronization challenge

SyncML is the next generation in synchronization protocols. It is an open standard for universal data synchronization, which supports a variety of transport protocols and arbitrary networked data. It is easily expandable and based on eXtensible Markup Language (XML), which is widely considered a future-proof choice. SyncML effectively addresses the resource limitations of mobile devices and has been optimized to work with a variety of applications and devices. SyncML is the only available synchronization protocol, which can deliver this kind of universal connectivity.

SyncML technology has firm market acceptance and is becoming the protocol of choice for universal data synchronization. At the moment, there are over 650 companies supporting SyncML technology worldwide. Ericsson, IBM, Lotus, Matsushita, Motorola, Nokia, Openwave, Starfish Software and Symbian are actively contributing to the development of the technology. In addition, major communications industry bodies - Third Generation Partnership Project (3GPP) and WAP Forum - have adopted SyncML as the preferred remote synchronization protocol.

SyncML Technology

SyncML is based on the server-client architecture, where a SyncML server handles the synchronization session between two databases. As illustrated in the following figure, a consumer connects with a SyncML enabled mobile device to a SyncML synchronization server. The SyncML server is connected to the remote database, which may also be updated via a Web-calendar by other consumers. The SyncML server handles the resolution of conflicting entries and updates the information both on the mobile device, and the remote database. The information in the handset and the remote database become synchronized and, thus, identical.

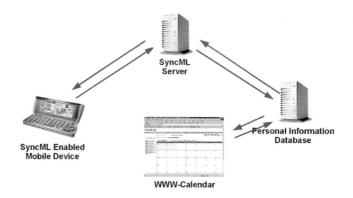

A SyncML synchronization session in practice

SyncML is transport independent and, thus, capable of synchronizing data over many different data protocols used by wireless and networked applications. It works smoothly and efficiently over the Internet (HTTP), the Wireless Session Protocol (WSP), Object Exchange (OBEX), e-mail standards (e.g., Simple Mail Transfer Protocol (SMTP), Post Office Protocol 3 (POP3) and Internet Message Access Protocol (IMAP)), Transmission Control Protocol/Internet Protocol (TCP/IP) networks and other wireless and wireline communication protocols. The most commonly used protocols are HTTP, WSP and OBEX.

SyncML		
HTTP	WSP	OBEX
Internet/Intranet	WAP	IrDA, USB, Bluetooth

Current SyncML transport bindings

In short, SyncML offers a versatile tool for developing data synchronization capabilities for any service concept. The protocol does not mandate how data is represented or structured on the device or within the networked data repository after synchronization is complete. Since SyncML enables mobile devices to communicate with a vast range of data, the protocol also supports simultaneous synchronization with multiple network data stores.

In the first phase, SyncML devices synchronize Personal Information Management (PIM) data (e.g., contacts/phone numbers, calendars, to-do lists). This provides carriers and service providers with an opportunity to offer new sticky services, where the personal information of consumers is stored and synchronized with their system. By combining SyncML and time-critical content, calendaring can be turned into a valuable service where interesting events and time-critical information are fed into mobile devices. The consumers will receive the information they need and want at any time and in any place.

Security

The SyncML specification uses the security mechanisms of the underlying transports. Thus, no restricting security schemes have been defined. SyncML authentication is based on basic or digest access authentication defined by Internet Engineering Task Force (IETF): Base64 coding and MD5 are used, respectively. Authentication can be performed on several levels: between the synchronization server and a client, on the database level or on operation levels. Other authentication schemes can be specified by prior agreement between the originator and the recipient.

Interoperability

To ensure interoperability, the protocol describes how common data formats are represented over the network. To ensure extensibility, the protocol permits the definition of new data formats as needed. In addition, the protocol allows implementers to define experimental, non-standard synchronization protocol primitives. The most commonly used data formats, which the protocol supports from the outset include vCard for contact information and vCalendar for calendar, to-do list and journal information.

Interoperability is ensured by a specific interoperability testing procedure determined by the SyncML Initiative, a non-profit organization developing SyncML technology. Currently, the SyncML Initiative has made available a self-test for testing a specification conformance and holds SyncFest events for interoperability testing with the products of other vendors and manufacturers. The First SyncFest was held at the end of February 2001 in Europe, and it has been followed by tri-monthly events in different geographical locations.

Conclusions

SyncML is rapidly evolving as the de facto standard in universal data synchronization. The SyncML Initiative, an open industry organization for SyncML technology has firm industry backing.

The market is likely to experience explosive growth. In June 2001, Nokia introduced the first commercially available SyncML enabled mobile device, the Nokia 9210 Communicator. Since then, other SyncML enabled mobile devices and servers have also appeared on the market. Several carriers, service providers and corporates are introducing their first SyncML enabled services.

Mobile Domain Technologies – Mobile Internet Layer

The first services will concentrate on offering personal information synchronization, and combining the information with interesting content. In the future, SyncML is going to be expanded to support new data types (e.g., images, files, and database objects). New expansions will be used to build and offer vertical applications for corporate users (e.g., company files, sales data, SCM, and fleet management) and consumers (e.g., information services and interactive games).

SyncML enabled services

Currently the SyncML initiative is also looking into opportunities to leverage the forthcoming SyncML device base in remote device management, which would, for example, allow carriers to configure mobile devices over the air.

Middleware in the Mobile Domain

The current trend in developing future mobile networks is to utilize Internet protocols, with the immediate implication that IP is the layer 3 protocol and addresses are IP addresses. This is not sufficient, however. Other solutions both above and below the IP protocol are also needed to meet the requirements for the next generations of mobile networks. Issues under study in the Internet community and in various standardization bodies, forums and consortia include mobility, Quality of Service (QoS), security, management of networks and services, discovery, ad-hoc networking, dynamic configuration and location.

Another significant trend is the requirement of ever-faster service development and deployment. The immediate implication has been the introduction of various service/application frameworks platforms. Middleware is a widely-used term denoting a set of generic services above the operating system. Although the term is popular, there is no consensus for a definition [1].

Typical middleware services include directory, trading and brokerage services for discovery, transactions, persistent repositories and different transparencies (e.g., location transparency and failure transparency). Examples of middleware include Common Object Request Broker Architecture (CORBA) [2], Java 2™ Enterprise Edition and Java 2™ Micro Edition (J2EE™ and J2ME™) [3, 4], Distributed Component Object Model (DCOM) [5], and Wireless Application Environment (WAE) [6]. Characteristically, the competing middleware specifications provide many similar but slightly different services. In order to overcome the problems due to different specifications, the Parlay Group [7] has specified a set of Unified Modeling Language (UML) models and corresponding Application Programming Interfaces (API) that can be implemented in CORBA, Java and DCOM environments.

This section examines ways to incorporate existing middleware solutions and Internet protocols in the Mobile Internet.

Application Requirements

The next figure depicts a highly abstracted vision of how a Service Application is distributed among various application servers, network elements and mobile devices. It should be noted that, for simplicity, the figure only shows a single mobile device although multi-party applications will be much more important and challenging than one-party applications. In addition, we must also be ready to cope with solutions based on proximity and home communication systems.

The execution environments or the service stratum consist of middleware, operating systems and protocol stacks, which should support fast service development and deployment.

The service stratum should make it easy to divide the application logic into cooperating parts or components, to distribute and configure these components as well as to redistribute and reconfigure them. Additional requirements for future mobile applications include adaptability to changes in the execution and communication capabilities, efficient use of available communication resources, dynamic configuration of consumer devices as well as ultimate robustness, high availability and stringent fault-tolerance.

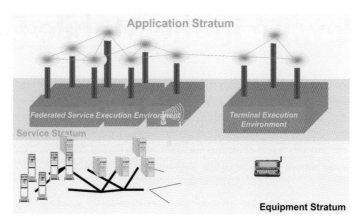

Mobile Distributed Application

The requirements for data accessed by these applications are quite similar. The execution environment should provide a consistent, efficiently accessible, reliable and highly-available information base. This implies a distributed and replicated world-wide file system which also supports intelligent synchronization of data after disconnections.

Internet Protocols and Middleware

The following figure decomposes the abstract notion of the execution environment in the previous figure into a layered architecture. It should be noted that the operating system is not shown in this figure because it would introduce another dimension. In most operating systems the layers below the middleware are encapsulated into program libraries which are available e.g., through system calls, and file descriptors. The main stream in the industry seems to be the "Internet Protocol (IP)-Transmission Control Protocol (TCP)-HyperText Transfer Protocol (HTTP)-eXtensible Markup Language (XML)" protocol stack, despite the fact that:

- o TCP has performance challenges, [8, 9, 10],
- o HTTP uses TCP capabilities inefficiently, and
- o XML is resource consuming to process.

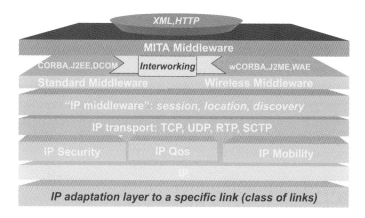

Layered Internet Protocol Architecture for Middleware

In TCP transport over a wireless path, there have been significant enhancements due to the proposals of the IETF working groups on transport area (*tsvwg* [11]), performance implications of link characteristics (*pilc* [12]), and endpoint congestion management (*ecm* [13]). The HTTP transport over a wireless link has not yet been addressed in IETF, but practical proposals are available [14, 15]. For XML the situation is even worse. The WAP forum has proposed a binary XML format [16], but this only addresses the efficient presentation of tags. The real problem is how to present tag values efficiently on the wire. The XML Protocol Activity in W3C [17] has laid the basis for a solution but has not yet solved this problem.

Programming Models

The programming model of the Internet is based on sockets, protocol specific APIs and the state machines of application protocols, which need to be implemented separately for each protocol. Although the Internet approach is efficient for a single protocol, it is very cumbersome and expensive for application development involving several operating systems and protocols. Moreover, many Internet protocols do not explicitly define the presentation format on the wire, which has introduced serious interoperability problems between implementations. Another concern is that the Internet approach is prone to feature interactions since many protocols are developed independently of each other.

In middleware solutions, particularly in CORBA, a considerable amount of effort has been put into enhancing interoperability in heterogeneous operating environments. This has resulted in a uniform baseline protocol which also specifies the presentation format. In the case of CORBA, the unifying protocol is the Generic Inter-Object Request Broker (ORB) Protocol (GIOP) and the presentation format is the Common Data Representation (CDR), which specifies how different data types and complex data structures are put on the wire. Therefore, many details that need to be handled by applications in the Internet solution are handled by the platform in the middleware solution.

When the CORBA approach - as well as the Java and DCOM approach - is compared to the Internet approach, the conclusion is that the middleware solutions provide a more convenient programming model than the Internet solution. The middleware solutions are usually based on object-oriented programming and method invocations. These invocations are based on strongly typed interfaces which provide both compile and run time error checking. They also hide many implementation details. Therefore, middleware-based application development is much faster than Internet-based development.

The fundamental problem of the current middleware specifications is that they only take advantage of a narrow subset of useful Internet protocols. The current middleware specifications were born in a time when the Internet protocols were a synonym of TCP/IP transport. Later they have developed solutions of their own for Quality-of-Service, directory, discovery, and so on, independently of each other and IETF specifications.

Next Steps

From the software architecture point of view, the fundamental challenge is how to harmonize developments in Internet protocols and different middleware solutions. This requires both an overall vision on a broad frontier of standardization activities and deep insight into various specifications. In particular, we do not want to discard existing middleware solutions since they are well-established and provide obvious benefits for application development.

By harmonization we mean two things. Firstly, we need to solve the problem of incorporating evolving Internet solutions of QoS, mobility, discovery and security into the existing middleware specifications without breaking those specifications. Secondly, we need to discover how different middleware solutions can become interoperable, in the sense that the components of an application can be executed on different middleware platforms.

In middleware standardization, we have focused on interoperability and interworking, particularly for aspects of mobile or nomadic users and wireless links. By active involvement we try to reduce the footprint of run-time software as well as the memory requirements and computational complexity of system software and application components in the consumer device.

The increasing diversity of devices - terminals, network elements and application servers - imply that different middleware solutions will also be used in the future. This heterogeneity requires interoperability on two levels: between middleware platforms and between parts of an application running on different middleware platforms.

Interoperability between different platforms is quite mature. In contrast, interoperability between parts of an application running on different middleware platforms is still immature. There are practically no tools available to support this kind of interoperability, and the burden of interoperability is left totally to application developers.

The Object Management Group (OMG), however, has started comprehensive development of a new architecture called the Model Driven Architecture (MDA) [19]. The objective is to interrelate Interactive Data Language (IDL) specifications, UML modeling [20], Meta-Object Facility [21], and XML Metadata Interchange (XMI) [22]. The forthcoming MDA might provide a useful starting point for tools supporting interoperability between parts of an application running on different middleware platforms.

Conclusions

The second (or third) generation of middleware products begins to be mature enough for deployment in mobile networks. However, opportunities for improvements still exist and ORB, Java and Web Services communities are working towards developing specifications, which are still missing.

The main implication of Internet protocols on the middleware is the increasing richness of available functionality and features. This implies that the middleware standardization bodies need to take another look at their solutions for Quality of Service, mobility, discovery, and security. They should also consider alternative transport mechanisms to TCP/IP, which is a good protocol for bulk data transfer but not for messaging. When the middleware specifications are based on the IP solutions for QoS, mobility, discovery and security, interoperability and interworking will be much easier to arrange.

References

[1] Network Policy and Services: A Report of a Workshop on Middleware. RFC 2768, February 2000.

[2] OMG CORBA v2.4.2 Specification. OMG document formal/2001-02-01, February 2001.

[3] Sun Microsystems. Java™ 2 Platform, Enterprise Edition Specification, v1.3, February 2001.

[4] Sun Microsystems. Java™ 2 Platform, Micro Edition Specification: http://java.sun.com/j2me/

[5] Eddon, G. and Eddon, H. Inside Distributed Com (Microsoft Press, 1998).

[6] WAP Forum. Wireless Application Environment Overview. Document id 195-WAEOverview-20000329-a. March 29, 2000.

[7] Parlay Group. Parlay 2.1 Specification. 2000-2001: http://www.parlay.org/

[8] Alanko, T.; Kojo, M.; Laamanen, H.; Liljeberg, M.; Moilanen, M.; and Raatikainen, K.: "Measured Performance of Data Transmission over Cellular Telephone Networks," Computer Communications Review 24(5): 24-44, October 1994.

[9] Cáceres, R. and Iftode, L.: "Improving the Performance of Reliable Transport Protocols in Mobile Computing Environments," IEEE Journal on Selected Areas in Communications 13(5):850-857, June 1995.

[10] Kojo, M.; Raatikainen, K.; Liljeberg, M.; Kiiskinen, J.; and Alanko, T.: "An Efficient Transport Service for Slow Wireless Telephone Links," IEEE Journal on Selected Areas in Communications 15, 7 (Sep. 1997), pp. 1337-1348.

[11] IETF TSVWG home page. http://www.ietf.org/html.charters/tsvwg-charter.html

[12] IETF PILC WG home page. http://pilc.grc.nasa.gov/pilc/.

[13] IETF ECM WG home page. http://www.ietf.org/html.charters/ecm-charter.html.

[14] Liljeberg, M; Helin, H; Kojo, M and Raatikainen, K: Mowgli WWW software: Improved usability of WWW in mobile WAN environments. In Proc. IEEE Global Internet 1996 Conference, 1996, 33-37.

[15] Hausel, B. and Lindquist, D. B.: "WebExpress: A System for Optimizing Web Browsing in a Wireless Environment," in Proceedings of MobiCom'96, ACM SigMobile, 1996, 108-116.

[16] WAP Forum. WAP Binary XML Content Format. Document id WAP-192-WBXML-20000515, Version 1.3, May 15, 2000

[17] W3C. XML Protocol Activity: http://www.w3.org/

[18] Campadello, S., Helin, H., Koskimies, O., and Raatikainen, K. Wireless Java RMI. http://www.cs.Helsinki.Fl/research/monads/papers/edoc2000/rmi.html

[19] OMG. Model Driven Architecture, A Technical Perspective. OMG Document Number ab/2001-02-01, February 14, 2001.

[20] OMG. UML Specification. OMG document formal/2000-03-01, March 2000.

[21] OMG. MOF Specification. OMG document formal/2000-04-03, April 2000.

[22] OMG. XMI Specification. OMG document formal/2000-11-02, November 2000.

Authentication Methods and Technologies

Authentication, i.e., building trust between participating identities, can be done by leveraging several different methods or combinations of methods. Depending on these methods or combinations and whether the identity is real, i.e., linked to a physical person or device or whether it is a pseudonym, a different level of authenticity is required.

The most common high-level way to define authentication methods is to divide them into three different prime categories: *What you know* authentication is based on the idea that a consumer knows something that verifies her authentication. Normal User Identification (ID) - password authentication is a good example of this category. *What you have* authentication proves an identity if some physical or virtual identification item is used. Examples of this category include a corporate ID card, a driver's license or a normal key. *What you are* authentication proves an identity based on a unique physical characteristic (e.g., a fingerprint).

If combinations of these methods are required at the same time, the certainty of authenticity is increased. The disadvantage, however, is that a system's usability becomes worse, and consumers then try to find ways to bypass security. The term *two-factor authentication* is used if two methods are required to prove identity. In practice, this means a person has to hold some physical identity object like a smart card and in addition know a password or Personal Identification Number (PIN). The term *multi-factor* is used if more than two methods are required in one authentication transaction. This method increases security, but may decrease usability and lead to attempts to bypass the security system. Such problems can be avoided by paying special attention to system design.

One challenge when designing the authentication infrastructure is to build reliable, functional and secure key distribution and management. As described above, authentication is based on shared secret or private/public keys. The shared secret can be a password or symmetric key known by both communication parties. An important factor is how to deliver and upgrade secrets or keys in a cost-effective and reliable way. This can be done, for example, by physical means (e.g., posting passwords or smart cards containing the secret), or over the network, by utilizing some authentication protocol.

These methods are typically implemented in computing systems using one or a combination of the following authentication technologies:

- o   Password authentication
- o   One-time passwords
- o   Hardware token (e.g., RSA SecurID™, Radio Frequency ID (RFID))
- o   Symmetric key based solutions (e.g., Kerberos, Global System for Mobile Communications (GSM) Authentication)

- Public Key Center based solutions: Digital Certificates, Digital Signatures, Public Key Infrastructure (PKI) and Certificate Authorities, Public Key Encryption, Wireless Identity Module (WIM)
- Biometric authentication
- Remote Authentication Dial-In User Service (RADIUS), Diameter
- Single Sign-on

These technologies are not equal in scope, but parts of different technologies are often needed to build a complete authentication system.

Password Authentication

In private and public computer networks, including the Internet, authentication is commonly done through the use of log in passwords. Log in is the procedure used to get access to an operating system or application, usually on a remote computer. Log in almost always requires the consumer to have a user ID and password. Often, the user ID must conform to a limited length (e.g., eight characters) and the password must contain at least one digit and not match a natural language word. The user ID can be freely known and is visible when entered at a keyboard or other input device. The password, however, must be kept secret and is not displayed as it is entered.

Knowledge of the password is assumed to guarantee that the consumer has access privileges. Each consumer registers initially or is registered by someone else, by using an assigned or self-declared password. On each subsequent log in, the consumer must know and use the previously declared password. Some Web sites require consumers to register in order to use the site; registered consumers can then enter the site by logging in.

In the simplest password authentication systems, the user ID and password are transferred openly over data networks, but this enables eavesdropping crimes. In more advanced systems, challenge and response schemes are followed, and secret information is encrypted with symmetric or public/private keys utilizing either a software or a hardware solution.

The weakness of a password authentication system, especially for transactions which are significant in value (e.g., the exchange of money), is that passwords can be stolen, revealed accidentally, or forgotten.

Onetime Passwords

Onetime passwords protect against eavesdropping, as the password is invalidated after one use. In off-line implementations, the most common way of using onetime passwords is sending a list of passwords to the consumer beforehand. Every time the consumer accesses the protected resource, one password from the list is used. A more modern way to utilize onetime passwords is to send one to the consumer when needed. This may happen, for example, by sending a password in a text message to a mobile device. The SecurID™ card, described in more detail under Hardware Token Based Authentication, also generates onetime passwords.

For onetime passwords, it is important that a password and/or the password list is never sent over the same communication channel at the same time as the user identification number or hardware security token.

Hardware Token Based Authentication

SecurID™ and RFID tokens represent the *what you have* security method. They can be combined with the *what you know* scheme, if a PIN is required to get final approval for access.

RSA SecurID™

Essentially, SecurID tokens are onetime passwords for user authentication and can be used to authenticate the right to access restricted services. SecurID can be used together with a RADIUS system to increase the security level from a simple user ID password approach. The time-synchronized SecurID card has a display that shows a string of numbers that changes every minute. When authenticating, the user is prompted for a user ID and the Passcode, which is generated dynamically by the SecurID card when the user enters a PIN via the keypad on the card. The authentication request is then sent over the network to the RSA ACE/Server™, which determines the validity of the Passcode. A validated Passcode will grant a network access.

Radio Frequency ID

RFID is a passive radio receiver and transmitter absorbing its energy from radio waves. Nowadays most implementations handle the basic challenge-response protocol with encryption support, which protects against the simplest man-in-the-middle attacks.

Symmetric Key Based Solutions

The most popular symmetric key authentication technology is the one used in Global System for Mobile communications (GSM) authentication, utilizing Subscriber Identity Module (SIM) - smart cards. Another popular example is Kerberos authentication. In addition to these, we will also briefly review smart cards used in PC log in. The disadvantage of symmetric key systems is that they cannot provide bulletproof non-repudiation.

Smart Card Authentication

Smart cards can store either digital certificates or just log in information. Certificates are described in the Public Key Cryptography section.

The beauty of using smart cards is that the log in process is transparent to the consumer. As with an Automatic Teller Machine (ATM), the consumer cannot access the system without a card. Once the card is entered, the consumer has to enter a Personal Identification Number (PIN) and the system takes over from there. Smart cards can also feature the ability to lock out a consumer after a set number of unsuccessful log in attempts. Once this occurs, an administrator must unlock the card before it can be used again.

Subscriber Identity Module Authentication

The basic idea of SIM Authentication, defined in the GSM specifications, is to encrypt all traffic between a mobile station (i.e., a mobile device) and a base station. However, the carrier network is considered to be secure. To reach this target the client and the Authentication Center, which are linked closely to the Home Location Register (HLR), share a common secret called Ki. To be more specific, this symmetric key, Ki, is stored in the consumer device SIM. When a device wants to authenticate itself with a network, the network generates a random challenge called RAND in the GSM standard. The secret, Ki, is used to calculate the response, called Signed RESponse (SRES), to the challenge. If SRES matches one calculated by the network, access is granted. The home carrier, as opposed to the visiting carrier, is always the one verifying the authenticity of the consumer. The visiting carrier acts as a mediator, passing requests from the visiting device to the home carrier's Authentication Center and vice versa.

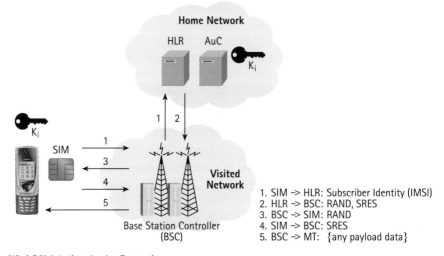

Simplified GSM Authentication Protocol

After successful authentication, the mobile device calculates a session key, Kc, from the random authentication challenge. The standard says that from this point on all communication must be encrypted. However, in certain situations, encryption might be bypassed, and indeed in some countries, encryption is not even allowed. Generally, if encryption is not used, it is considered a lack in the quality of service defined by the standard. Note that encryption is used only on the radio channel.

Kerberos Authentication

Kerberos is a network authentication protocol created in the early 1980s at the Massachusetts Institute of Technology. Kerberos is, by nature, a distributed authentication service that allows a process (a client) running on behalf of a principal (a user) to prove its identity to a verifier (an application server, or just a server) without sending data across the network that might allow an attacker or the verifier to subsequently impersonate the principal. Optionally, Kerberos provides integrity and confidentiality for data transmitted between the client and the server.

The Kerberos authentication system uses a series of messages encrypted with shared, i.e., symmetric keys to prove to a verifier (target service) that a client is running on behalf of a particular principal. Initially, both, the principal and the service are required to have keys registered with the Kerberos Authentication Server (AS). The user key is derived from a password that she chooses; the service key is a randomly selected key (since no consumer is required to type in a password).

When a client wants to access a service (verifier, v) on the network, the client requests authentication from the AS server. The server then creates two copies of the session keys, also known as the shared key (Kc,v). The session keys are accompanied with other data relevant to the protocol like a time stamp. Time stamps are used to prevent a hacker from capturing the session and playing it back. After this, one of the shared keys (Kc,v) with related data is encrypted with the client's secret key (Kc) and the other is encrypted with the verifier's secret key (Kv). Next, both encrypted packets are sent back to the client, which forwards the session key to the verifier. The verifier can now decrypt the message using its private key (Kv). Only an authentic client and verifier can decrypt the shared session key (Kc,v) enabling further communication between the parties. The protocol also enables mutual authentication. A simplified description of the protocol is shown in the figure below.

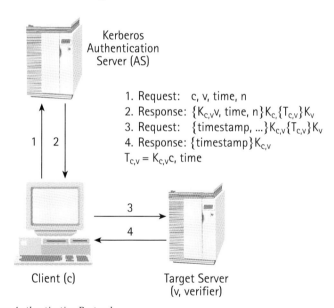

Simplified Kerberos Authentication Protocol

One of the main benefits of using Kerberos is that it provides single sign-on capability using the Security Service Application Programming Interface (SSPI) standard. This is realized with the help of an additional server system, part of the Kerberos server called a ticket-granting service. It is used to support subsequent authentication after the initial authentication without the need to re-enter a principal's password. Cross-realm authentication approaches also exist, where the authentication of a principal registered with a different authentication server is enabled.

There are several enhancements to Kerberos. For example, in the Windows 2000™ version of Kerberos, the initial user authentication is done using public key certificates instead of the standard shared secret keys.

The description of the Kerberos protocol in this section has been simplified for clarity. Readers should consult RFC 1510 for a more thorough description of the Kerberos protocol.

Public Key Cryptography

Weaknesses in password authentication and symmetric key systems for transactions have led to a need to develop something more advanced. Mobile commerce, Internet business and many other transactions require a more stringent authentication process. The use of digital certificates issued and verified by a Certification Authority (CA) as part of a public key infrastructure will become the standard way to perform authentication on the Mobile Internet.

The distinct advantage of public key-based remote authentication over mechanisms that mimic authentication in the local environment is that sensitive authenticating information (e.g., a password), is never sent over the network. In other authentication methods, if server Alice holds a copy of client Bob's password or thumbprint, Bob must authenticate himself by proving that he knows or has this information. This is typically accomplished by Bob conveying this information to Alice upon sign-on. During the process, the information is vulnerable to third-party attacks.

Essentially, when PKC is used in a mutual authentication, the mobile device and the destination server's service present their certificates to each other. The certificate that contains the name and public key of the holder are signed by a private key of the certificate authority contained in a root certificate. If the exchanged certificates are able to be decrypted with the public key of the CA, the authenticity and integrity of the certificate is proven. Now, if the initiating party of the communication is able to sign the challenge sent by the other party with its private key and the first party is able to decrypt it with a public key from a valid certificate, the authenticity of the signing party is proven. The same can also be done in the other direction. Instead of getting these certificates directly from the other communicating party, an alternative source is the central directory service.

The root certificates provided by a trusted authority are then stored on the consumer's device. Root certificates are the basis for certificate verification and should be supplied only by a system administrator. The certificate authority can also invalidate a certificate by marking it invalid on the revocation server, which is where the communication party should check a certificate's validity.

Biometric Authentication

Fingerprint readers, voice recognition software, facial identification cameras and retina scanners are among the options used in biometric authentication. They represent the *what you are* authentication principle.

Even with the promise of high accuracy and ease of use in biometric systems, they have not been widely used. Major concerns are their current high cost, unreliability and lack of accuracy. While biometric systems continue to develop, they still suffer from the effects of environmental factors (e.g., sweaty or wet fingers).

Remote Authentication Dial-In User Service

RADIUS is a client/server protocol and software, which enables remote access servers to communicate with a central server to authenticate dial-in users and authorize their access to the requested system or service. The primary functions of RADIUS are authentication, authorization and accounting. RADIUS allows a company or network service provider to maintain user profiles in a central database which all remote access servers can share. It provides a single point of administration for multiple remote access servers. This eases setting up a unified policy for clients independent of their access point to the network. Having a central service also means that it is easier to track usage for billing and to keep network statistics. RADIUS is the de facto industry standard and is a proposed Internet Engineering Task Force (IETF) standard.

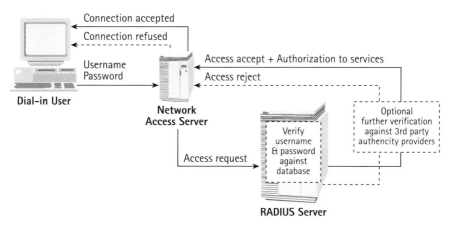

RADIUS authentication and authorization

A remote Network Access Server (NAS) operates as a client of RADIUS. The client is responsible for passing consumer information to designated RADIUS servers, and acting on the response that is returned. RADIUS servers are responsible for receiving consumer connection requests, authenticating the consumer, and returning all configuration information necessary for the client to deliver the service to the consumer. A RADIUS server can act as a proxy client to other RADIUS servers or other kinds of authentication servers, (e.g., ActivCard™, ActivEngine™ or RSA ACE/Server). The RADIUS server can support a variety of methods to authenticate a consumer.

Transactions between a client and a RADIUS server are authenticated through the use of a shared secret, which is never sent over the network. In addition, any user passwords are sent encrypted between the client and the RADIUS server, to eliminate the possibility that someone snooping on an unsecured network could determine a consumer's password.

When a client is configured to use RADIUS, any consumer of the client presents authentication information to the client. This might be with a customizable log in prompt, where the consumer is expected to enter his username and password. Alternatively, the consumer might use a link framing protocol (e.g., the Point-to-Point Protocol (PPP)), which has authentication packets to carry this information. Once the client has obtained such information, it may choose to authenticate using RADIUS. To do so, the client creates an access request containing the necessary authentication information.

If the correct authentication information (e.g., matching user ID and password pair), is provided to the RADIUS server, it responds with an *access-accept* message indicating an accepted connection. The server also sends authorization information about the services the consumer can access. Instead of providing immediate access, the server may prompt further proof of authenticity, e.g., in the form of a SecurID challenge. If the consumer is not able to provide correct authentication information, the RADIUS server sends an *access-reject* response indicating that this consumer request is invalid.

The previous figure shows the authentication and authorization process.

A more detailed description of RADIUS can be found in RFC 2865.

Diameter

Diameter is being specified in the Authentication, Authorization and Accounting working group of the IETF. Diameter can be considered as a next generation RADIUS, supporting authentication, authorization and accounting in a number of situations. Diameter is primarily designed as a Network-Node Interface (NNI), with Mobile IP extensions for key distribution purposes.

Diameter attempts to solve IP roaming issues between home and foreign networks, taking into consideration the authentication of a consumer, consumer access and accounting issues. Additionally, Diameter allows key distribution, where the AAA servers can act as a key distribution center that generates and distributes keys.

Single Sign-On

Single Sign-On (SSO) is a mechanism whereby a single action of user authentication and authorization can permit a consumer to access all computers and systems where he has access permission, without the need to authenticate explicitly to all systems by, for example, entering multiple user IDs and passwords. Single sign-on reduces human error, a major component of systems failure, and is therefore highly desirable. User friendly, secure and reliable single sign-on can be seen as one of the primary catalysts boosting mobile commerce.

Many current computer systems or platforms provide a single sign-on feature. However, none of these satisfy the need for real cross-platform single sign-on that is needed when multiple service providers utilizing varying technical platforms are involved. This happens easily on the Mobile Internet, as it is not likely that one vendor will run the same computing platforms. To meet the requirement of cross-platform single sign-on, cross industry activity is needed.

There are four basic ways to implement single-sign-on:

o   In the *Standard identity and secret* scheme, the same protocol, identity and secret (e.g., password) are used for all systems. This model is unrealistic, as consumers want to be identified differently in different situations. To reach its ultimate target, this scheme must be followed to some extent, as there is not enough memory and computing resources in mobile devices to build support for several different protocols.

o   The second way to implement single sign-on is to hide the differences between multiple protocols and authentication information in a mobile device. This is called the *Intelligence in terminal* scheme.

o   The third solution, *Intelligence in proxy*, hides the differences of multiple protocols and authentication information on a trusted server in the network. The advantage is that the differences are hidden from the mobile device, which allows single-sign-on regardless of whether the mobile device has been specifically designed for it or not. The disadvantage is that the consumer must trust the server and the communications path between the device and the server.

o   The fourth way is to set up a centralized authentication server or a distributed cluster of authentication servers, which are able to communicate securely with each other. Once authentication is accomplished, the server generates a security token, which proves the consumer's identity. The token may be called a ticket or a voucher and may include credentials describing the rights of the consumer in the destination system. Examples include HLR + Authentication server in GSM, Home Subscriber Server (HSS) and Diameter in the Mobile domain. The server or the cluster can be called an *Identity broker*. The advantage of this approach is that the broker does not become a bottleneck in the system as easily as in the proxy case, because the broker is consulted only in the authentication phase and is not needed in other communication.

None of these models work in their pure incarnation and a functioning system usually has elements from all of them.

A SSO solution clearly has merits and risks. The three main merits are:

1. **Convenience**: Consumers need only identify/authenticate themselves once to the primary domain and not for every platform or application.

2. **Security**: Because there is only one sign-on, the risks inherent in remembering multiple login/password combinations are eliminated.

3. **Trust**: Two-factor single sign-on, where access is granted based on the combination of what a consumer knows (e.g., a password or PIN) and what they possess (e.g., a biometric device), is very secure.

The three main risks of SSO are:

1. **Security**: If an intruder compromises a consumer's account or password, that intruder could have extensive and easier access to far more resources.
2. **Cost**: Single sign-on implementations can be expensive, both in the cost to purchase and in the manpower to deploy.
3. **Errors**: Accidentally typing the combination user ID and password of one service when authenticating to another service gives the service provider an opportunity for fraud. Even worse risks follow from the typical habit of using the same user ID and password for multiple services.

Conclusions

Identification and authentication are fundamental elements on the mobile Internet. They are needed for many important use cases, enabling the setting up of trust relationships between communicating parties. Globally functional interoperable solutions are required for Mobile and Web domains that support the consumer's preferred set of identities on a proprietary security level. The robustness of the complete system must be taken into account, as the mobile device and frequency spectrum are limited. Meeting all these requirements requires applying many of these authentication technologies, for one size does not fit all. The complexity of the system needs to be hidden from consumers with proper user interface implementation. Single sign-on is an essential target in reaching an enjoyable consumer experience.

References

[1] http://whatis.techtarget.com

[2] Security Engineering, Ross Anderson (ISBN 0-471-38922-6)

[3] RFC 2865 - Remote Authentication Dial In User Service (RADIUS), C. Rigney, S. Willens, A. Rubens, W. Simpson. June 2000. (Status: DRAFT STANDARD), http://www.ietf.org

[4] http://www.livingston.com/marketing/whitepapers/radius_paper.html

[5] http://web.mit.edu/is/help/kerberos/

[6] RFC 1510 - The Kerberos Network Authentication Service (V5). J. Kohl, C. Neuman. September 1993. (Status: PROPOSED STANDARD), http://www.ietf.org

[7] http://www.activcard.com

[8] http://www.rsa.com

[9] http://www.microsoft.com

[10] http://www.baltimore.com

[11] http://www.verisign.com

[12] http://www.projectliberty.org

Part 4.3

Application Layer

- Browsing in the Mobile Domain
- Messaging in the Mobile Domain
- Multimedia Messaging
- Instant Messaging and Presence
- Wireless Village Interoperability Framework
- Public Key Infrastructure
- Digital Rights Management
- Mobile Payment
- Location Technologies in the Mobile Domain

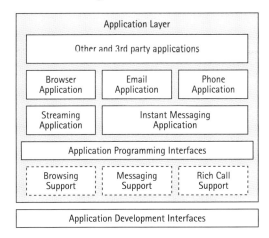

Browsing in the Mobile Domain

For mobile devices, the future of Wireless Application Protocol (WAP) lies in its close alignment with widely accepted Internet standards. The WAP Forum and the World Wide Web Consortium (W3C) have defined mobile Internet standards over the past several years. Most recently, the WAP Forum has moved to adopt the Extensible HyperText Markup Language (XHTML) Basic standard from W3C as the basis for WAP 2.0. This transition to XHTML will strengthen the browsing solutions on the Mobile Internet and allow for a far greater range of presentation design and formatting possibilities than previously possible. Consequently, consumers will enjoy a wider array of services, more intuitive user interfaces, and a generally more useful experience. At the same time, it will also enable carriers to align and customize the look and feel of the services they provide through their mobile portal.

Overview

In its simplest form, a mobile browser is an application in a mobile device that fetches content from servers and presents it to the viewer so that the viewer can interact with the content and generate new requests for content to be presented. The main type of content to be supported by mobile browsers in the near future is XHTML, which is the basis for all browser-based services. A wide variety of other content formats can also be supported (e.g., images, audio, and video) in the form of either static (added at implementation time) or dynamic (added dynamically at runtime) content handler plug-ins. However, the main body of services will be implemented by using XHTML. Other content formats can either be embedded inside or be pointed to by the XHTML document.

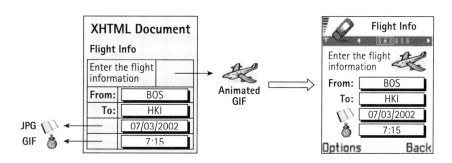

Rendering of an XHTML Document with links to images

In its basic structure and behavior, the mobile browser is very similar to any other browser. The differences become evident when we look at the individual parts of a mobile browser-based service solution and compare them with the corresponding parts of a fixed solution. Starting from the device, it is obvious that the size of the display is limited because of the small size of the device itself, input mechanisms are limited, memory size must be kept to the smallest possible, and the life of a small battery is not very long. Thus, there are limitations on how much content can be presented at once, and what kind of content can be supported.

The next challenge is the optimal use of mobile network bandwidth so that services are fast to use and response times are not too long. Finally, services themselves must be tailored to the liking of a consumer. They must be fast to find, easy to use and provide the service with no extra baggage, in order to keep consumer expenses down. All these special requirements have been the leading guidelines for the work done in several standardization forums to come up with technologies best suited to the Mobile domain.

XHTML Basic

According to the W3C specification, XHTML Basic defines a document type that is rich enough to be used for content authoring and precise document layout, yet which can be shared across different classes of devices - desktop, Personal Digital Assistant (PDA), TV and mobile device. XHTML Basic is the mobile adaptation of XHTML 1.0, and includes everything in XHTML 1.0 except those capabilities (e.g., frames), which are not appropriate for devices with small screens. XHTML itself, according to the W3C, is the first major change to HyperText Markup Language (HTML) since HTML 4.0 was released in 1997. XHTML brings clean structure to Web pages, which is especially important given the small screens and limited power of mobile devices. W3C is recommending XHTML for all future Web development for desktops as well as all other devices.

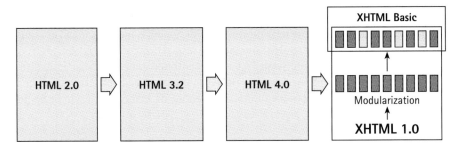

From HTML to XHTML Basic

XHTML is the keystone in W3C's effort to create standards providing richer Web content on an ever-increasing range of platforms. Using XHTML, content providers will find it easier to produce content for a broad set of platforms and with better assurance of how the content is rendered. In contrast, the initial WAP markup language Wireless Markup Language (WML) 1.x does not ensure consistent layout across different devices. This is of particular concern to carriers, which are managing mobile devices from a number of providers and with different user interfaces from one device model to another.

XHTML Mobile Profile

The WAP Forum has defined the XHTML Mobile Profile (XHTML MP) as a superset of XHTML Basic. The XHTML MP adds a few additional elements and attributes from full XHTML 1.1, so it is a subset of XHTML 1.1, and is thus fully compatible with XHTML Basic. The additional items are: <hr>, <b>, <i>, <big>, <small>, <style> and style="..."

With the XHTML MP, a document can be presented on the maximum number of Web clients, including a variety of mobile devices featuring different display formats and presentation capabilities. And since XHTML MP is sanctioned by both the W3C and the WAP Forum, it is an integral part of the overall Internet standards set, which will ensure its widespread acceptance now and continued viability over the long term. This section describes the key features of the XHTML MP for mobile browsers and its benefits for content developers, carriers and consumers. It will also present examples demonstrating the display opportunities and presentation flexibility enabled by the XHTML MP.

Key Features and Capabilities

Support for the XHTML MP will bring a number of positive changes to the mobile browser. The XHTML is a stricter formulation of an established markup language, since it is based on the same HTML tags used in every Web page today. This means that existing Web design and presentation tools can work with it. Similarly, a developer can view an XHTML MP document in a standard desktop Web browser and the layout will be consistent. There are also significant differences from HTML.

The same XHTML page, viewed using the Nokia Mobile Browser (left) and Microsoft Internet Explorer (right)

Well-Formed Content

From a technical standpoint, the XHTML MP strictly enforces the rules for well-formed, valid eXtensible Markup Language (XML). The XML is a W3C standard for specifying well-structured documents that can be processed by computers (e.g., a browser). A well-formed XML document conforms to all the rules for the XML syntax. For example, every document must have an identifying XML header that is written properly, each tag (instruction) is matched by a closing tag and is correctly nested, and attribute values are always quoted. By following all the XML syntax rules, the XHTML MP enables a layout that is easier to process and can be consistently displayed on different browsers. By comparison, HTML need not be well formed, so the resulting display might contain elements that do not appear as intended on some browsers.

HTML

```
<p>This is a paragraph.
<p>This is another.
. . .
```

XHTML

```
<p>This is a paragraph.</p>
<p>This is another.</p>
. . .
```

HTML tags are not always closed but XHTML tags must be closed

Valid Content and the Document Type Definition

Valid XML simply means that the document uses the language correctly - that it includes only defined tags, and each tag is used in the correct context. For example, in XHTML, a table row tag is only valid inside a table tag, but a table row would be senseless outside of a table. The Document Type Definition (DTD) defines the rules for XHTML (or any valid XML language). The DTD (or the address at which to access the DTD) appears before the head element in each XHTML document. The head forms the beginning section of a Web document and contains information that is not part of the actual page content. It also contains controls for the overall document.

Document.xhtml

```
<?xml version="1.0" ?>
<!DOCTYPE html PUBLIC "-//WAPFORUM//DTD XHTML Mobile 1.0//EN"
 "http://www.wapforum.org/DTD/xhtml-mobile10.dtd">
<head>
. . .
```

An example DTD specifier

According to the W3C, the DTD facilitates hassle-free production and presentation. The DTD identifies what each tag means and specifies how it should be treated. In short, it provides the rules used by editors and other authoring and transformation tools to validate the online document, which ensures that the document can be displayed on any XHTML browser. This is particularly important on the Mobile Internet, where devices offer a wide range of display formats and capabilities.

By strictly following XHTML rules, the memory cache in mobile browsers can be reduced, allowing for a smaller memory footprint and greater performance efficiency, which is essential in mobile devices. If you look at the size of standard desktop Web browsers, much of the complexity comes from handling the ambiguities of HTML.

Cascading Style Sheets

Central to the XHTML MP is its support for Cascading Style Sheets (CSS). CSS describes how documents are presented on screen in the browser. The W3C has actively promoted the use of CSS on the Web with all browsers, desktop and mobile. Through the use of CSS, document creators can control the presentation of documents without sacrificing device independence or adding new markup language tags, as was done with WML 1.x. The use of well-known standard HTML tags and CSS properties will reduce content development costs by eliminating the need for developers to learn new tags, to store multiple versions of content, or to master different tools. In addition, servers will not experience the added overhead that results from the expensive transcoding required to prepare content for an array of different markup languages.

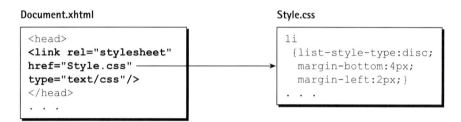

An XHTML document referencing a style sheet

Through CSS a document creator can specify the presentation of a Web application in one place, the style sheet. If the presentation needs to be changed at any time, the change is made once in the style sheet and the modification is dynamically reflected throughout all the pages on the site.

CSS separates the content of the document or application from the presentation, thus allowing developers to easily create browser-specific versions of the same content simply by creating the appropriate style sheet. For example, when a consumer requests a Web page, the content server identifies the requesting device and returns the content with a link to the appropriate style sheet. The style sheet is downloaded once and cached by the browser for use with subsequent pages, which speeds the rendering of all pages on the site.

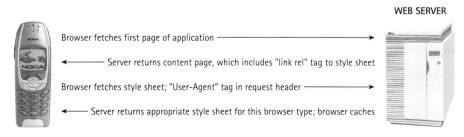

All pages on the web site can be the same for any class of device and screen size; only the style sheet is tuned for different needs.

The power of CSS lies in the precise control it offers document creators and the ease with which they can optimize content for presentation on any device. Every aspect of the document's formatting (e.g., positioning, fonts, text attributes, borders, margin alignment, flow) can be defined in the style sheet. A change to any aspect of the document needs to be made only once.

CSS also gives carriers greater control over the look and feel of the services they provide through their mobile portal. A carrier can define a default style sheet for all devices it supports and require all content developers to include a link to their style sheet, which will ensure a basic look and feel consistent for all devices, even those from different manufacturers. For example, a carrier might specify that all links are underlined or all headings are bold and centered. Of course, content developers could choose to modify that style through their own use of CSS if different behavior is required, but the default style sheet determines the overall look of all applications for a particular browser.

Browser Examples

The precise control provided by CSS makes it easy to reformat content for different mobile devices. One of the previous figures shows the same XHTML MP page displayed for different browsers. Through the power of the XHTML MP, the presentation, style and navigation are easily reformatted for each browser. The following figures show other examples of the same page with different formatting applied with different CSS style sheets.

Precise layout control using CSS and XHTML

Note how CSS allows padding and paragraph indentation to achieve precise positioning. Even sophisticated typographic treatments (e.g., drop-caps) can be used. Flexible alignment (e.g., center, left, right), borders of varying thickness and margins can also be used to create advanced effects (e.g., floating paragraphs).

Identical files, but different CSS style sheet changes appearance.

The different bullet styles are not defined in the XHTML page.

Transformations

As an XML-based standard, the XHTML MP allows for the automatic parsing and transcoding of content through the use of eXtensible Stylesheet Language Transformations (XSLT). A part of XSL and a W3C standard, XSLT provides a language for transforming XML documents into other XML documents.

Note that XSLT is completely different from CSS. Although both names contain the words *style sheet*, CSS controls the display of a document by decorating it, whereas XSLT specifies rules for transforming or rewriting a document from one format to another.

Using XSLT, a system can automatically transform the same XML content into multiple markup languages depending on the browser. Through such transformations, content can be created for one device and automatically transformed to appear on another device. For example, a news service could create its news feed once in XML and use XSLT to dynamically convert it to XHTML MP and HTML for presentation on mobile or other devices.

Better still, as XHTML becomes the standard for all Web browsers, a Web site could use XHTML for all its content. In this way, developers would create content once for display on different devices, eliminating the need to write different versions of the same content for different devices.

XHTML is the Future

The adoption of the XHTML MP within the mobile industry as the standard browser language brings mobile communications into alignment with the latest advances in Internet content development and presentation. Supported by the WAP Forum and the W3C, XHTML MP establishes a global standard for all browsers.

By adopting the XHTML MP as their browser standard, mobile device manufacturers, carriers, developers and consumers alike can get a head start on experiencing the many benefits its offers. Foremost among these benefits is compliance with the global standard for mobile Internet computing going forward.

Messaging in the Mobile Domain

Short messaging has become a surprisingly successful mobile service. It is worth remembering that the Short Message Service (SMS) was optional in the first Global System for Mobile Communications (GSM) networks. Initially, some carriers even went so far as to disable the feature on devices for fear of detracting from voice revenue. However, these fears proved unfounded and SMS became a success, increasing carrier revenues dramatically. Similarly, on the legacy Internet, Instant Messaging solutions coupled with presence have become very popular. Consumers use Instant Messaging to keep in contact with peers in a fun and inexpensive way. These two examples illustrate the popularity and importance of flexible interpersonal communication services. In the same way, a full range of mobile communication services will be developed for the Mobile Internet.

In the Mobile Internet, messaging is the first rich service having a unified set of multiple services (voice coming later). The SMS evolution leading to MMS integrated with Web domain SIP based Instant Messaging are the cornerstones of Mobile Messaging. SIP provides a technology for collecting and provisioning of presence information and for developing presence based applications. Combined with the previously described messaging solutions, these create a new paradigm of Mobile Messaging.

Messaging in the Mobile Domain consists of multiple technologies including SMS, the Multimedia Messaging Service (MMS), Instant Messaging (IM) and Presence. Also, the interoperability between messaging systems is essential. These messaging technologies are covered in a separate section.

The first section introduces the Multimedia Messaging Service. We will demonstrate how MMS relates to SMS from the consumer's perspective and will also present the major aspects of this new service: standardization, network architecture and inter-network routing, encapsulation of messages, MMS applications, and interoperability.

The second section presents an Instant Messaging and Presence solution based on the latest generation of Internet protocols, (e.g., Session Initiation Protocol (SIP)). It outlines the advantages of such a solution, clarifies what consumer services will be supported, and discusses its likely evolution.

The last section introduces the Wireless Village interoperability framework. It demonstrates how the Wireless Village defines and promotes a set of universal standards and specifications for mobile instant messaging and presence services.

Multimedia Messaging

The Short Message Service (SMS) has been a great success story, and the service provides a great source of revenue for carriers and service providers. However, the content that can be transmitted over SMS is rather limited: text, ringing tones and small monochrome BitMap pictures. With recent advances in technology, creating rich multimedia content has become easier and more affordable. For instance, digital cameras are offering a new means of instant photography: point, shoot and the picture is ready to be transmitted. At the moment, however, there is no application, which would allow consumers to instantly share those images. Today, consumers would typically first have to transfer their images to a PC and then send them to a friend through e-mail, hoping that the message will be read promptly. But very often the magic of the moment has passed by the time that happens.

It would be much better if the image could be sent directly and displayed instantly on the color screen of the friend's mobile device. It would allow moments of life to be shared when they occur, when they are still magical.

The Multimedia Messaging Service (MMS) is the next evolutionary step from the SMS. For a consumer, it works similarly to SMS. However, MMS provides the opportunity to utilize a wide variety of richer content types than SMS. These include color pictures, audio, music and video clips.

This section presents various aspects of the new Multimedia Messaging Service, from the technology and challenges to the applications MMS will allow.

Short Message Service

Originally launched in 1992, SMS has become the most successful mobile data service. SMS has mainly been used for mobile device to mobile device messaging. In SMS, a consumer typically composes a text message on his mobile device and sends it to another person using the recipient's phone number as the address. The message is received by a Short Message Service Center (SMSC), which acts as a store-and-forward unit. It is responsible for delivering the message to the destination device. It may have to wait until the receiving device becomes reachable (powered on or within network coverage) for that to be possible. The message is delivered without any intervention from the recipient; it appears naturally on his device screen and a sound alerts him to the arrival of the new message.

Another important aspect of SMS, from a usability point of view, is that once the destination device has received the message, a delivery confirmation is conveyed to the sender. Since most consumers carry their mobile devices with them, it provides confidence that the message has reached or will soon reach the recipient.

On the other hand, SMS does have several limitations. Each SMS message is limited to 160 characters, a rather short length. Methods exist to concatenate multiple SMS messages to create longer messages. There are also ways to convey other types of (binary) content (e.g., monochrome BitMaps and ringtones) using SMS. However, SMS is transmitted over the signaling channels of a mobile network. This makes it impractical to use SMS to convey multimedia messages, which are several kilobytes in size.

Multimedia Messaging Service

The concept of Multimedia Messaging Service is built on the familiar and successful SMS service. From the consumer's perspective, the service's logic and behavior are similar to SMS.

Like SMS, MMS is based on a store and forward service model. Receiving multimedia messages in a mobile device works similarly to SMS: messages are delivered directly to the mobile device with no need for consumer interaction. Furthermore, the sender receives delivery confirmation.

However, there are important differences between SMS and MMS. A major one is that MMS provides the opportunity to utilize richer content types than SMS. The content will include pictures in formats such as the Joint Photographic Experts Group (JPEG) and the Graphics Interchange Format (GIF) as well as audio, music and video clips. Because MMS uses data traffic channels and not signaling, it is well suited to deliver much larger multimedia content. The following figure shows the evolution of mobile messaging.

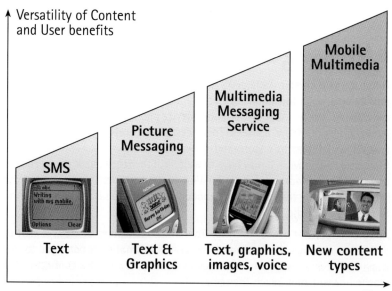

Multimedia messaging migration path

SMS was mainly used for messaging between mobile devices. In addition, MMS will be used for messaging from Web applications to mobile devices and between the Internet and mobiles devices as shown in the figure below. These exchanges will lead to the development of numerous media-rich applications, e.g., picture messages, electronic postcards, instant pictures, video clips and audio messages.

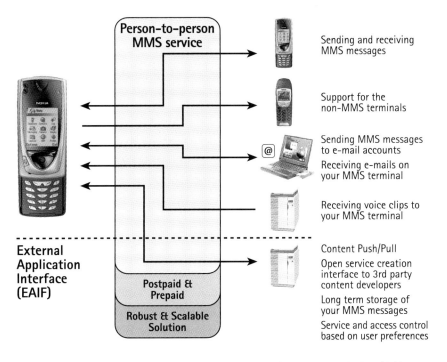

Exchange of MMS messages between mobiles devices, the Internet (e-mail, instant messaging), and Web applications

In sending multimedia messages, the address used can be either the recipient's phone number or the Mobile Station Integrated Services Digital Network Number (MSISDN) familiar from SMS, or an e-mail address [RFC822]. Messages addressed by MSISDN are routed to the recipient device whereas e-mail addressed messages are routed directly to the Simple Mail Transfer Protocol (SMTP) network which handles the delivery.

MMS messages are routed via both the sender's and recipient's Multimedia Messaging Service Centers (MMSC). The routing mechanism enables fast introduction of services, and, more importantly, messaging between networks of different technologies, (e.g., the Global System for Mobile Communications (GSM), Time Division Multiple Access (TDMA), Code Division Multiple Access (CDMA) and Third Generation (3G)). This interoperability makes MMS a truly global service.

Standards

The MMS architecture and the overall concepts have been standardized in the Third Generation Partnership Project (3GPP). Based on the work and requirements of 3GPP, the Wireless Application Protocol (WAP)-based implementation specifications are the responsibility of the WAP Forum. WAP 2.0, released in July 2001, contains the specifications for MMS version 1.0 [WAP-205], [WAP-206], [WAP-209].

These specifications address the MMS concepts and requirements presented in 3GPP Release 99 [TS22.140][TS23.140].

Work is underway in the WAP Forum and 3GPP to address future MMS functionalities. WAP MMS version 1.1, based on 3GPP Release 4 and backward compatible with WAP MMS version 1.0, will offer new features such as:

o   Forwarding: a feature allowing a recipient to forward a message to other recipients without having to first retrieve the message on his device.

o   Reply-charging: a feature allowing the sender of a message to accept the fees for the reply to his message.

Network Architecture

The overall architecture of the MMS implementation in mobile networks is shown in the following figure. The role of the Multimedia Messaging Service Center (MMSC) is analogous to the role of the SMSC: it provides access to the mobile network with WAP, store and forward service logic and an External Application Interface (EAIF) towards applications that provide value-added services on top of MMS.

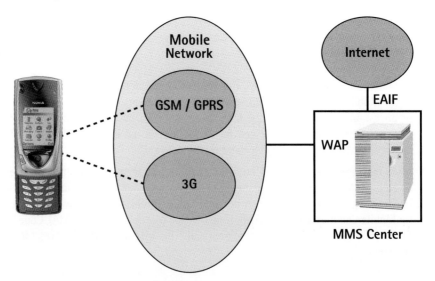

Multimedia Messaging Network Architecture

Multimedia Messaging

The mechanisms employed to deliver messages in the mobile network are different from the ones used in SMS. In MMS, the message sizes in the first phase are estimated to be around or under 30 kbytes, depending on the content types. This is considerably more than the maximum allowed of an SMS message. The WAP protocol suite was selected as the mechanism because it provides data transport services that are optimized for mobile networks. WAP also provides uniform transport services regardless of the underlying network.

The Wireless Session Protocol (WSP) layer [WAP-203] is used as the basis of the transport mechanism. WSP can be viewed as a binary version of the HyperText Transfer Protocol (HTTP) [HTTP1.1]. The following figure shows the message flow for multimedia messages delivery over a General Packet Radio Service (GPRS) network. Message notification at the receiving device uses the WAP push mechanism. A WAP push Service Indication (SI) is delivered to the recipient's device on top of the SMS bearer. This notifies the recipient's device that a new message has arrived. The service indication triggers a WSP/GET operation from the receiving device that fetches the message to the device. After the message has been fetched, a delivery notification will be issued to the sender.

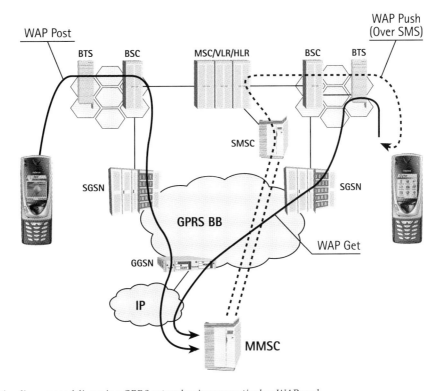

Multimedia message delivery in a GPRS network using connectionless WAP push

Inter-Network Routing

Delivery between networks in MMS is based on architecture different from SMS. Short Messages are delivered from the sender's SMSC directly to the recipient whereas in MMS the messages are routed via both the sender's and recipient's MMSCs.

The reasons for this architecture and routing mechanism are:

o   Introducing recipient services is easy and fast - no need for agreement between carriers regarding the services.

o   Messaging between different network technologies (GSM, 3G, TDMA and CDMA) is made feasible. This interoperability makes MMS a truly global service.

The MMS network architecture in a two-network case is illustrated in the following figure.

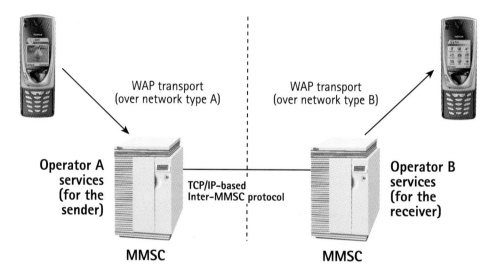

Inter-MMSC Architecture

Being able to route messages between networks of different technology (e.g., between GPRS and CDMA networks), is a step forward from the SMS world where messages are transmitted only on similar networks. This means that the messaging volume potential with MMS exceeds that of SMS. The benefits, however, do not materialize free of charge: MSISDN-based (or other) routing requires routing tables to be set up between MMSCs.

An added complexity is handling Mobile Number Portability, which requires routing decisions to be based on individual MSISDNs instead of prefixes. However, the benefits brought by the network architecture adopted outweigh the disadvantages.

Inter-MMSC protocol standardization work is progressing in 3GPP.

Message Encapsulation

The multimedia message is encapsulated in a WSP Protocol Data Unit (PDU) consisting of MMS headers and a message body. These two entities form a PDU. This is illustrated in the following figure.

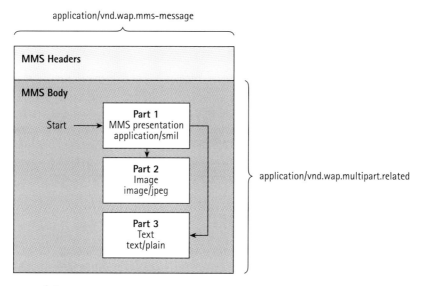

Message encapsulation

The MMS headers contain information related to the transfer of the message from the sender to the recipient. Some headers originate in RFC 822 (e.g., from, to, date, subject) while others are MMS-specific (e.g., X-Mms-Transaction-ID and X-Mms-Expiry). In MMS, the message body typically has the Multi-purpose Internet Mail Extensions (MIME) multipart/related structure of RFC 2387 [RFC2387]. It is composed of distinct multimedia objects and an optional presentation part. The format of choice for the presentation part is the Synchronized Multimedia Integration Language (SMIL) [SMIL2.0], but other formats (e.g., the Wireless Markup Language (WML) [WAP-191]) can be used.

The body can also follow the MIME multipart standard [RFC2045-7], providing compatibility with e-mail systems. In any case, conversion from/to multipart mixed/related and multipart can be easily performed.

The message body is present only for posting and retrieving messages. For other transactions (e.g., message notification and delivery reporting), only headers are sent.

Applications

MMS introduces a generic mechanism to encapsulate and transport multimedia content without restricting the formats used. Because of this, MMS can be the foundation of numerous and very diverse applications:

o   Mobile device to mobile device: sending/receiving photos, voice mail, business cards.

o   Web applications to mobile device: electronic postcards, greeting cards, advertisement, news of the day (video/audio clips), screen savers, animations, maps, photo albums.

o   Internet to/from mobile device: receive selected e-mails, send e-mails.

For instance, although postcards are very popular, it is troublesome to search for a post office or a mailbox or to find correct stamps. It would be more convenient to take a picture with a camera attached to or integrated in the mobile device and send it to a friend using MMS. The message reaches the destination almost instantly and can be personalized (e.g., a picture of yourself in Paris). Indeed, several new mobile devices will have a built-in camera.

Message Adaptation

Multimedia Message Adaptation (MMA) will be a key technology to ensure the success of multimedia messaging. Its role is to seamlessly link different services to MMS for the customer and ensure interoperability between MMS devices having diverse capabilities. MMA is achieved by modifying the multimedia message in the network to make it suitable for the destination device and application. This section presents the motivation behind MMA.

Interoperability With Internet Applications

Interoperability with Internet applications is important for the success of multimedia messaging especially in the early stage as the service is introduced. Initially, MMS customers will be relatively low in number and will start using the service only if it allows them to reach many consumers. The way to access those consumers is to link MMS with existing technologies (e.g., e-mail, SMS and the Web).

To support these scenarios, the messages will need to be adapted within the delivery system. Such adaptation includes message and media format conversion, e.g., WAP MMS to e-mail message format or the Portable Network Graphics (PNG) to JPEG and media modification (e.g., image scaling, file size reduction) to match the mobile device characteristics. Also, adaptation and routing should be hidden from consumers. When sending a message, the consumer should not have to consider the destination application of the message (e.g., MMS, Web or e-mail). Consumer should only need to provide the recipient's address and the message delivery system should handle the rest, automatically performing the required adaptations.

Interoperability Between Mobile Devices

Pervasive Mobile Device Environment

Most mobile devices support the Short Message Service (SMS). Similarly, it is expected that mobile devices of several different types will include MMS. However, MMS defines the message exchange mechanisms over which a wide variety of content can be transmitted without restricting them. This is an important aspect since it allows the evolution of the service and its support by a wide variety of mobile devices. MMS will operate and evolve in a very pervasive environment.

Both the composition and presentation of multimedia messages will involve mobile devices with very different capabilities (e.g., display size, color or gray scale display, processing power, support of content formats, available memory to store messages, regional differences). The technological advancement in mobile device technology further adds to the diversity of devices that will soon be involved in MMS.

MMS content will evolve from text and images, to audio and motion picture clips (video and vector graphics). Therefore, support for previously deployed mobile devices will always remain an issue as MMS content and services evolve. Since the MMS content needs to remain flexible and still ensure interoperability, message adaptation will play a key role in the service.

Profiles or interoperability guidelines on how to use MMS are very useful as long as they only define minimum requirements that can be exceeded. Otherwise, product and service differentiation and evolution may be severely compromised.

For instance, some high-end mobile devices may support Video Graphics Array (VGA) (640x480 pixels) resolution images while lower-end devices may support images closer to 160x120 pixels. It would be an unnecessary restriction to limit mobile devices to sending images with a resolution corresponding to the lowest common denominator. Content adaptation in the network would enable the exchange of VGA resolution images between high-end devices while reducing the resolution when the destination is a low-end device. Network-side content adaptation encourages devices to send messages with the best quality possible while still ensuring interoperability.

Automatic Adaptation

It is not an acceptable option to have multimedia messaging only work satisfactorily between similarly capable devices. Nor it is an option that the sender of a message needs to tailor the message so that it fits the receiver's device capabilities. In general, the consumer will not even know how to do that and he will probably not be able to predict which device the recipient will use to receive the messages. When the service works seamlessly for him, using MMS will be a much better consumer experience because he can concentrate on his communication needs. By contrast, requiring the consumer to consider such technical constraints will decrease his experience and using MMS will not be perceived as enjoyable.

Automatic network-side message adaptation will improve the consumer experience. Without adaptation, delivering the multimedia message may even fail completely in some cases (e.g., if the size of the message is too large to fit into the receiving device).

Message Adaptation Use Cases

Here are some examples of message adaptation scenarios:

o   MMS to e-mail: when sent to an e-mail recipient, the message delivery system automatically converts the multimedia message into a multi-body mail message. It also converts any content in formats only specific to the Mobile domain (e.g., Wireless BitMap (WBMP), WML and Adaptive Multi-Rate (AMR)) to formats widely used on the Internet.

o   E-mail to MMS: when sent from an e-mail user, the message delivery system automatically converts the multi-body mail message into an MMS multi-part message. Also, it converts any content in formats not supported by the recipient's device to one it does support (e.g., PNG to JPEG or GIF). Finally, the layout is adapted to the characteristics of the display. However, it is not realistic to expect that all content suitable for a PC can automatically be adapted successfully for any mobile device.

o   Web publishing: publishing messages on the Web requires that the message be translated to both common Web formats and to WAP formats. This allows supporting the cases where the recipient uses either a Web browser or a WAP browser from a mobile device. This includes changing the SMIL presentation to WML for a WAP browsing-capable mobile device.

All messages targeted to MMS devices should be transformed to best fit the specific capabilities of the receiving device. The original message content should be kept unchanged as much as possible. The capabilities of the receiving device that mainly need to be considered when performing adaptation are:

o   Supported content formats,

o   Maximum accepted message size,

o   Maximum image resolution supported, and

o   Display properties such as resolution.

Mobile Device Capabilities Negotiation

In order to properly adapt messages for the destination device, its capabilities and characteristics must be reported to the server. There are several ways to achieve this. First, the mobile device, upon requesting the download of the message, may indicate its device type in the User-Agent header field of the WSP transaction. It may also use the User Agent Profile (UAProf) specification [WAP-174]. Servers could also maintain databases containing the device characteristics corresponding to each mobile number (a user profile). These solutions will help the server perform proper adaptation.

Conclusions

The new Multimedia Messaging Service has been presented, and major differences between SMS and the Multimedia Messaging Service described:

o   Use of data traffic channels for increased message size.

o   Generic encapsulation mechanism allowing usage of rich multimedia content.

o   Inter-network traffic architecture providing global messaging service between different access networks.

o   Need for content adaptation to ensure interoperability between applications and mobile devices of different capabilities, and to enhance the consumer experience.

References

[WAP-205]   "Wireless Application Protocol, WAP MMS Architecture Overview Specification." WAP-205-MMSArchOverview. WAP Forum. URL: http://www.wapforum.org.

[WAP-206]   "Wireless Application Protocol, WAP MMS Client Transactions Specification," WAP-206-MMSCTR. WAP Forum. URL: http://www.wapforum.org.

[WAP-209]   "Wireless Application Protocol, WAP MMS Encapsulation Protocol Specification," WAP-209-MMSEncapsulation. WAP Forum. URL: http://www.wapforum.org.

[WAP-203]   "Wireless Application Protocol, Wireless Session Protocol Specification," WAP-203-WSP. WAP Forum. URL: http://www.wapforum.org.

[TS22.140] "Multimedia Messaging Service: Service aspects; Stage 1," 3rd Generation Partnership Project TS 22.140 Release 1999. URL: http://www.3gpp.org/ftp/Specs.

[TS23.140] "Multimedia Messaging Service: Functional description; Stage 2," 3rd Generation Partnership Project TS 23.140 Release 1999. URL: http://www.3gpp.org/ftp/Specs.

[RFC2045]   "Multipurpose Internet Mail Extensions (MIME) Part One: Format of Internet Message Bodies," Freed N., November 1996. URL: ftp://ftp.isi.edu/in-notes/rfc2045.txt.

[RFC2046]   "Multipurpose Internet Mail Extensions (MIME) Part Two: Media Types," Freed N., November 1996. URL: ftp://ftp.isi.edu/in-notes/rfc2046.txt.

[RFC2047]   "MIME (Multipurpose Internet Mail Extensions) Part Three: Message Header Extensions for Non-ASCII Text," Moore K., November 1996. URL: ftp://ftp.isi.edu/in-notes/rfc2047.txt.

[RFC2387]   "The MIME Multipart/related content type," Levinson E., August 1998. URL: ftp://ftp.isi.edu/in-notes/rfc2387.txt.

[HTTP1.1]	"Hypertext Transfer Protocol HTTP/1.1," Fielding R., Gettys J., Mogul J., Frystyk H., Masinter L., Leach P., Berners-Lee T., June 1999. URL: ftp://ftp.isi.edu/in-notes/rfc2616.txt.
[RFC822]	"Standard for the Format of ARPA Internet Text Messages," Crocker D., August 1982. URL: ftp://ftp.isi.edu/in-notes/rfc822.txt.
[SMIL2.0]	"Synchronized Multimedia Integration Language (SMIL) 2.0 Specification," W3C Recommendation, August 2001. URI: http://www.w3.org/TR/smil20.
[WAP-191]	"Wireless Application Protocol, Wireless Markup Language Specification, Version 1.3," WAP-191-WML, WAP Forum, 19-February-2000. URL: http://www.wapforum.org.
[WAP-174]	"WAP User Agent Profile," WAP-174-UAProf, WAP Forum, URL: http://www.wapforum.org.

Instant Messaging and Presence

Instant Messaging and Presence (IM/P) is a great success on the Internet. In the Mobile World, people on the like have to keep in contact with their peers, and IM/P will help them to maintain their essential social links when on the move. The Session Initiation Protocol (SIP) is a perfect for these communication services.

Session Initiation Protocol

SIP is going to be the standard signaling protocol/mechanism supporting Voice over IP (VoIP) for third generation mobile networks, a choice ensuring that future VoIP services will be truly Internet-based. Defined by the Internet Engineering Task Force (IETF) Session Initiation Protocol workgroup in RFC 2543, SIP is very different from traditional signaling methods used in telecommunication networks. It shares a lot with existing Internet protocols like HyperText Transfer Protocol (HTTP) [3] and Simple Mail Transfer Protocol (SMTP) [4]. SIP was designed to establish, modify and terminate multimedia sessions or calls. It is text based which makes SIP relatively easy to implement, easy to debug, extensible and modular. The modularity helps to implement SIP clients in different device categories from small devices with limited capabilities to desktop computers with full features.

SIP uses the Transmission Control Protocol (TCP) [11], User Datagram Protocol (UDP) [12], or Stream Control Transmission Protocol (SCTP) as a transport mechanism. Actually, SIP is independent of the lower layer transport protocol. SIP has its own reliability system and can be run over any protocol, which provides byte stream or datagram services. A SIP message is composed of headers and an optional payload, e.g., Multi-purpose Internet Mail Extensions (MIME) type. The default payload type is Session Description Protocol (SDP) [6], which can be used to describe multimedia sessions.

```
INVITE sip:jari.kinnunen@nokia.com SIP/2.0
Via: SIP/2.0/UDP midas.research.nokia.com:5060;branch=z9hG4bk38a7c76
Max-Forwards: 70
From: "Bouret Chris" <sip:chris.Bouret@nokia.com>;tag=c39854
To "Kinnunen Jari" <sip:jari.Kinnunen@nokia.com>
CSeq: 1 INVITE
Call-ID: a58b6c67682e0@midas.research.nokia.com
Content-Type: application/sdp
Content-Length: 143
Subject: Project Questions
Accept: application/sdp
Contact: <sip:chris.Bouret@midas.research.nokia.com:5060>

V=0
o=username 0 0 IN IP4 midas.research.nokia.com
c=IN IP midas.research.nokia.com
t=0 0
m=audio 1038 RTP/AVP 0
a=rtpmap:0 PCMU/8000
a=ptime:20
```

SIP Message

SIP deliberately utilizes addressing based on global Uniform Resource Identifiers (URI) [2], which is similar to Internet e-mail (sip:<user>@<domain>). SIP addresses personal mobility using a personal, unique identity. A consumer may originate, receive and access services from any device and any location.

Instant Messaging and Presence Demonstrator implemented on the Nokia 9210 Communicator

Services

Voice is not the only SIP based service available on the Mobile Internet. Mail, Web browsing, Web services, Instant Messaging and Presence will be the first services available and voice over IP will come as a complement to existing services (i.e., SIP allows initiation of any interactive session between any kind of network endpoint).

The Mobile Internet offers a way to combine these services and SIP offers a perfect evolutionary path by supporting:

- o Instant Messaging (exchange of content between a set of participants in real time)
- o Presence (subscription to and notification of changes in the communication state of a consumer).
- o Voice can be introduced seamlessly when the required Quality of Service (QoS) becomes available on the network. The voice path is established with suitable protocols not part of SIP (e.g., Real-time Transport Protocol (RTP)) [5].
- o Click-to-Call from an electronic document (sip:chris.bouret@nokia.com). This means that the document includes a SIP link as a tag.
- o Transport of MIME types (reference to Web/Wireless Application Protocol (WAP) pages, mail, pictures) in SIP signaling
- o Interactive games with SIP signaling
- o Signaling to any kind of SIP-enabled elements in the network (e.g., servers, intelligent appliances).

Instant Messaging and Presence

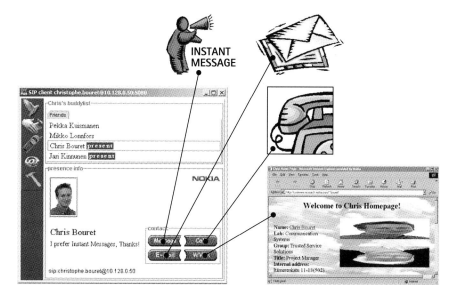

SIP Services

Flow Examples

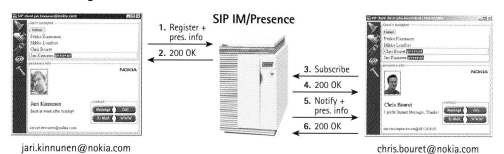

Presence Call Flow Example

We can describe SIP presence functionality as follows: First, Jari sends a registration request to his Presence server (message 1 and 2). The register message includes his presence information coded in XML format. Presence information can also be delivered to the presence server using some other mechanism (e.g., HTTP/Simple Object Access Protocol (SOAP)). Chris, interested in Jari's presence information, subscribes by sending a message to Jari's presence server (3 and 4). After authorization from Jari (not shown here), Chris gets a NOTIFY message including Jari's presence information.

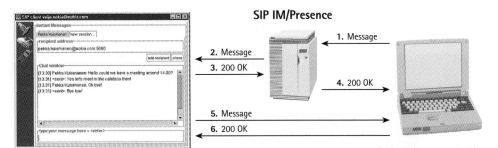

Instant Messaging Call Flow Example

Simple Instant Messaging message flow: The first time Pekka sends a message to Veijo (message 1) it goes to the IM/P server that forwards or redirects it to Veijo's actual address (message 2). Subsequent messages (message 5 and 6) can be sent directly from peer to peer but using proxies to allow messages to go through firewalls.

Another alternative is to establish a standard SIP session between endpoints and to deliver messages in a separate negotiated media channel, though standardization work on this issue is still in an early phase.

Service Creation Model

SIP services are taking advantage of the Internet structure:

o   The core network is transporting and routing the signaling
o   Consumer services are executed on application servers or directly on the device

The Internet architecture opens possibilities for third party service providers (e.g., the Application Service Provider (ASP), corporate or even consumers), to create and control services. On the application server side, SIP service creation uses existing well known interfaces (e.g., the Servlet converged Application Programming Interface (API, including HTTP/SIP) [10], the SIP Common Gateway Interface (CGI) [7,8] the Call Processing Language (CPL) [9], and SIP Hypertext Preprocessor (PHP) [14]) to ensure a large base of potential service creators. These standard interfaces also help to create services starting with a HTTP request (e.g., a click to call), that starts a SIP service setting up the actual call. On the device, the service is integrated in the SIP client, making a secure environment (e.g., Java™) very attractive. Thus, MIDlet-like service logic may be executed on the mobile device.

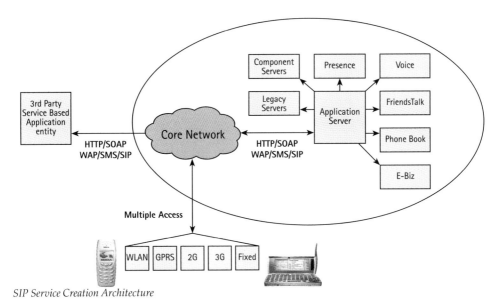
SIP Service Creation Architecture

Current Status and Future of the Technology

SIP IM/P in IETF

SIP is currently being extended by the SIMPLE (SIP for Instant Messaging and Presence Leveraging Extensions) workgroup of IETF to provide SIP with presence information transferring capabilities. In SIMPLE, presence is defined as subscription to and notification of changes in the communication state of a user. This communication state consists of a set of communication means, a communication address, and the status of that user. A presence protocol provides such a service over the Internet or any IP network.

Providing information about consumer presence is particularly well suited to SIP. Since SIP registrars and location servers already hold consumer presence information (through REGISTER messages coming from the UAs) and SIP networks already route INVITE messages from any consumer on the network to the proxy that holds the registration state, SIP can also provide presence services to consumers. This means that SIP networks can be reused to establish global connectivity for presence subscriptions and notifications. In order to handle these kinds of event notifications with SIP, it was extended with two new methods, the SUBSCRIBE and NOTIFY methods. The data format enabling presence and instant messaging (CPIM) interoperability is defined in IETF by the Instant Messaging and Presence Protocol (IMPP) Working Group.

```
<presence xmlns="urn:ietf:params:cpim-presence:">
    <presentity id="chris.bouret@nokia.com"/>
    <tuple id="mobile">
        <status>
```

```
        <value>OPEN</value>
      </status>
      <note>I feel good!!</note>
    </tuple>
    <tuple id="e-mail">
      <status>
      <value>OPEN</value>
      </status>
      <contact priority="1">mailto:chris.bouret@nokia.com</contact>
      <note>My mail!</note>
    </address>
  </tuple>
<tuple id="web">
    <status>
    <value>OPEN</value>
    </status>
    <contact priority="1"> http://www.nokia.com</contact>
    <note>My Web page</note>
  </address>
 </tuple>
</presence>
```

A Presence Document Example

Since all Presence and Instant Messaging solutions defined by the IETF will support this format, it will be easy to translate from one transport protocol to another.

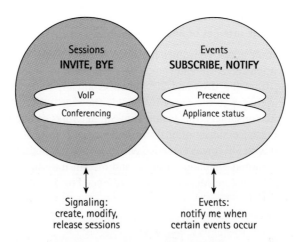

Addition of the event notification framework to basic session initiation

SIP IM/P in 3GPP

Since in third generation mobile networks the Call Control and the Service Creation Environment is based on Internet protocols including the SIP, it seems feasible to introduce SIP based presence services in 3G mobile networks as well. The proposed solution is to place a Presence Server in the IP Multimedia Subsystem [13] of the Release 5 architecture [12] as part of the Application & Services cloud. Since 3G architecture definition started before the IETF SIMPLE and Wireless Village standardization effort, the basic SIP solution will support them as much as possible.

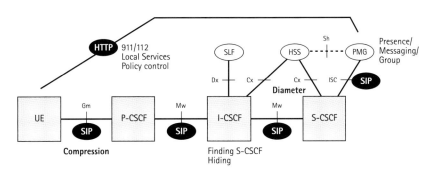

3GPP Architecture extended with presence service

Conclusions

On the Mobile Internet, Instant Messaging and Presence is a key feature. As we have seen in this section, SIP is a perfect fit to implement IM/P. SIP-based services are integrated with existing Internet services and SIP service creation is easy and simple. Finally, the SIP solution is defined in different standardization bodies working in co-operation. SIP Instant Messaging and Presence is a key solution in the Mobile Internet, opening the door to rich multimedia communication services.

References

[1] RFC2543, Session Initiation Protocol, http://www.ietf.org

[2] RFC2396, Uniform Resource Identifiers, http://www.ietf.org

[3] RFC2616, HyperText Transfer Protocol - HTTP/1.1, http://www.ietf.org

[4] RFC821, Simple Mail Transfer protocol, http://www.ietf.org

[5] RFC1889, RTP: A Transport for Real-Time Applications, http://www.ietf.org

[6] RFC2327, Session Description Protocol, http://www.ietf.org

[7] Internet Draft, draft-lennox-sip-cgi-02.txt: Common Gateway Interface for SIP

[8] Common Gateway Interface, http://www.w3.org/CGI/

[9] Internet Draft, draft-ietf-iptel-cpl-framework-01.txt: Call Processing Language Framework and Requirements

[10] JAIN SIP Servlet API, http://jcp.org/jsr/detail/116.jsp

[11] Transmission Control Protocol, http://www.ibiblio.org/pub/docs/rfc/rfc793.txt

[12] User Datagram Protocol, http://www.rfc-editor.org/rfc/rfc768.txt

[13] W3C Note Simple Object Access Protocol (SOAP), http://www.w3c.org/TR/SOAP/

[14] HyperText Preprocessor, http://www.php.net

Wireless Village Interoperability Framework

In the Wireless Village model, Presence takes on a richer meaning, including mobile device availability (e.g., device is on/off, or in a call), consumer status (e.g., available, unavailable, or in a meeting), location, mobile device capabilities (e.g., voice, text, packet data, multimedia) and searchable personal statuses, (e.g., mood – happy, angry) and hobbies (e.g., football, fishing, computing, and dancing). Since presence information is personal, it is only made available according to the consumer's wishes - access control features put the control of the consumer's presence information in the consumer's hands.

The Wireless Village defines and promotes a set of universal standards and specifications for mobile instant messaging and presence services. The standards and specifications will be used for exchanging messages and presence information between mobile devices, mobile services and Internet-based instant messaging services. All will be fully interoperable and will leverage existing Web technologies.

The Wireless Village interoperability framework includes the Wireless Village system architecture and an open protocol suite at the Instant Messaging and Presence (IMPS) Service application level to provide interoperable mobile IMPS services among workstations, network application servers and mobile devices.

System Architecture

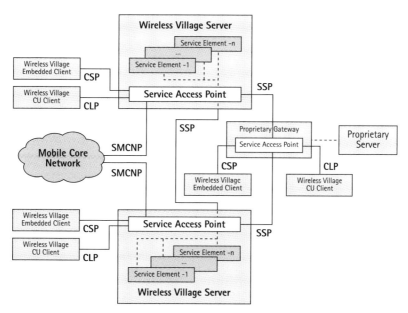

The Wireless Village system architecture

The Wireless Village System Architecture, as shown in previous figure, describes the IMPS system and its relation to mobile networking and the Internet. This is a client-server-based system, where the server is the IMPS server and the clients can be either mobile devices, other services/applications or Web domain clients. For interoperability, the IMPS servers and Gateways are connected with a Server-to-Server Protocol (SSP).

The architecture gives implementers more choices in Wireless Village Servers or Gateways, but with the Wireless Village brand and technology. The Wireless Village Server is the central point in this system. It is composed of four Application Service Elements that are accessible via the Service Access Point. The Application Service Elements are:

- o Presence Service Element
- o Instant Messaging Service Element
- o Group Service Element
- o Content Service Element

The Wireless Village Client consists of an Embedded Client and a Command-Line Interface (CLI) Client. It communicates with the Wireless Village Server to accomplish IMPS features and functions and to provide consumers with IMPS services.

The Wireless Village System Architecture is consistent with the Third Generation Partnership Project (3GPP) TS 22.121 Virtual Home Environment and the 3GPP TS 23.127 Open Service Architecture. The interoperability between Wireless Village Servers and Clients, and between Wireless Village Servers, is achieved through the Wireless Village Protocol Suite.

Protocol Suite

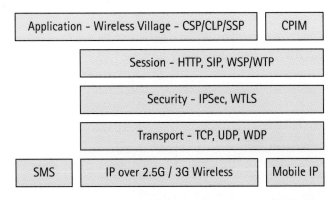

The Wireless Village protocol suite

The Wireless Village Protocol Suite consists of the Client-Server Protocol (CSP), SSP and Common Line Protocol (CLP). The protocol stack is shown in the previous figure.

CSP is designed to provide Embedded Clients in mobile devices and Web domain clients access to the Wireless Village Server.

SSP is designed to provide the communication and interaction means among the Wireless Village Servers and the SSP Gateways. SSP allows the Wireless Village clients to subscribe to the IMPS services provided by different servers which are distributed across the network. SSP allows Wireless Village clients to communicate with existing proprietary Instant Messaging networks through the SSP Gateway.

CLP is designed to provide the Wireless Village server and the CLI client with the means to communicate and interact with each other to support the IMPS services in a legacy CLI client.

The Wireless Village Protocol Suite runs at the application level, and is compliant with Internet Engineering Task Force (IETF) Request For Comments (RFC) 2778, RFC 2779 and the *impp* working group CPIM model. The Wireless Village Protocol Suite may run independently over different transport layer and bearer protocols.

Reference

www.wireless-village.org

Public Key Infrastructure

Public Key Infrastructure (PKI) is a security technology, which can be used to build trust into the Mobile Internet. When accompanied with a set of cryptographic mechanisms and policies PKI provides the foundation for essential security services (e.g., authentication, confidentiality, integrity and non-repudiation).

PKI is not an application on its own. Rather, it is an enabling technology with a huge amount of potential applications ranging from secure e-mail to Digital Rights Management (DRM). PKI builds on top of Public Key Cryptography (PKC), also known as asymmetric cryptography.

Motivation

Consumers are used to conducting business in the physical world. Security mechanisms that protect consumers' daily life and businesses against misuse are features such as handwritten signatures, hard-to-forge holograms on banknotes and micro printing features in identification documents. Unfortunately, none of these mechanisms can be used in the Mobile Internet because their security is based on physical characteristics that do not exist there. Bits are easy to copy and modify. Every copied bit is just as perfect as the original one. Scanning your handwritten signature and attaching it to an e-mail message does not prove anything about the origin of the message. The natural laws in the digital world are simply different from what consumers are used to. However, the security requirements for conducting business remain the same; consumers just need to replace the old security mechanisms with new ones.

Cryptography, the science of secret writing, is the ultimate technology for replacing the old security mechanisms. Actually, viable alternatives for cryptography do not exist. Without cryptography it is difficult, if not impossible, to do business on the Internet.

However, cryptography on its own is just a collection of mathematical algorithms, and a key management infrastructure needs to be deployed on top of these algorithms. It is often claimed that implementing the cryptographical algorithms is 20% of the work and managing the keys is the other 80%. PKI is one such key management infrastructure.

Security Services

The four fundamental security services essential for both the physical and digital world are:

- o Confidentiality
- o Authentication
- o Integrity
- o Non-repudiation

Access control and authorization are also important security services and could be added to the list, but the four mentioned above are the most fundamental. The lesson is to discover that there are many mechanisms to choose from, each of which has its own security and other characteristics.

Confidentiality means that information is available for authorized parties only. Example mechanisms for confidentiality include a sealed envelope, storing a document in a safe or using an encryption algorithm to encrypt an e-mail message.

Authentication is the process of assuring the identity of a peer or the origin of information. In the physical world, authentication can be performed with a passport or some other identification document. In cyberspace, the most commonly used authentication mechanisms are based on usernames and passwords. There are also several security protocols that support authentication based on cryptography and PKI, using, for example, digital signatures.

Integrity-protected information cannot be undetectably modified by unauthorized parties. In the physical world, a sealed envelope with a see-through window is a good example of protecting the integrity of a message. If the seal is intact, the integrity of the message is preserved. In the digital world, mechanisms such as Message Authentication Codes (MAC) and digital signatures accompanied with PKI certificates can be used.

Non-repudiation is the process of gathering some evidence that prevents an entity from later falsely denying that a transaction took place. The evidence can be presented to an independent arbitrator who can make a fair judgment based on the evidence.

An example from the every day world is a handwritten signature, which commits the signer to a transaction. In cyberspace, usernames and passwords or other shared secret schemes are problematic from the non-repudiation point of view because an independent arbitrator can not determine which one of the two parties who shared the same secret produced the evidence. PKI with digital signatures provides better technical grounds for non-repudiation. However, non-repudiation is not just a technical concept but requires a legal framework and appropriate policies in place.

Cryptography

As discussed earlier, cryptography is a viable technology for providing the four fundamental security services listed above. Cryptography is based on four building blocks: encryption, cryptographic hash algorithms, message authentication codes and digital signatures. These building blocks can be used to implement security services.

Cryptography is based on an algorithm and a key. The fundamental principle is that algorithms are not secret - only keys are. This is known as Kerckhoffs' principle: cryptographic systems should be designed to remain secure, even for attackers who know every detail of the design, except the values of secret keys. At first it may seem that disclosing the details of the algorithm weakens its security but history has proven that security through obscurity is not the approach to take.

Encryption

Encryption is used to provide information confidentiality by transforming plaintext information into seemingly random ciphertext. Decryption is the reverse transformation that takes the ciphertext as input and produces the original plaintext. In both encryption and decryption a key is used to control the transformation. Encryption algorithms can be classified into symmetric and asymmetric algorithms.

Symmetric Encryption

In symmetric encryption algorithms, the same secret key is used for both encryption and decryption as shown in the figure below.

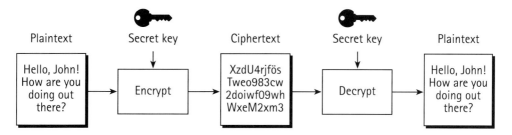

Symmetric encryption process

Actually, some symmetric algorithms use different keys for encryption and decryption. However, the keys are expanded from the same base key, making one reveal the other. The implication of using the same key for encrypting and decrypting is that the same secret key must be shared between both communicating parties. Sharing must be done in a secure way without risking the confidentiality of the shared secret. That is a difficult task. In addition, the number of keys required to be managed by each peer increases quadratically with the number of peers.

Typical symmetric algorithms which are widely used are Data Encryption Standard (DES), Triple DES (3DES), Rivest Cipher 5 (RC5) and Advanced Encryption Standard (AES).

Asymmetric Encryption

Asymmetric (a.k.a. public key) encryption algorithms provide significant improvements over symmetric algorithms. In asymmetric algorithms, two different but mathematically related keys are used. A public key is used for encryption and a private key is used for decryption as shown in the following figure.

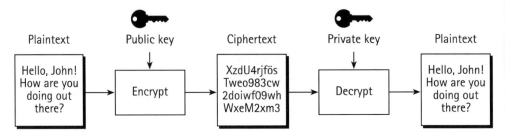
Asymmetric encryption process

The basic idea in public key encryption is that the public key, as the name implies, can be shared with everyone. The confidentiality of the private key must be protected with the same care as the secret key in the symmetric case. Everybody can encrypt a message with my publicly available public key, but only I can do the decryption with my private key. From the point of view of key management, the benefits are obvious. There are no confidentiality requirements for sharing public keys. In addition, the growth in the number of keys is linear rather than quadratic in terms of the number of peers.

The assumption behind public key cryptography is that it is computationally infeasible to deduce the private key from the public key. In most algorithms that infeasibility is based on a hard mathematical problem (e.g. factoring large numbers into prime factors).

Typical widely used asymmetric algorithms are Rivest, Shamir, Adleman (RSA) and Diffie-Hellman. Diffie-Hellman is actually not a public key encryption algorithm but a key exchange algorithm used to share a symmetric key between two entities. However, it is based on the same principles as public key encryption and is widely used in security protocols.

In terms of performance, asymmetric algorithms are several orders of magnitude less efficient than symmetric ones. In practice, most applications use hybrid encryption which combines the efficiency of symmetric encryption and the key management benefits of asymmetric encryption. The potentially large message is encrypted with an efficient symmetric encryption algorithm with a randomly chosen secret key that is then encrypted using the public key of the recipient. This minimizes the amount of data encrypted and decrypted with the asymmetric algorithm.

Cryptographical Hash Algorithms

Hash algorithms are useful for both MACs and digital signatures. They are used to calculate a fixed size fingerprint (typically 16-20 bytes) of a potentially large message. The output of a hash algorithm is called a hash value, fingerprint or a message digest. Hash algorithms are very efficient one-way functions. The two main requirements for a cryptographic hash algorithm are that

- it is infeasible to reverse it and
- it is very difficult to find two different messages which produce the same hash value.

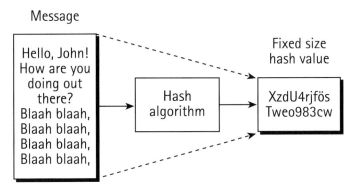

Hash algorithm

Typical hash algorithms are Secure Hash Algorithm 1 (SHA-1), Message Digest 5 (MD5) and RIPEMD-160.

Message Authentication Codes

Message Authentication Code is a cryptographic checksum, which protects the integrity of a message. MACs are based on a secret key and a symmetric encryption algorithm or a hash algorithm. In the latter case the term HMAC is commonly used.

Using a MAC is a two-step process. In the first step, (1) MAC calculation, the MAC is calculated over a message with the MAC algorithm and a secret key. The MAC value is attached to the original message and sent to the recipient. In the second step, (2) MAC verification, the recipient performs the same MAC calculation over the received message and compares the attached MAC value and the MAC value he calculated himself. If they are equal, the message has not been tampered with. If they are different, somebody has modified either the message or the MAC value.

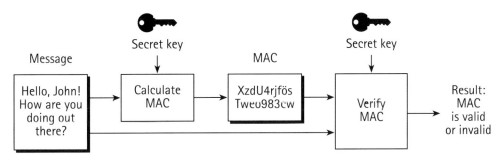

MAC process

The security of the MAC is based on the confidentiality of the secret key used in the MAC calculation and verification. From the perspective of key management, this is identical to the symmetric encryption case and also inherits the same problems.

In addition to data integrity, MACs can also be used for authentication purposes.

Digital Signatures

Digital signatures provide authentication, data integrity and non-repudiation services. The signing process is a mirror image of the asymmetric encryption process consisting of two steps: (1) creating a signature with the private key and (2) verifying the signature with the public key. The process is identical to the MAC calculation and verification process. If somebody modifies a single bit of either the message or the digital signature the verification will fail.

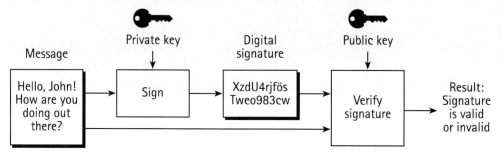

Digital signature process

The basic idea in digital signatures is that only one person can do the signing but anyone can verify his signature. The security is based on the confidentiality of the private key and the integrity and authenticity of the public key. The key management benefits when compared to MACs are similar to the benefits of asymmetric encryption over symmetric encryption.

Some of the encryption algorithms (e.g., RSA), can also be used for digital signatures. Other typical algorithms are the Digital Signature Algorithm (DSA) and the Elliptic Curve DSA (ECDSA).

Digital Signatures With Hash Algorithms

Signing a large document of several megabytes with an asymmetric signature algorithm is a computationally expensive operation. However, an optimization similar to hybrid encryption can be used to make signing large documents feasible. The optimization takes advantage of the characteristics of hash algorithms, as a small fingerprint of the document is signed instead of the whole document. The figure below describes the signature process utilizing a hash algorithm.

Public Key Infrastructure

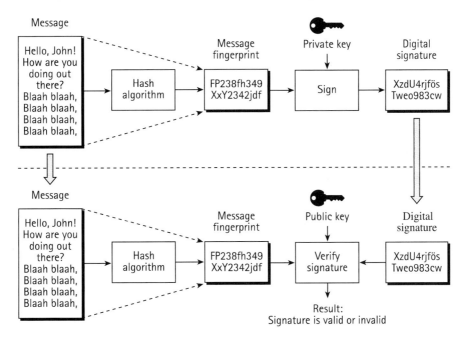

Digital signature process with a hash algorithm

A Few Hard Assumptions

The previous section described the building blocks and algorithms for implementing security services. Most commonly used algorithms and protocols are available for free as open source implementations. It is easy to digitally sign a message with a private key or to encrypt it with a public key, for you can do this by writing a few hundred lines of source code. For example, the Java™ programming language provides excellent Application Programming Interfaces (APIs) for cryptographical algorithms.

However, the security of the mechanisms described in the previous section is based on two quite hard assumptions:

1. **The private key can only be used by the correct person**
    - Need to protect the confidentiality of the private key
    - Need to take care of the private key access control
2. **The public keys are authentic and not tampered with**
    - Need to protect the integrity of the public key (and the association between the public key and the identity of the consumer)

If either one of these assumptions fail the whole security mechanism will fall apart, rendering the whole system insecure.

The first assumption can be solved in some threat models by using, e.g., a smart card or other hardware-based security technologies in combination with, e.g., Personal Identification Number (PIN) codes and trustworthy operating environments.

The second assumption requires a security infrastructure for managing the public keys, i.e., a public key infrastructure. The fundamentals of PKI will be introduced in the next section.

PKI Technology

PKI is about binding names or other attributes to public keys and distributing these bindings to the consumers of the PKI.

The History of PKI

Public Key Cryptography was introduced to the scientific world in the mid-1970s by Whitfield Diffie and Martin Hellman [DH76], although according to current knowledge individuals in the British intelligence agency knew the basic concepts of PKC already in the late 1960s.

The first proposals for distributing the public keys were based on the telephone directory concept that stores consumers' public keys instead of telephone numbers. Consumers could browse the directory and search for their friend's public key using his name as the search criteria. The obvious problem is how to protect the integrity of the database and the information within it, and how the consumer can be sure that the downloaded public key is really the correct one. We could use digital signatures to protect the integrity of the public key, but the chicken and egg problem remains - what public key should be used to verify that digital signature and how to protect the integrity of that public key. At best, the problem reduces to a single ultimate public key, a root key that could be used to verify the integrity of other public keys. This is the fundamental concept of PKI.

The evolution in key management is remarkable. In the symmetric cryptosystems there are a lot of keys and key distribution is really difficult because of the confidentiality requirements for shared secrets. By introducing PKC the number of keys is reduced significantly and the confidentiality requirement for public key distribution disappears. As the last step in the evolution, PKI takes care of the integrity protection of public keys by relying on a single, or a small set of, root keys that are used to verify the integrity of other public keys.

Digital Certificates

PKI is about managing digital certificates binding together public keys and names. The name refers to the subject of certification, the entity that owns the private key corresponding to the public key. Certificates are digitally signed to protect their integrity and authenticity. The party who issues the certificate, a Certificate Authority (CA), sometimes called a Trusted Third Party (TTP), does the signing.

Usually, certificates are made publicly available by storing them in a directory service on the Internet. The entities actually using certificates are called relying parties, and the relying party verifies the integrity and authenticity of the certificate and makes actions based on the information inside the certificate, e.g., extracts the public key from the certificate and uses it to encrypt an e-mail message or to verify a digital signature in a mobile commerce transaction.

Issuing a Certificate

The party issuing the certificate has many responsibilities. First of all, the information within the certificate must be accurate. The name and the public key must be correct; otherwise the certificate fails to do its main job of binding these two together. The two processes guaranteeing correctness are Proof Of Identity (POI) and Proof Of Possession (POP).

Proof of identity is the process during which the CA authenticates the identity of the subject of certification, which can be done by using an official identity document (e.g., a passport).

Proof of possession is the process during which the CA verifies that the subject of certification is really the correct owner of the key pair. Requesting the subject of certification to perform a cryptographic operation with the private key usually does that.

After the proof of identity and proof of possession are successfully performed the CA issues the certificate, i.e., binds the public key and the name together by digitally signing them with the CA's private key.

Using a Certificate

Every PKI-based transaction consists of two separate steps, one performed by the subject of certification and the other by the relying party. The subject of certification uses his private key to perform a private key operation, either to digitally sign something or to decrypt an encrypted message. The relying party uses information from the certificate to either verify a digital signature or to encrypt a message.

From the point of view of PKI, relying party activities are more interesting than the private key activities done by the subject of certification. The relying party must always ask herself a few questions before proceeding with the transaction at hand, e.g., who issued this certificate or do I trust that CA in general, more specifically, do I trust that CA to issue the certificates being used for the transaction at hand. To answer these questions and proceed with the transaction the relying party must have a trust policy.

If the trust policy says that the issuing CA is trusted for these types of transactions the relying party validates the authenticity of the certificate by verifying the digital signature on the certificate by using the public key of the CA. If the signature is authentic, some additional checks are done, including validity period checking. If everything is OK, the public key is extracted from the certificate and the appropriate cryptographical operation is performed with the key.

Certificate Details

A widely used specification defining the syntax and semantics of the certificate object is RFC 2459 written by the Internet Engineering Task Force (IETF) *pkix* working group [RFC2459]. It is based on International Telecommunication Union (ITU) standard X.509 [X.509]. The syntax is defined using Abstract Syntax Notation One (ASN.1) [X.208] and can be represented unambiguously as a string of bytes using Distinguished Encoding Rules (DER).

The most important elements of a certificate are:

- o   Public key
- o   Subject name - who owns the key pair
- o   Issuer name - which CA issued this certificate
- o   Certificate serial number - unique within the domain of a CA
- o   Validity period
- o   CA's digital signature
- o   Extensions - possibility to introduce more semantics, e.g., key usage information, certificate policies, information about revocation status services

A typical size for a certificate varies between 500 - 1000 bytes, depending mostly on the complexity of the naming scheme and the size of the public key and CA's digital signature.

Certificate Revocation

CA needs to have some mechanism to withdraw an issued certificate, should some information in the certificate become invalid. Examples of such events include people getting married and changing their names, resigning from the company or the suspicion that the private key has been compromised. Certificate revocation is the responsibility of the CA, but the revocation status information is of utmost importance to the relying party. If the certificate at hand is revoked, the transaction that the relying party is pursuing should not be continued.

The first revocation mechanism that was introduced in X.509 was based on periodically issued Certificate Revocation Lists (CRLs) that contain the serial numbers of the revoked certificates. The CRL is digitally signed by the CA to protect its integrity. The whole concept is similar to the black lists used in the credit card world.

The downside of this CRL mechanism is that the CRL may grow to be quite large in size. In addition, most of the information in a CRL is not relevant for a single relying party who is only interested in the status of a specific certificate. However, it is possible to divide the CRL into smaller partitions and download only the relevant partition based on the information inside the certificate. If we continue that line of thought all the way down to the CRL partition size of a single certificate, we have introduced another concept for revocation - online certificate status checking.

pkix group in the Internet Engineering Task Force (IETF) has already defined the Online Certificate Status Protocol (OCSP), which a relying party can use to query the status of a specific certificate [RFC2560]. The OCSP responder protects the integrity of its response by signing it digitally.

Both CRLs and OCSP have their merits. The latency times for OCSP are small when compared to periodically issued CRLs. The OCSP requires online connections whereas CRLs can be locally cached, although the cache needs to be refreshed quite often. The features and characteristics of the relying party's device should determine which revocation mechanism to choose, if not both.

Transitive Trust

In X.509 based PKI, trust is transitive, meaning that a CA can certify another CA to issue certificates for consumers. This can also be delegated further on to another CA, should the intermediate CA decide to do so. Certificate validation actually becomes validating a chain of certificates starting from a trusted root CA certificate and optionally going through intermediate CA certificates and finally reaching the consumer certificate. That chain is usually called a certificate path.

Support for certificate paths calls for two functions: (1) certificate path construction and (2) certificate path validation. The path construction process takes as inputs the consumer certificate and a set of trusted root CA certificates and builds a certificate path from one of the roots to the consumer certificate by optionally retrieving intermediate certificates from various online directory services. Path validation is the process of checking the validity of a constructed path, including signature verification and revocation status checks for the certificates in the certificate path.

The transitive trust concept enables CAs to set up different hierarchies of CAs limiting the damage introduced by, e.g., the situation when some CA's private key is compromised.

CA certificates can be revoked using similar mechanisms as are used for revoking consumer certificates. CRLs containing revocations of CA certificates are called Authority Revocation Lists (ARLs).

Traditionally, CAs certified only subordinate CAs within the same organization. However, technically nothing precludes CAs from cross-certifying CAs in other organizations as well. In these cases, the hierarchical CA topology becomes a network of independent CAs.

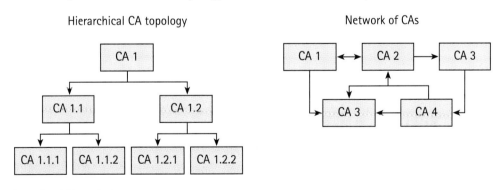

Examples of CA topologies

In a hierarchical CA topology, the CA at the top of the hierarchy is in typical deployments the only trusted root CA for the consumers of that PKI.

Attribute Certificates

A certificate binding a name to a public key is usually called an identity certificate. However, instead of a name we could bind all kinds of other attributes to the public key, for example date of birth, gender, nationality or a title. Going even further, we could bind the rights to access a company's Intranet into a public key. Such certificates are called authorization certificates or attribute certificates. Usually attribute certificates complement an existing identity certificate.

Even though attribute certificates have interesting applications, most of the issued certificates today are identity certificates, identifying a natural person, e-mail address or the domain name of a HyperText Transfer Protocol (HTTP) server.

Applications Using PKI

As discussed previously, PKI is an application-independent enabling technology and does not have any value on its own. The next sections, however, describe some existing PKI based applications.

Secure Web Applications

Transport Layer Security (TLS) is a transport layer security protocol for establishing confidential, integrity protected and authenticated connections between a client and a server [TLS]. It is used in thousands of Web based services, for example Internet banking. TLS is based on and backwards compatible with Secure Sockets Layer (SSL) 3.0.

A TLS server certificate binds the Web server's domain name to its public key. The server certificate is used for server authentication and to set up a secure communication channel between the client and the server.

Secure E-Mail

Secure/Multi-purpose Internet Mail Extensions (S/MIME) is an Internet standard defining how to encrypt and digitally sign e-mail messages [S/MIME]. PKI plays a central role in the key management of the S/MIME standard.

Pretty Good Privacy (PGP) is another very popular secure e-mail application. It is also based on public key cryptography and certificates, although its trust model and syntax deviates from X.509.

Virtual Private Networks

Virtual Private Network (VPN) is a technology for building secure communication links over insecure networks by tunneling information that is exchanged between peers. Security is achieved by using cryptography. PKI is one of the technologies which can be used for key management in VPNs.

IP Security, as defined by Internet Engineering Task Force (IETF) [IPSec], is the most well known standard specifying the details of tunneling and cryptography.

Wireless Public Key Infrastructure

The WAP Forum has specified a wireless PKI solution [WPKI] building on top of a Wireless Identity Module [WIM] and the X.509 standard. The Wireless Transport Layer Security (WTLS) protocol also supports PKI technology for both server and client authentication. In addition, the WAP Forum has specified a WIM based digital signature mechanism for non-repudiation services.

Conclusions

The technical merits of PKI are undisputable. If high-end security is required, PKI is the state of the art solution for building security mechanisms in the Mobile Internet. The status of standardization is reasonably good, although interoperability between PKI vendors has been a problem in the past. However, vendors have recognized this and are currently working together to avoid unnecessary fragmentation of the PKI market. The PKI Forum and the European Forum for Electronic Business (EEMA) PKI Challenge are examples of this on-going effort.

The infrastructure of PKI is one of its problems. Deploying PKI is expensive, especially for the consumer markets. Actually, there is an important distinction which needs to be made: all consumers using a modern PC are already using PKI in the role of a relying party when, e.g., accessing secure Web applications using the TLS protocol. The PKI is part of the underlying operating system or the Web browser and mostly invisible to the consumer.

Attempts to issue certificates for normal consumers have not yet reached the critical mass that would encourage a large amount of service providers to support the consumer PKI. The typical substitute methods for authenticating consumers are based on usernames and passwords exchanged over a server authenticated secure connection. PKI is an integral part of setting up the secure connection but the consumer is only in the role of the relying party.

PKI should be as invisible to consumer as possible. However, this is not the case for most PKI-based applications, which in fact degrade usability by requiring that the consumer understand the technical details of PKI. There is always a trade-off between security and usability - and finding the right balance is hard.

PKC and PKI provide the grounds for non-repudiation services. In addition, current developments in the digital signature legislation in the US and EU are facilitating the progress towards a world where ordinary consumers can make legally binding commitments in the Mobile Internet.

References

[SecretsAndLies]	"Secrets and Lies," Bruce Schneier, 2000
[S/MIME]	S/MIME RFC
[TLS]	TLS RFC
[IPSec]	IPSec RFC
[WTLS]	WTLS spec
[DH76]	"New Directions in Cryptography," Witfield Diffie, Martin Hellman, IEEE Transactions on Information Theory 22, 1976
[RFC2459]	"Internet X.509 Public Key Infrastructure Certificate and CRL Profile," Russ Housley et al., RFC 2459
[RFC2560]	"X.509 Internet Public Key Infrastructure Online Certificate Status Protocol - OCSP," Mike Myers et al., RFC 2560
[X.509]	ITU-T Recommendation X.509 (1997 E): Information Technology - Open Systems Interconnection - The Directory: Authentication Framework, June 1997.
[X.208]	CCITT Recommendation X.208: Specification of Abstract Syntax Notation One (ASN.1), 1988.
[WIM]	"Wireless Identity Module," WAP-260-WIM-20010712a
[WPKI]	"Wireless Application Protocol - Public Key Infrastructure Definition," WAP-217-PKI, 2001

Digital Rights Management

Digital Rights Management systems aim to enable commercial digital distribution of content and enable new business models by protecting the rights of content creators (artists), content publishers (labels), content aggregators and retailers against illegal copying of digital content, while protecting privacy and fair use rights of consumers. Protection of media is nothing new, as we are all familiar with legal restrictions on copying books, limited access to pay channels on cable or satellite TVs, warning messages at the beginning of rental video tapes and area restrictions on DVDs. Less visible media protection practices are levy charges (i.e., taxes paid for empty audio- and videocassettes), more recently also for empty recordable compact discs (CD-R) and even hard discs.

Digital and networking technologies have together acted as a catalyst for the explosive growth of seemingly free content and its distribution on the Internet. First, digital media may be easily and perfectly reproduced by consumers. Second, the distribution of these copies is practically instantaneous through the Internet, and is facilitated by such technologies as distributed computing. Worse, this distribution mechanism is practically anonymous and the fundamentally international nature of the Internet makes it difficult to impose legislation from one jurisdiction to another. Added to these enablers for copyright infringement is the added complication of the social and legal environment of the Internet. In the early rush of companies to build a business on the Internet, content was given away freely to attract visitors to sites. Consumers are faced with mixed messages from society, legal and social domains, which has lead to the current situation in the Web domain where there is a culture of accepted and widespread copying and distribution of copyrighted works. One of the major labels studied consumer reaction to Napster and found that, on average, after a consumer has downloaded less than a half dozen songs without paying, he will expect this to be fair and normal. Unfortunately, this also naturally leads to a situation where artists and other participants in the value chain are no longer paid for their work and are thus less motivated to provide content for consumers.

Contrary to consumer's experience in the Web domain, mobile service subscribers have always paid for services rendered. Whether it is paying for talk time or ring tones, logos, or alerts, consumers have become accustomed to paying for media in addition to paying for access to the network for voice and data calls. Not only do they view this practice as normal, consumers are encouraged to make spur of the moment purchases as the content is low in price and integrated with a simple and pervasive billing mechanism: your phone bill.

An interesting example of consumer attitudes towards paid content in the Mobile and Web domains can be seen in the following figure from [Jupiter2002]. Even if the per item value of ring tones is much lower than music tracks both in money and user evaluation, the market value of mobile content was more than twice the market value of total PC Internet content in Europe in 2001.

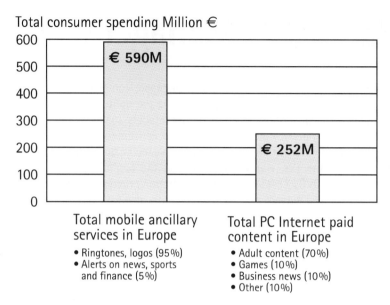

Revenues of paid content in Europe

The content industry and other actors in the content distribution value chain are still trying to establish new rules for the new game and DRM is seen as a critical enabling technology essential to protect the rights of all participants. DRM does not only enable protection of content, but also allows defining new sets of rules for usage (e.g., preview, play X times or location-based consumption). DRM can also enable new, cost-effective content discovery and delivery methods (e.g., superdistribution). This also leads to new business and pricing model opportunities, which still have to be matured. Consumers might prefer to buy only a few of the songs instead of a whole album when new delivery methods enable cost-effective ways to do so. Paying for services must be easy. Providing a credit card for payment on the Internet, for example, is both overly complex and the transaction costs are proportionally too high for just one song. For content owners, subscription-based services combined with content protection and appropriate time out rules create a market. Mobile payment technology makes it pay to pay for content.

Technologically, DRM is interestingly different from most other security-related technology challenges. DRM aims to protect content mainly from device owner attacks, whereas most of the other security technologies concentrate on protecting the device owner and her content against attacks from others. Thus, the security model is practically turned on its head.

System Models

The target of DRM systems is simple. It is built to protect the content owner's rights against illegal copying and other use not respecting the agreement between the content provider and the content purchaser. This is typically achieved by encrypting the content and related rights before delivering them to the consumer. Competition, lack of standards and unclear market and business models have unfortunately created several incompatible, vertical DRM solutions. But even if this is the case in the Web domain, the situation can and should be avoided by early standardization in the Mobile domain.

Basic System Model

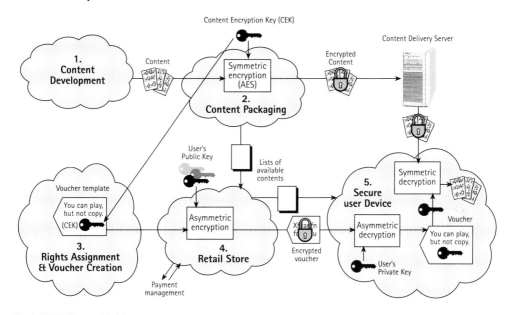

Basic DRM System Model

A typical DRM solution consists of the following logical parts:

1) Content development, creation of a digital content file (e.g., music, image, movie, e-book, ringing tone, game, multimedia application, Java applet).

2) Content packaging, typically using encryption with symmetric key technology (e.g., the Advanced Encryption Standard (AES)). After content has been encrypted it is usually considered safe even for free distribution, since only the Content Encryption Key (CEK) has to be carefully protected.

3) Rights assignment and voucher creation. An entity containing CEK, related usage rights (e.g., use once, use many times, copy, modify) and reference to encrypted content package is usually called a voucher. Usage rights are typically expressed either with Extensible Markup Language (XML) based Rights Expression Languages (REL) or with some proprietary methods. Vouchers can also include metadata about the content.

4) Retail store, which lists available content and delivers the voucher to the consumer. The voucher is typically encrypted using the consumer's public key, based on asymmetric encryption to achieve a higher level of security. The retail store is typically also responsible for ensuring that consumers pay for the content before gaining the right to open the content package.

5) Consumer consumes the content. In order to be able to open the content package, a consumer has to have both the content package and the voucher containing the rights and the CEK. The consumer's device uses the private key of the consumer to open the encrypted voucher. Since the private key enables the consumer to open all vouchers delivered to him and provides access to both usage rights and CEK, it has to be strongly protected. It should be noted, however, that even if one consumer would be able to get access to his private key, vouchers targeted to other devices have not been breached.

System Model Variations

There are several variations of the basic DRM system model described above. Some of the main possibilities are discussed in this section mainly to provide a glimpse to the wide range of technical solutions and to help us understand why so many vertical DRM systems exist on the Internet and why standardization is so important.

One of the many reasons for having so many different solutions is that all content types are not created equal. Some have more value from the very beginning of their lifecycle and the value of some content last longer than others. One view of this is presented below.

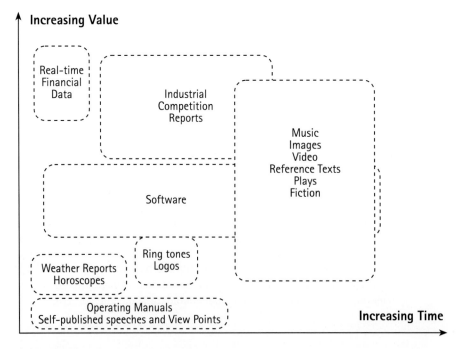

Time value of content

Content Delivery and Encryption

There are several ways how content encryption can differ from the basic DRM system model. One of the most usual is encrypting the content using the device's public key or session-related keys instead of a content-specific symmetric key. This solution offers somewhat higher security since there is no risk of somebody illegally distributing the encryption key to others, since all devices use different keys for decryption. The problems with this approach relate to the scalability of the service (i.e. all content have to be encrypted per device) and it does not offer the same flexibility in content distribution as the basic model.

Encrypting the content with CEK and allowing the encrypted content package to be freely distributed offers many possibilities for delivering content to the consumer. Content can be downloaded from Web sites with HyperText Transfer Protocol (HTTP) or Wireless Application Protocol (WAP) connections, sent to devices using WAP push, broadcasted on digital television networks, distributed on CDs or memory cards or forwarded from a device to another using Bluetooth or infrared (IR) connections. In these cases, the content package usually includes some reference to the place from where vouchers can be obtained and often also some metadata of the content itself. This also enables so-called superdistribution, in which people forward content they like, or they believe their friends would like, to their friends using Multimedia messaging, Bluetooth or other distribution technologies. When their friends receive the content packets all they have to do is to approve the charge for getting the voucher for the content. Superdistribution is expected to be a big success, since letting friends propose content to each other minimizes the difficulty of content discovery.

Some implementations also allow and/or require that a used encryption algorithm or key be changed during the end-to-end delivery process. This can happen, for example, on a PC before forwarding protected content to a DRM-enabled portable music player or from a Web server supporting several DRM solutions. In some cases, content might be provided over a secure channel to a mobile device even in unencrypted form, if the target device is not capable of copying or forwarding the content further. It should be noted that transcryption (decrypting content and encrypting it using a different algorithm or key) does not affect the content quality as transcoding (changing content format, for example, between different audio codecs) usually does.

Voucher Creation and Rights Assignment

In DRM solutions, various entities can be responsible for voucher creation. Depending on the value of the content, trust between participants of the value chain and business models, rights can be determined and defined in digital form by the content creator, publisher, aggregator or retailer. Generally, it can be stated that the closer to the content creator rights are defined and vouchers and content are encrypted, the more secure and complex the solution becomes. For example, allowing the retailer to define the rights and encrypt content and thus be the first actor in the DRM solution chain reduces the complexity of the system significantly, but it also requires a high level of trust between the retailer and content providers and is typically used only for content with low per entity value (e.g., ring tones).

Some of the existing end-to-end content management and distribution solutions aim to support several DRM solutions simultaneously. One possible approach for achieving this is to build a system supporting different vertical DRM solutions under one general content distribution solution. This approach usually requires that content and vouchers are stored in multiple ways and the consumer can download content packages in the format supported by his device. Another approach is to decrypt content and rights stored for different DRM solutions and encrypt them again to a format specified for the particular target system. The latter approach, while possibly more attractive for the consumer (only one DRM solution is needed in the device), is problematic since DRM solution providers and content owners are usually not willing to give other solution providers the right to open the packages before they reach their consumer and thus break their vertical end-to-end business case. The place where all content packages and vouchers are opened and re-encrypted is also quite an attractive place for attacks. Lack of standardized rights expression makes it difficult to translate rights expressions between different DRM solutions.

Variations in Usage Right Definitions

Existing DRM solutions offer a varying selection of usage right definitions. It is quite easy to define fancy usage rights but ensuring that those rights are enforced in devices can be difficult. However, the majority of the most common use cases can be covered with a relatively small set of usage rights.

Usage rights are typically expressed by intents and restrictions. Typical intents are preview, save, modify and print rights, but many others have also been defined. Typical restrictions are validity period (e.g., until a specified date or for specified time after the content is used for first time), count (e.g., play or copy 3 times) and location (e.g., only within a specific cell, building or geographical area).

Complex rights expressions are not difficult only for devices (demanding, e.g., accurate and tamper resistant clocks, support for international time zones and the possibility to securely verify locations) but also for consumers and legislation authorities, while providing only questionable added value for content providers. Other problems are related to backing up and restoring rights, the ability to move content between different devices, e.g., in case of buying a new mobile device, and different responsibility, fair use and privacy issues when consumers and content are moving internationally. DRM technologies can be used to enforce these rules, but updates to legislation, international and industry wide agreements and standards are also needed.

Technologies Related to DRM

DRM is developed to enable secure distribution and usage of digital content. As discussed earlier the solution has to protect the rights of all participants in the distribution chain. Most of the technologies used for enabling this have not been created specifically for DRM purposes; the solutions are typically built by combining several security technologies together, possibly with small modifications. This enables the use of standardized and proven technologies and also allows memory optimization through reuse of software components, important especially in mobile devices.

Security Technologies

Access Control
Access control is probably the first form of protecting digital content against illegal use. Protected content is stored on computers, and consumers would need to provide valid usernames and passwords to access the server, directories and even specific files. Even in early versions of file systems different levels of access permissions (rights) were used (e.g., read, write, modify). From the DRM point of view, access control rights are not enough, though. Even if a consumer only has permission to read the content, he can still download it and then copy and modify it freely. Write permissions refer only to the permission on the specific server in the specific directory or file, not to the content itself.

Session or Connection Protection
Connection protection solutions were originally developed to protect consumers against eavesdropping, i.e., somebody illegally tapping into a connection and copying the bits transferred during a session. This is typically achieved by encrypting the bits with keys agreed at the beginning of the session.

Session protection is used in current DRM solutions mainly in communication between different processes either between two computers over a network or within one device, to enable, for example, secure exchange of encryption keys, user rights or identification information.

Symmetric Encryption
Symmetric encryption is used for encryption of large content files, because the encryption and decryption algorithms are much faster than asymmetric encryption algorithms. In symmetric encryption, the same key is used for both encryption and decryption.

If each content file is encrypted individually for each consumer, n*n keys are needed. Management of all the keys quickly becomes a problem in large-scale implementations. If one key is used for encrypting all contents for a specific consumer, the consumer can get access to all contents if she manages to get control of the key. Also, this approach demands encrypting all content files separately for all consumers, a solution which is difficult to scale. If each content file is encrypted only once, the solution becomes more scalable, but sharing keys with consumers still remains a problem.

Asymmetric Encryption
Asymmetric encryption is based on a solution using a key pair. A public key can be distributed freely and it can be used to encrypt content targeted to the owner of the public key. The owner uses the pair of the public key, the private key, for decrypting the content. When used in DRM solutions, the private key also has to be protected against attacks by its owner, mainly done with tamper resistance technologies used in the devices.

Tamper Resistance

Tamper resistance is a term used to describe solutions aiming to protect systems against different kinds of security attacks. It is generally agreed that none of the DRM solutions are fully tamper resistant, as all solutions can be broken with enough time and effort. Content encrypted with symmetric methods is considered sufficiently safe for free distribution, as are vouchers encrypted with asymmetric methods. In these cases, the effort required to open the packages is considered to exceed the value of the protected content. The biggest threat for DRM solutions are usually attacks to acquire access to the keys or rights in the consumer device. Another risk area is the handling of content in the device after the content has been decrypted.

Key and rights protection are typically achieved by storing them in a protected memory area in the device and utilizing processes in secure execution environments for reading the keys and decrypting the content. This is reasonably easy in closed, limited purpose devices with proprietary operating systems, but becomes an increasing challenge with open, multipurpose operating systems used on PCs and increasingly also on mobile devices. In an open platform supporting 3rd party applications and codecs, the DRM implementation has to be able to be sure that all applications requiring its services respect the restriction of usage rights. This is usually achieved by using digitally signed applications approved by a trusted party. Also the data streams between the processes have to be secured, typically achieved by the session protection methods discussed above.

The more complicated and expensive the resulting implementation, the higher the level of tamper resistance in devices and security of the whole DRM system needs to be. In order to match the security level of the implementation and the level of security demanded based on the value of the content, different models of security levels have been developed. One example is presented in the following figure.

Level	Attack needed to crack DRM	Attacker expertize needed	Potential attackers
0	(Just bad usability)	Anybody	70%
1	Configuration (e.g. time/date)	Anybody interested	40%
2	Legal SW download (e.g. file browser)	Anybody with knowledge	20%
3	Illegal SW download	Hobbyist	7%
4	Illegal SW, expert knowledge needed	IT-expert	3%
5	SW engineering	SW engineer	1%
6	Demanding SW engineering/tools	Experienced SW engineer	0.1%
7	HW based/laborious SW attack	Security expert	0.01%
8	HW based very laborious attack	Professional cracker	.001%
9	Military espionage level attack	Research lab	-

Security levels

Watermarking

Technology for adding either visible or hidden information to the content to enable tracing the original copy is called watermarking. It cannot alone protect content against copying, but together with legal tools, it makes illegal copying much more risky. Unfortunately, watermarking also usually decreases the quality of the content and can either be removed or changed with suitable tools. In some cases, transcoding of the content (e.g., when optimizing images for small screens in mobile devices) can also remove or distort watermarks. Generally, the fewer bits there are in the content package the more difficult watermarking becomes.

An everyday example of visible watermark usage is channel logos on top of TV programs. While not disabling copying, they reduce the value of content for illegal use.

Digital Fingerprints

As with humans, digital fingerprints are used for identifying content files. Digital fingerprints are created by using a hash algorithm to create a small file (typically 16-20 bytes) from a content file. The hash algorithms are designed so that it is very unusual that two even little different content files produce the same fingerprint. The technology has been used, for example, for archiving and finding music files and ensuring that they have not been modified. Digital fingerprints can also be used together with DRM systems to make sure that the content file which a voucher refers to is not modified or changed.

Fingerprinting is a method for adding personalized watermarks to a content file, enabling identification of devices or consumers that have touched the content file.

Content Distribution Methods

As discussed above, many DRM solutions support several content and voucher delivery mechanisms. After content has been encrypted, most of the solutions are actually delivery media agnostic, and voucher delivery usually demands some kind of direct connection between the server and device.

Physical Media

Typical physical media distribution methods are based on CDs and memory cards. Contents delivered on CDs are not typically encrypted, though lately there have been several efforts to protect music with watermarks. These still rather small scale trials by leading music labels, have raised an opposition. In addition to problems with consumer organizations, there have also been technical difficulties and objections from other industry players [CNETApril2002]. Computer games and other software distributed on CDs also use various copy protection mechanisms. One of the most typical is delivering a serial number needed during installation within the package of the CD. There is no technical reason why encrypted content could not be distributed on CDs, though it is difficult to develop economically sensible ways of delivering vouchers encrypted with a consumer device's public key on CDs.

Memory cards are at least not yet being used for mass distribution because of their relatively high cost per megabyte. They are, however, increasingly used for device to device or filling station to device delivery. Currently, most of the contents stored on memory cards are unprotected, but solutions for tying the content to a specific memory card exist for all widely used memory card types. Tying content to a card means usually encrypting that content with a key which is derived from the memory card's ID number. Again, there is no reason why content encrypted with CEK as described in the typical DRM model could not be distributed on memory cards. Delivery of public key encrypted vouchers from device to device is technically possible, but certification management and revocation can become a problem especially with devices which are not always connected to a network.

Legacy Internet

The legacy Internet is often quite a cost-effective way for delivering content from content providers to consumers. Using the legacy Internet, however, demands connections that are not always available for mobile devices or are not supported by devices at all. Even in those cases, content can be delivered to, e.g., PCs and then downloaded to the devices. As discussed above, several vertical DRM solutions developed for the Web domain also restrict their usability for delivering content to mobile devices.

Over the Air

Delivering content directly Over The Air (OTA) to mobile devices is an increasingly attractive option for consumers as connection speeds grow with 2.5 and 3G mobile networks. Connection costs will likely be a restraint for large scale downloading of music files for at least some time, however. Downloading of small media files, starting with ringing tones and carrier logos and continuing with Java™ midlets, Musical Instrument Digital Interface (MIDI) songs and e-books, is already feasible.

Peer-to-Peer

Peer-to-peer or device-to-device content delivery using a Bluetooth or IR connection offers a convenient and cost-effective way to distribute content between friends close to each other. If content is encrypted, it can be distributed freely as mentioned above. Distributing vouchers is more difficult for the same reasons as discussed in the case of physical media distribution. This problem is reasonably easy to overcome if devices support OTA delivery of vouchers leading to attractive superdistribution models.

Streaming

Streamed content has become increasingly popular on the Internet. Streamed media services include Internet radio stations, news or event multicast video services or on-demand video or audio streaming. Streamed media services will also be available in mobile networks. So far streamed content files have usually been large, or with short-term value so that encryption has not been needed. Falling memory prices are also bringing hard discs and other mass memory solutions to mobile and consumer electronic devices, which will also make it necessary to encrypt streamed content. The basic DRM model is also applicable to streamed services, possibly with a few optimizations. It is possible that packet losses will demand error resilient content encryption methods and, since most of the streamed services are based on multi- or broadcast solutions, voucher concepts have to be adapted to these scenarios.

Broadcast

Broadcast delivery is often the most cost-effective way of distributing content. Interesting concepts of over-night delivery combined with online or subscription-based ordering have been developed to further optimize distribution. Traditional access protection (even if technically based on encryption) used in television networks is not alone enough to protect content against illegal copying. Usage rights are not yet delivered together with broadcasted content in television networks, but standardization work is going on in the Digital Video Broadcasting (DVB) organization, for example, to address this problem.

Standardization

If we agree that the typical DRM model can be used as a starting point, we can begin by identifying the technology areas used. Based on the discussion above we can conclude that standards or widely-accepted de facto standards exist for both symmetric and asymmetric encryption techniques. Since tamper resistant solutions are typically strongly protected trade secrets and since they are only interacting with modules within a device they do not need to or cannot be standardized.

There is a clear need to standardize a way for expressing usage rights. As most of the industry players agree that rights could and should be expressed using an XML-based language. W3C has decided that it is not the right place to standardize the exact expressions used, even if the use of XML was strongly recommended. Rights Expression language standardization efforts were then moved to MPEG-21 for general content and to 3rd Generation Partnership Project (3GPP) and the WAP Forum for Mobile domain optimized solutions.

Coming to an agreement on the technologies used at different levels is not enough, however - a standardized protocol for acquiring vouchers and content files is also needed. Standardization of these solutions for the mobile domain has therefore also been started in the WAP Forum and 3GPP.

The need for creating an interoperable widely-used and scalable DRM solution is clearly understood by all the main players in the Mobile domain. Content providers want to ensure that market development, which has started so promisingly with ring tones, will continue to grow when networks and new types of mobile devices enable the use of richer content. Consumers want high quality content and services for their mobile devices and device manufacturers and solution providers want to have consumers who are able and willing to pay for their offerings.

The mobile device sets hard requirements on DRM implementation. The small memory footprint does not allow support for several DRM engines, which also makes standardization more attractive for device manufacturers. Key management and tamper resistance will also become more challenging issues when mobile devices start to support downloadable applications on open operating systems. But mobile devices also offer big opportunities; devices are always with the consumer and are always connected. This offers new, very attractive possibilities for superdistribution and peer-to-peer networking. In the long run, interoperability with other consumer equipment will also become essential for sharing content with other devices that the consumer owns.

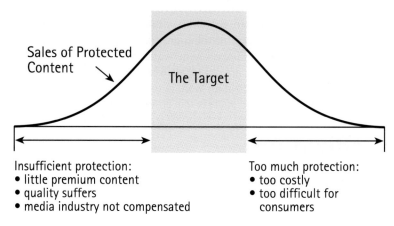

Sweet Spot analysis

Mobile environment standardization is proceeding according to a clear, step-wise approach based on the sweet spot analysis of content value, market size and implementation and usage complexity figure above. Standardization starts from building protection for contents that are most valuable in the Mobile domain today (i.e., ring tones and operator logos). As the per entity value of these content is rather low, the first level of Mobile DRM can be quite simple. The first implementation will most probably be based on encrypting the content but delivering vouchers through protected channels in unencrypted form. The next level, targeting mobile games and richer mobile media content, will introduce the possibility of acquiring the voucher independently of the content. An even higher level of protection will be reached in the next step, with voucher encryption based on a public key infrastructure.

References

Jupiter2002   "Paid content in Europe," Jupiter MMXI, Entertainment and Media, Volume 1, 2000.

CNETApril2002   "Dion disc could bring PCs to a standstill," CNET News.com, April 4, 2002.

Mobile Payment

Mobile payment is a subset of a bigger entity called mobile commerce. Mobile commerce is defined as any electronic transaction or information interaction conducted using a mobile device and mobile network that leads to transfer of real or perceived value in exchange for information, services or goods. As consumable content increases, there is a growing need for different mobile charging and payment methods. Mobile devices will soon be seen as the ultimate digital wallets used for a wide range of mobile transactions in the remote, local and personal environments. In other words, mobile transactions will be done over the digital mobile network, in the physical proximity of and in conjunction with other devices belonging to the same consumer (e.g., a PC or set-top box). As a device used to reliably conduct transactions (e.g., banking, trading, payment and ticketing), the mobile device is becoming a Personal Trusted Device (PTD).

The mobile transactions market has begun with remote transactions on the Mobile Internet. Soon, mobile transactions will also be conducted in the local environment, like at a point of sale, in a fast, easy and secure way. This opens up huge opportunities for the ultimate digital wallet.

Up to now the payment method used in most cases has been the phone bill. However, there is a growing need for carriers to have independent micro payment methods and credit and debit card payments. No matter what payment methods are used, ease of use and security will be the key factors driving the consumption of mobile commerce services as well as the adoption of mobile payments. Overly complex and time-consuming applications and services create barriers that may discourage customers from going mobile. The challenge is to implement a secure payment scheme that is convenient and simple to use.

Mobile Commerce Services

There are numerous alternatives for using a mobile device and/or network for buying, delivering and payment, as can be seen in the following figure:

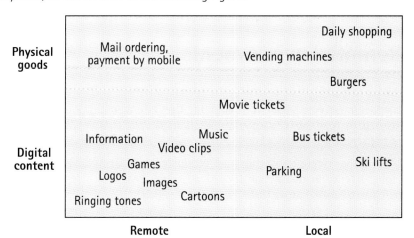

Mobile commerce enables buying, delivering and payment of various kinds of content and goods.

A major part of mobile commerce consists of the purchase of different types of digital content. Consumers are personalizing their mobile devices with ringing tones, graphics and picture messages. Soon games, downloadable phone applications, as well as music and video feeds will be purchased and consumed via mobile devices. In addition, physical goods have become part of mobile commerce. Mobile extensions of existing Internet services enable consumers to order and buy goods (e.g., books, CDs and tickets), via their mobile devices. One typical example of mobile access to existing services is mobile banking, where consumers can conduct money transfers and trade stocks using their mobile device. Essentially, the available services are the same as the bank's Internet services.

Currently, the successful e-commerce sites offer personalized user interfaces, and this should also be applicable to mobile sites. The more consumers have to scroll through the information layers to find what they need, the more likely they are to give up before getting there. Ultimately, service personalization means offering products or services that consumers require, at the right time, in the right place. A typical example is displaying the lunch menu of a nearby restaurant for a hungry tourist.

The launch of Wireless Application Protocol (WAP) services has provided valuable experiences that should be taken into consideration when launching mobile commerce services. First of all, there must be a lot of value-added mobile commerce services available so that consumers get into the habit of using mobile devices as payment devices. Secondly, it is not enough to make mobile extensions of current Internet services, since not all of them offer value added services in the Mobile domain. The mobile device does have its limitations (e.g., a small display).

Security solutions providing higher session security and client authentication methods for mobile payments will give mobile devices an undisputed advantage as a commercial platform for conducting e-business.

Transaction Environments

Mobile commerce transactions can be performed in three environments: a remote, local and personal environment. Each environment has its own mobile commerce services and characteristics requiring specific technologies. A mobile device will be the ultimate transaction device, since it can combine all three environments.

Remote Environment

In the remote environment, transactions are conducted via mobile networks and the location of the participants is not very relevant to this specific transaction. The effects of the mobile transaction become apparent through the User Interface (UI). The implications are that the UI needs to relay all the relevant information to maintain consumer trust and ensure the usability of the services.

Most of the remote transactions are conducted with menu driven applications meaning that higher-latency in transaction time is accepted. Examples of remote transactions are book purchases, banking, stock trading and downloading of digital content that will be consumed on the device. The technology used for remote transactions is in most of the cases WAP over a carrier provided access (e.g., Circuit Switched Data (CSD) or General Packet Radio Service (GPRS)).

Local Environment

In the local environment, mobile transactions are usually initiated over a short-range access technology. In the local environment, the consumer is in direct interaction with the other-end-device, so the mobile transaction has an effect on the consumer's immediate surroundings. The device's UI complements the UI of the other-end-device in keeping the consumer up-to-date when proceeding through the mobile transaction flow. Security (not privacy) is slightly less of an issue, because the consumer gets immediate feedback on the success of each transaction.

Since the consumer needs to wait at the vicinity of the other end for the completion of the transaction, most likely blocking other consumers from executing similar transactions, the performance requirements for local transactions are high. The transactions in the local environment have to be fast and convenient swipe transactions with extremely low transaction-latency. Purchases can be both low-cost, impulse transactions, like buying from vending machines, ticket use or parking, or mid to high cost, non-impulse transactions, like ticket purchases, shopping for groceries or paying at restaurants. Radio Frequency (RF) contactless is one suitable option for a local transaction interface meeting the requirements of speed and ease-of-use. For menu driven applications used in a local environment, a higher latency in transaction time is accepted, so that WAP over Bluetooth (with or without RF contactless) is an example of a usable technology.

Personal Environment

In a personal environment, the mobile transactions are (mainly) conducted on the UI of another device while the device's (security) functionality augments the functionality of the other device. The other-end-device is the one that the consumer is actually interacting with. Implications are that the consumer's interaction with the PTD is limited to the security-essential functions, like Personal Identification Number (PIN) entry, since the primary interaction is with the other-end-device.

An example of a transaction in the personal environment is the mobile device enabling secure transactions with another communications device (e.g., a PC). The mobile device participates in authentication and authorization by making available service certificates and performing any signature operations required by its security element. The consumer may thereby use another communications device without having to personalize it.

Payment Concepts

Today, prepaid/post-paid account based billing is the dominant remote micro payment model and will likely be so in the near future. The first signs indicating the harmonization of billing interfaces are visible. The carriers are in the good position of having an existing functional billing infrastructure and customer relationship management tools for this purpose. In the macro payment field, credit and debit payments are dominating and are more in the interest of financial institutions. A promising payment concept for mobile remote macro payment is wallet-based Mail Order/Telephone Order (MOTO) credit card payment. Digital signature Wireless Identity Module (WIM) functionality gives an additional possibility to make authenticated credit card payments. Given that the magnetic stripe is a payment concept commonly used in the current payment world, the wallet-supporting magnetic stripe card image could be a logical next step for mobile payments, especially for local payments.

The EuropayMastercardVisa payment protocol (EMV) standard is an emerging payment standard that has been developed to improve the security of existing magnetic stripe card based credit and debit payment schemes. The EMV specifications describe the interface between payment chip cards and Point-Of-Sale (POS) -devices and the payment application running on the chip and in the POS-device. In the future, the mobile optimized version of EMV might play an important role in the area of local payment, depending on how well EMV is adopted.

Security in Mobile Transactions

The desired security level in mobile commerce should be considered a business issue and it should be adjusted according to the value of the transaction. The following figure shows how the level of needed security varies depending on the value of the transaction.

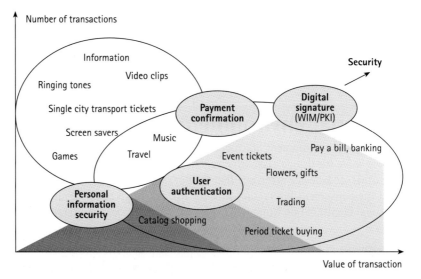

The level of needed security varies depending on the value of the transaction

The highest volumes in mobile commerce will probably be the purchase of low value digital content, therefore extra attention should be put in to the area of micropayment. The frequency of buying this kind of content will be high and potential loss for consumers in fraud situations is quite moderate, so a smooth buying process would be something to aim for. One example of user-friendly authentication would be Mobile Station Integrated Services Digital Network Number (MSISDN) -based authentication instead of usernames and passwords. If additional security is needed, MSISDN could be combined with a PIN code.

However, various mobile commerce services in the macropayment area need a higher security level than only MSISDN-based authentication. Most banking and corporate services need secure access with strong user authentication and session security. Also, stock trading, betting and high value purchases on the Mobile Internet will require that the consumer be strongly authenticated.

The usability of security methods is one of the key success factors for mobile services. Complex security solutions may create barriers which discourage consumers from going mobile. The challenge is to implement the security scheme so that it remains convenient and simple to use. From the usability point of view, a good solution would be Wireless Public Key Infrastructure (WPKI), which makes high security, combined with a convenient user experience, possible.

Security in Macro Payment Transactions

The WAP Forum has specified some security components which can be used to provide a high security level when making mobile commerce transactions in a remote environment. Wireless Transport Layer Security (WTLS) enables server authentication and data encryption. It invisibly encrypts and decrypts information sent between a WAP client and a WAP gateway aiming to prevent any third party from deciphering the communication between these two parties. The protocol also protects the integrity of communications, providing the recipient information to verify that content has not been altered in transit.

The extension to the Mobile domain, called Wireless PKI, Mobile PKI or WAP PKI, covers the infrastructure and the procedures required to enable the trust provisioning needed for authentication and digital signatures for servers and clients. The WPKI specification will address certificate enrollment and lifecycle management, and more specifically, the creation, distribution, verification and revocation of the certificates. The key elements of the WPKI system are:

- Mobile devices with WIM support
- WAP Gateway enhanced with certificate-based identity validation capability
- Registration Authority (RA) for certificate enrollment
- Back-end PKI infrastructure with access to Certification Authority infrastructure

The back-end PKI infrastructure is almost the same for both wired PKI and wireless PKI, with the main difference being certificate enrollment. The process of enrolling a Certification Authority certificate into the mobile device needs to be re-worked to accommodate an efficient mass-market rollout.

Wireless Identity Module

The Wireless Identity Module is a security element, which stores and processes the information and security keys needed for user identification and authentication. Additionally, it has the ability to perform cryptographic operations, though its main function is to enable the digital signatures required for authenticating mobile transactions. The digital signature is executed in the WAP application security layer.

WIM has been defined as a separate tamper-resistant hardware device, such as a smart card or a Subscriber Identity Module (SIM) card. Technically, a WIM can be implemented in several ways, where the main differences are in business and usage considerations.

WIM functionality can be stored on an SIM-card issued by the carrier. However, the mobile network subscription functionality is separated from the other applications which require authentication and signature capabilities.

WIM functionality can also be stored on the second smart card in the device. The Dual Chip approach means that a second smart card is placed semi-permanently in the mobile device. It is removable and issued independently of the SIM card (e.g., by a mobile commerce service provider or financial institution). The strength of such system is that it separates the network subscription from the other applications creating more latitude in terms of business models. Naturally, this requires device designs with the possibility of storing two smart cards.

Eventually, consumer requirements and demand will have an impact on which of the proposed WIM solutions will be taken into use on the market. However, it is good to bear in mind that from a service provider's technical perspective, the implementation of WIM is irrelevant.

Location Technologies in the Mobile Domain

Mobile device location means in its simplest form defining the geographical coordinates of a mobile device. The motivations for this are the various services and applications, which can be based on such information.

There are various ways to group location information based applications, for example:

1) *Government applications*
   o The most important example is to locate mobile devices which have an on-going emergency call. In the United States, the Federal Communication Commission (FCC) has stated that all emergency calls should be located with 67% accuracy of 50 m or better, when location methods requiring mobile device changes are used. Correspondingly, this 67% accuracy should be 100 m or better, when methods not requiring mobile device changes are used. This requirement came into effect in October 2001.
   o According to the legislation of different countries, various electronic surveillance applications may be used.

2) *Carrier applications*
   o Home zone calls. Carriers can offer reduced tariffs when a call is made in certain areas (e.g., at home).
   o Mobile devices and mobile traffic can be located for network planning purposes (e.g., so-called hot spots can be detected).
   o Handover decisions may be done based on location information.

3) *Commercial applications*
   o Fleet management. For example, a taxi company can use location information to track their vehicles.
   o Valuable packages can be tracked.
   o Car navigation.
   o Emergency roadside services.
   o Information services.
   o Search for stolen property.

These are just some examples of location applications, and there could be many more.

The richness of location applications sets numerous requirements for a good location method:

- o The method should give good accuracy. This is perhaps the most important feature of a positioning method. For applications such as car navigation or emergency call location, good accuracy is important.
- o The area where the mobile device can be located, should be as large as feasible. It should also be possible to define in advance where a device can be located; ideally, location should be possible within the whole coverage area of the mobile network.
- o The location method should be fast, i.e., the time it takes to locate a mobile device should be as short as possible. This is again important for applications like car navigation and emergency call location.
- o The location method should not generate too much signaling load within the mobile network.
- o The effects of location method on the mobile device should be minimum, i.e., the power consumption, size and price should not increase. This is naturally important for mobile device manufacturers and consumers.
- o The location method should have a minimum impact on the mobile network. This is important for carriers.
- o Consumer privacy should be ensured, for example, by allowing a consumer to disable the location feature.
- o It should be possible to locate all mobile devices.
- o For certain applications it is necessary to locate a large number of mobile devices at the same time.

Some of the requirements mentioned above are clearly contradictory. This means that different location methods are more suitable for certain purposes than others, and vice versa. No single method is suitable for all applications. Thus, different location methods have been studied and adopted.

Mobile Location Methods

Cell Coverage

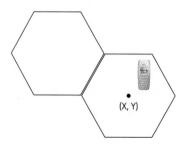

Cell coverage location

The basic idea of a mobile network is to divide the coverage area of the network into cells. Thus, the most straightforward location method is to use the identity of the serving cell to give the location estimate, which can be, for example, the calculated center of the coverage area.

In most cases, only minor software changes in the network are required. Thus, cell coverage is a low cost method. A drawback, however, is that there are applications for which the achieved accuracy is not sufficient. For example, in Global System for Mobile Communications (GSM) the cell radius can be 35 km or even larger (extended cells). Cell coverage may provide an easy way to locate the mobile device in some environments, e.g., in city centers (micro and pico cells).

In some systems, there might be identifiers describing areas consisting of more than one cell. For example, in the third generation Universal Mobile Telecommunication System (UMTS) the Service Area Identifier (SAI) can consist of one or more cells. In such cases, the center of all cells can be used as the location estimate.

Received Signal Levels

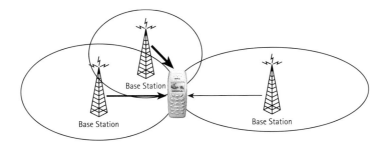

Received signal level location

The strength of a signal the mobile device receives from a base station depends on the distance between them. There exist propagation models, which predict how the signal level depends on the distance and other parameters. Thus, by using measured signal strengths, it is possible to estimate the distance that defines a circle in which the mobile device should be. When at least three such circles are available, the location estimate can be found at their intersection.

In systems such as GSM, the network already knows the received signal levels during a call. Thus, this method is low in cost, and its implementation is straightforward. Drawbacks here are that changing propagation conditions can cause problems, and accuracy can vary a great deal.

The received signal level location method is already in use in commercially available solutions (e.g., in GSM).

Angle of Arrival

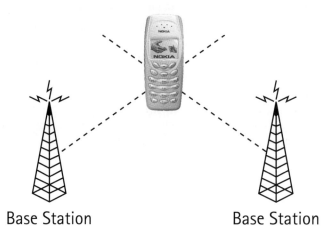

Angle of Arrival location

Angle Of Arrival (AOA) has widely been used for locating radio transmitters. Also, in a mobile network the angle of arrival of the signal from the mobile device can be measured, for example, in a base station by directional antennas. The direction together with the coordinates of the measuring antenna defines a line. When at least two lines are obtained, their intersection can be determined and used as a location estimate.

In mobile networks, coordination is required so that the base stations know which mobile device to measure and when. This generates extra signaling load. Also, the capacity is limited, since one antenna/receiver can locate only one mobile device at a time. A benefit of AOA is that it can often be implemented so that no changes are needed in the mobile device. AOA is relatively accurate in good conditions, but possible problems can be reflections and Non Line Of Sight (NLOS) conditions, i.e., situations where the radio signal cannot propagate a direct route due to obstacles. The accuracy also depends on the distance between the base station and the mobile device. AOA might also require hardware (e.g., antennas) and software changes to base stations. Thus, the cost can be high.

Timing Advance

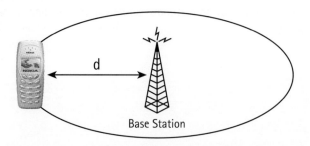

Timing Advance location

In GSM, the time delay between the mobile device and the serving base station, Timing Advance (TA), must be known to avoid overlapping time slots. TA gives the distance d, which defines a circle around the base station:

```
d = (TA * c) / 2,
```

where c is the speed of radio waves, and TA is in time units.

The accuracy of TA is not very good since the resolution of TA corresponds to 550 m. In GSM standards, only one TA circle (to serving cell) is used. This together with serving cell coverage area can be used as a back-up location method. Since both serving cell identity and TA are already available in the GSM network, this is a low cost method.

The first GSM location systems using cell identities and Timing Advance values are already available.

Round Trip Time

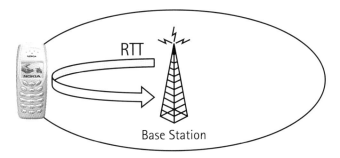

Round Trip Time location

In UMTS standards, Round Trip Time (RTT) measurement has been introduced. RTT is similar to Timing Advance in GSM, but is used only for location purposes.

Accuracy is expected to be much better than in GSM, since the time duration of one UMTS chip, i.e., the smallest part of the spread spectrum code, (appr. 80 m when multiplied by the speed of radio waves) is much shorter than the time duration of a GSM data bit (appr. 1100 m when multiplied by the speed of radio waves).

In the case of a soft handover situation, when a mobile device or User Equipment (UE), as it is called in UMTS, has active communication with more than one base station, accuracy can be improved by using more than one RTT circle, as shown in the following figure.

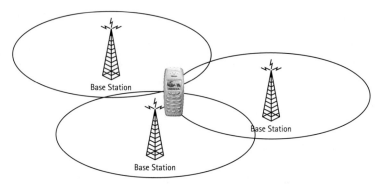

Round Trip Time location with 3 circles

Global Positioning System

The Global Positioning System (GPS) is a satellite-based, widely-used location system. It is based on a fleet of Medium Earth Orbit (MEO) satellites transmitting two L-band spread spectrum signals. A straightforward way to use GPS for mobile device location is to integrate a GPS receiver into a mobile device. GPS has good accuracy, up to 2 m with differential corrections. Also, the removal of Selective Availability (SA) disturbance feature in GPS signals means that basic GPS can have an accuracy of up to 10 m.

GPS needs Line of Sight (LOS) to at least 3 satellites. This means that problems can occur indoors and in city environments. Integrating a GPS receiver into a mobile device increases the manufacturing price and power consumption, and antenna integration can also be a problem.

A mobile network can provide assistance information to the mobile device and its GPS receiver in order to improve its performance and to reduce some of these problems. Assistance data can consist of reference position and time, differential corrections, ephemeris data, ionosphere models, almanac information, GPS time transfer data, Universal Time Coordinates (UTC) models, satellite acquisition and health data. Assistance data fastens acquisition, decreases power consumption, increases sensitivity and expands coverage area (e.g., urban environment).

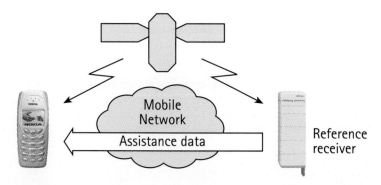

Assisted GPS

Assisted GPS (A-GPS) has been standardized both for GSM and UMTS.

Time of Arrival

In this method, the Time Of Arrival (TOA) of a signal from the mobile device is measured at different places in the network. In GSM, a mobile device is made to transmit random access bursts by forcing it to make an asynchronous handover. Normally, measurements are done at base station sites.

The difference in TOA values (TOA1, TOA2) measured between two places determines a hyperbola:

```
c*(TOA1-TOA2) = d(MD,BS1) - d(MD,BS2),
```

where c is the speed of radio waves, and d(MD,BSx) denotes the distance from the mobile device to a base station. When at least two hyperbolas are available, the position estimate of the mobile device can be determined in most cases. Three hyperbolas are needed for a unique solution.

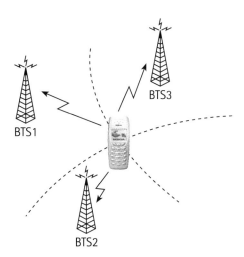

Time Of Arrival location

TOA is relatively accurate. Simulations performed in the T1P1 standardization organization (GSM 05.50) show that TOA accuracy varies depending on the environment (e.g. number of hyperbolas). For example, in urban areas, 67% accuracy is appr. 200 m with 2 hyperbolas, and appr. 100 meters in suburban areas. TOA can be implemented so that no changes in the mobile device are required (in GSM). TOA requires a common clock at measurement devices, and coordination between base stations, which means that extra signaling load is generated. Also, the capacity is limited, and TOA may require hardware and software changes to the mobile network.

Enhanced Observed Time Difference

Enhanced Observed Time Difference (E-OTD) is actually a mirror image of the Time Of Arrival location method. In E-OTD, the mobile device measures the time difference between bursts received from the serving and neighbor base stations (Observed Time Difference (OTD)).

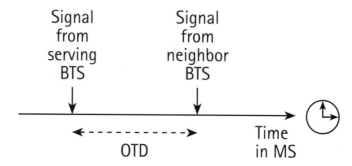

Observed Time Difference

The basic equation for E-OTD location is:

```
OTD = RTD + GTD.
```

- OTD Observed Time Difference between the receptions of bursts from two base stations measured by the mobile device.
- RTD Real Time Difference is the synchronization difference between base stations, i.e., the relative difference in transmission times of their bursts.
- GTD Geometric Time Difference is due to different propagation times (different distances) between the mobile device and the two base stations. GTD includes actual information about location:

```
GTD = [d(MD,BS1) - d(MD,BS2)] / c
```

where d(MD,BTSx) is the distance between the mobile device and the base station x, and c is the speed of radio waves. Thus, the possible location for the mobile device observing a constant OTD value between two base stations is a hyperbola. When at least 2 hyperbolas are obtained, the location estimate can be found at their intersection. A unique solution requires 3 hyperbolas.

E-OTD requires mobile device to support the needed measurement functionality. E-OTD also requires a synchronized mobile network, or the timing differences of the base stations have to be known. According to GSM standards, RTD values can be measured using Location Measurement Units (LMU). RTD values can be calculated from the equation:

```
RTD = OTD - GTD,
```

where OTD is measured by the LMU, and GTD can be calculated from the known coordinates of the LMU and base stations.

E-OTD is relatively accurate. Simulations performed in T1P1 (GSM 05.50) show that E-OTD accuracy varies depending on environment, number of hyperbolas etc. For example, in urban areas 67% accuracy is appr. 200 m with 2 hyperbolas, and appr. 100 m in suburban areas. Some field tests indicate 67% accuracy up to 50-150 m, depending on the environment, network, procedures etc. E-OTD measurements can be performed relatively quickly, e.g., 1 or 2 second measurements should normally be sufficient. Since each mobile device performs the necessary measurements itself, location capacity is not limited by the measurements (as in the case of e.g., Time Of Arrival or Angle Of Arrival location methods).

E-OTD is included in GSM location standards, and the first commercial solutions are being adopted.

Observed Time Difference of Arrival – Idle Period Down Link

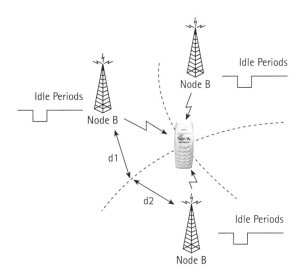

Observed Time Difference of Arrival - Idle Period Down Link location

Observed Time Difference Of Arrival -Idle Period Down Link (OTDOA-IPDL) is similar to Enhanced Observed Time Difference (E-OTD), i.e., the mobile device measures Observed Time Difference values between reception times of signals from base stations (called Node B in UMTS). It is going to be used for UMTS.

Simulations indicate that OTDOA-IPDL should be rather accurate. It is expected to be more accurate than E-OTD due to the larger bandwidth of UMTS.

V MOBILE INTERNET STANDARDIZATION

- o Standardization in the Mobile World
- o Open Mobile Alliance™
- o Wireless Application Protocol Forum
- o Third Generation Partnership Project
- o The Internet Engineering Task Force
- o International Telecommunications Union
- o European Telecommunications Standards Institute
- o The Institute of Electrical and Electronics Engineers
- o Wireless Ethernet Compatibility Alliance
- o Bluetooth SIG
- o Open Source Development Lab and Carrier Grade Linux Work Group
- o Service Availability Forum
- o Java™ Community Process
- o World Wide Web Consortium
- o Web Services Fora
- o Object Management Group
- o Liberty Alliance
- o Location Inter-Operability Forum
- o Mobile Commerce Fora
- o Wireless Village

Standardization in the Mobile World

Historically, there have been separate standardization activities in the telecom and datacom industries. Both of these have had multiple organizations established for different reasons or objectives, (e.g., organizations with a regional focus or industry forums for technology harmonization). Even today, there are a number of standardization organizations, each with its own objectives. In many cases, they are making similar or even almost overlapping efforts, with only minor differences. This multitude of standardization activities has led to other associated needs (e.g., how to manage a growing number of implementation options or guarantee acceptable interoperability between products in a multivendor environment).

The next few years will see the convergence of mobile communications and the Internet resulting in new technologies, new business models and business opportunities. We are moving towards a Web-based business model where mobility and the Internet are unified. This will require new competencies from all parties involved in the industry, including standardization organizations. They will also face new challenges and requirements.

Understanding mobility and the unique characteristics of mobile business will be vital to building the networks and services of the future. Success in the Mobile World will be about speed. It will be about how well services and content are controlled, and how well common requirements and drivers for standardization actions are implemented in the Mobile Internet era.

Standardization arose out of the desire to reduce dependence on individual vendors and to allow development of systems composed of smaller parts. While these basic objectives are still valid, some new challenges can be identified. There are number of reasons for these challenges, but perhaps the most important are the need to introduce new end-to-end services more quickly, the disappearance of the division between fixed and mobile technologies and the requirement to standardize issues which have not been standardized previously.

Challenges for standardization in the Mobile Internet include:

o   How to extend vertical network standardization with end-to-end aspects
o   How to focus standardization work items and manage increasing complexity
o   How to reduce the number of different standardization options
o   How to enable evolution speed for different technologies on different layers
o   How to create more added value to consumers
o   How to support application development in standardization

As can be seen from the list, these challenges involve a wide range of issues. Some of the topics are visible directly to consumers. Others are seen only by communications professionals. However, it is essential to remember that standards are made to bring some benefit to the industry and they are only important if they address the right issues at the right time.

Standardization Challenges

How to Extend Vertical Network Standardization with End-to-End Aspects

Many telecom standards are based on a model where vertical interfaces are standardized, enabling independent implementations and interoperability testing on agreed test cases using automated test suites. The following figure illustrates a few vertical interfaces in the MITA end-point-to-end-point interconnection model.

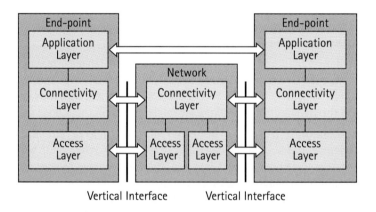

A few vertical interfaces in the MITA end-point-to-end-point interconnection model

Another approach in standardization is to focus on specifications for User-Network interfaces (UNIs) and for Network Node Interfaces (NNIs). Typically UNI specifications cover consumer device interfaces for related access network(s), whereas NNI specifications cover interfaces between carriers. The following figure illustrates UNI and NNI interfaces in the MITA end-point-to-end-point interconnection model.

Standardization in the Mobile World

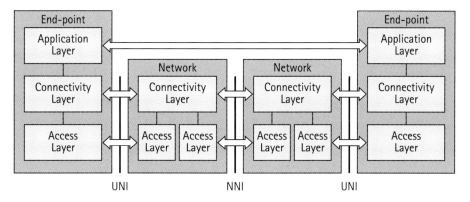

UNI and NNI interfaces in the MITA end-point-to-end-point interconnection model

As illustrated in the above figure, both UNI and NNI interfaces are peer-to-peer interfaces at either the Connectivity and/or Access layers. Interconnection signaling between end-points is provided at the Application layer. In the MITA model, the Connectivity layer is an end-to-end transport service provider for Application layer information. The Connectivity layer should also provide end-to-end services for end-to-end communication (e.g., Quality of Service, Security). These end-to-end aspects are illustrated in the following figure.

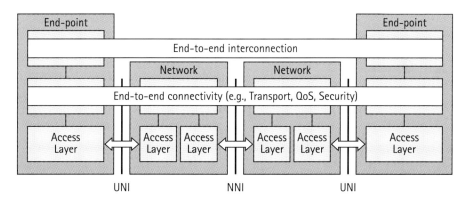

End-to-end aspects between end-points

The above figure, however, does not address all the end-to-end aspects of the user plane that must be taken into account in the Mobile Internet. The generic content delivery model in MITA means that content creation and content consumption processes are needed at the end-points. Internal architectures at both end-points and network entities need to be structured to meet the end-to-end requirements (e.g., delay and jitter). The following figure shows end-to-end user plane flow between end-points.

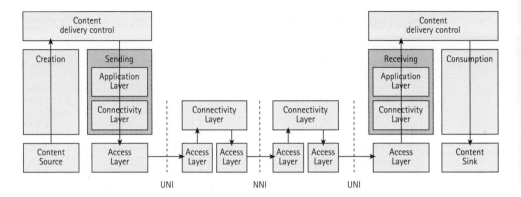

End-to-end user plane flow between end-points

The comparison of the cases described above clearly reveals that standardization work must take into account end-to-end aspects. However, this does not mean that standardization organizations should start to work on everything instead of looking at certain parts of networks. It is rather a question of bearing in mind the big picture and judging standardization results against their merits as part of the end-to-end communication system.

How to Focus Standardization Work Items and Manage Growing Complexity

Today, many standardization forums have a mixture of work items. They have an internal working structure and in some cases also an internal architecture that has been used for mapping the current work items and/or identifying candidates for new work items. It is not always a simple task to compare or map similar or almost equal work items from different standardization forums.

As addressed in other chapters, MITA is based on a layered approach in which there are layered models for architecture element, network, and identities. MITA also has a generic model for content delivery and access independent interface concepts. These architecture principles can also be used to map standardization work items.

The Mobile Internet environment uses and combines features of both the Mobile and Web domains. The evolution-based approach means that a certain number of backward compliant features are maintained in Mobile Internet standards. This backward compatibility itself increases the complexity of standards. If standardization of these necessary and combined features is not managed well, the Mobile Internet may become an over-specified environment.

One of the key ways to manage complexity is to be aware of its threats and to maintain a clear view of the overall concept. In practice, this leads to some very simple, yet fundamental questions: Is this standard necessary? Do we really need it? Does it provide more benefits than increases in complexity?

How to Reduce Standardization Options

The previous section pointed out that it will be impossible to avoid the growth of complexity in Mobile Internet standardization; however, such growth can be managed. One way to exercise influence on this phenomenon is to reduce the optional features in standards.

Today, technology alternatives have been included in standards with equal implementation priorities, though these kinds of mandatory options add complexity to standards. Another part of the phenomenon is that one solution is selected as a mandatory feature, and one or more are included as optional features. This approach does enable competition between different implementations, but may lead to a lack of good interoperability between products.

In a Web domain, interoperability between products is not essential because at the installation phase, interoperability between entities can be tested and be provided from that point onwards. In a Mobile domain, however, the roaming requirement between networks invalidates the possibility of fixed settings. Without good interoperability between entities, roaming could not be provided.

In addition, product development costs tend to get higher with more standardization options. This also slows the introduction of new products and features.

It is already half a victory to realize that the increasing number of options directly decreases the value of standards. General frameworks with many options may be nice on paper, but they hardly constitute a practical standard.

How to Enable Different Technology Evolution Speeds at Different Layers

In the past, telecom networks essentially offered a few basic services. Standardization was often done so that the entire network and related services were standardized and upgraded at the same time. However, with the multitude of consumer services, standardization at different layers and parts of the network happens at different speeds.

The MITA Element has three layers: the Application Layer, the Mobile Internet Layer and the Platform Layer. The speed of technology evolution at different layers is very different today. At the Application Layer, it is relatively easy to invent new applications and solutions, even without requiring any substantial new technologies.

On the other hand, the progress of technology related to access technologies at the Platform layer needs more research, simulations, and prototyping than new applications, before new inventions have matured into standards. This is partly influenced by the restrictions of the physical media, legislative issues and the high investments needed to develop these technologies. At the same time, development and standardization of the Application Layer technologies may proceed considerably faster since they do not have similar restrictions.

The third layer, the Mobile Internet Layer, has two major roles to tackle. It has to provide solutions for the end-to-end aspects presented above, and also provide enough functionality so that new applications need not be vertically synchronized with the slower progress of Platform layer technologies. These challenges are illustrated in the following figure.

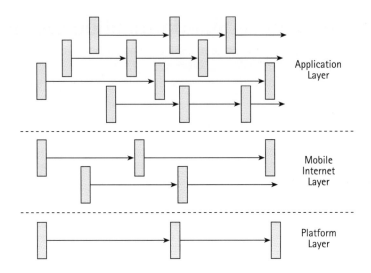

Different technology evolution speeds at different layers

It is relatively easy to conclude that one synchronized standardization release can be made for the Platform and the Mobile Internet Layers but the Application Layer needs its own releases.

How to Enable More Added Value to Consumers

In the past few years in the Mobile domain, network evolution from Global System for Mobile Communications (GSM) to General Packet Radio Service (GPRS) and then to 3G has led standardization actions. There have also been service improvements from Short Message Service (SMS) to Multimedia Messaging Service (MMS) and Wireless Application Protocol (WAP) 1.0 to WAP 2.0. The latest new dimension in this evolution is Java™ enabled application programming for mobile devices.

In parallel with these evolutionary tracks, more and more weight has been put on added value for consumers. This increases the role of services in standardization. In a similar way, application development requirements will impact standardization when the number of active developers increases.

In the Mobile World, added value for consumers is a combination of network capabilities, the number of attractive applications, and the service portfolio offered. In order to create more value, enough attention must be paid to standardizing all of these components. The last two items, in particular, are visible to consumers. For example, a few well-placed standards could lead to faster acceptance of new services and better usability of applications.

How to Support Application Development in Standardization

As we saw above, application development capabilities in Mobile Internet entities play an essential role in the development of the Mobile Internet services and added-value for consumers. The vertical interface model in standardization assumes that system integrators provide service integration and interoperability. Typically, the vertical model does not address or adequately take into account a software developer's requirements because the standardization focus is on interconnection interfaces.

Today, software interfaces are either vendor-specific proprietary interfaces or open interfaces via either standardization (e.g., Java Community Process (JCP)) or open source principles. In many cases, these application programming interfaces are not specified as part of the original standardization process that made the main technology selections, but are made in another and mostly independent, organizations.

In the Mobile Internet era, both technical specifications and service enablers based on these technical capabilities have equal importance for service creation. This requirement addresses a need to improve the link between technical standardization and software interface specification processes.

Principles for Standardization in the Mobile World Era

The challenges of standardization can be addressed by utilizing the MITA technical architecture modeling principles. Below is a list of the most significant standardization principles:

o   Layered models for elements, network and identities should be utilized to focus standardization work items.

o   In network-related work items, User, Control and Management planes should always be separated.

o   A vertical interface model (e.g., UNI and NNI specifications) will not suffice in the future. Actions taking into account end-to-end aspects (e.g., transport, Quality of Service, and security) at the Connectivity layer and the interconnection requirements at the Application layer should be included in standardization with its own end-to-end work items. The Interconnection model between subsystems is a tool for modeling and solving end-to-end aspects.

o   Work items specifying functionality for multiple layers should be avoided. When multilayer work is needed, work items should be divided into layer specific subsystem definitions with clearly specified primitives between subsystems.

o   Complexity of systems should be reduced by improving comparison analyses before technology selections are made for subsystems and then avoiding unnecessary redundant options in standards.

- Managing standardization processes with complete master releases causes extra delay and does not meet the challenges of the future. Framework models (e.g., applications and services, element platform and end-to-end connectivity) are needed. In these frameworks, results from multiple standardization forums are combined. The frameworks should have independent release schedules to enable attractive technology evolution speed and continuous growth of added value for consumers.
- The subsystem model expects that both technical solutions and related software and protocol interfaces are specified simultaneously. To support the growth of application development for Mobile Internet systems, subsystem modeling should be used widely in standardization.

Open Mobile Alliance™

The formation of the Open Mobile Alliance (OMA) was announced on June 12, 2002. The Open Mobile Architecture initiative and WAP Forum joined to form the foundation for this new organization and, as a result, nearly 200 of the world's leading carriers, device and network suppliers, information technology companies and content providers joined forces to achieve open standards and ensure interoperability. The work previously executed in the Open Mobile Architecture initiative and the WAP Forum continues in the Open Mobile Alliance.

At the time of the announcement, the Location Interoperability Forum (LIF), the MMS Interoperability Group (MMS-IOP), the SyncML Initiative Ltd. and the Wireless Village initiative expressed their intent to start discussions to consolidate with the Open Mobile Alliance. Other industry fora focusing on mobile service specifications are welcome to join.

The creation of the Open Mobile Alliance was to an extent an evolution of the Open Mobile Architecture initiative that was launched in November 2001 and received support from a number of industry-leading companies in the Mobile domain.

Scope of the Alliance

The members of the Open Mobile Alliance are working together to create a common industry architectural framework, drive service interoperability and promote adoption of open specifications and standards.

The Open Mobile Alliance will collect market requirements and define specifications designed to remove barriers to interoperability and accelerate the development and adoption of a variety of new, enhanced mobile information, communication and entertainment services and applications. The definition of these common specifications and the testing of interoperability will promote competition through innovation and differentiation, while ensuring the interoperability of mobile services across markets, devices and carriers, throughout the entire value chain.

Commitment to Interoperability

The charter for the Open Mobile Alliance is to:

- Deliver responsive and high-quality open standards and specifications based upon market and consumer requirements,
- Establish centers of excellence for best practices and conduct Interoperability Testing (IOT), including multi-standard interoperability to ensure a seamless user experience,
- Create and promote a common industry view on an architectural framework, and

- Be the catalyst for the consolidation of standards fora; working in conjunction with other existing standards organizations and groups, such as Internet Engineering Task Force (IETF), Third Generation Partnership Project (3GPP), Third Generation Partnership Project 2 (3GPP2), World Wide Web Consortium (W3C) and Java™ Community Process (JCP).

The principles of the Open Mobile Alliance are:

- Products and services are based on open, global standards, protocols and interfaces and are not locked to proprietary technologies,
- The application layer is bearer agnostic (examples: Global System for Mobile Communications (GSM), General Packet Radio Service (GPRS), Enhanced Data Rates for Global Evolution (EDGE), Code Division Multiple Access (CDMA) and Universal Mobile Telecommunication System (UMTS)),
- The architecture framework and service enablers are independent of Operating Systems (OS), and
- Applications and platforms are interoperable, providing seamless geographic and inter-generational roaming.

Benefits of the Open Mobile Alliance

The alliance will offer its members new avenues of growth and revenue by enabling a multi-vendor ecosystem, built on open industry standards.

The distinct benefit of the Open Mobile Alliance is a holistic approach to the value chain of mobile services and applications. This is reflected in the comprehensive inclusion of the specification work - no other industry organization has had such extensive participation from all parties involved in the entire mobile services value chain. The interoperability issues can be solved more quickly and efficiently when the industry specification fora are under the same working process within the same organization. Another benefit is that the companies involved in the Open Mobile Alliance will not need to invest in several, but one substantial specification body.

Furthermore, the Open Mobile Alliance was formed to enable the creation of mobile services designed to meet the needs of the consumer. To grow the mobile services market, members of the Open Mobile Alliance will work towards stimulating the fast and wide adoption of a variety of new, enhanced mobile information, communication and entertainment services.

Carriers, IT infrastructure providers, device and network vendors as well as content and media providers will all find benefits in a non-fragmented market where technical enablers are standards-based and open.

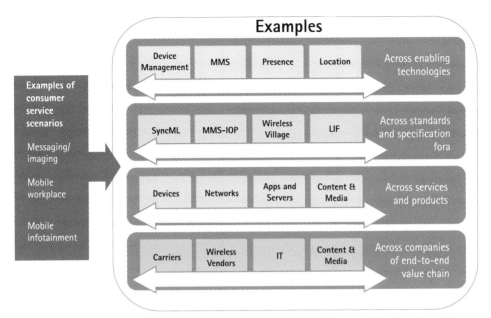

Open approach required throughout the industry: Examples of service scenarios, technologies and needs for interoperability.

Reference

http:// www.openmobilealliance.org

Wireless Application Protocol Forum

"Note: On June 12, 2002, a new global organization, the Open Mobile Alliance, was formed to foster worldwide growth in the mobile services market. The foundation of the Open Mobile Alliance was formed by joining the Open Mobile Architecture initiative and the WAP Forum."

The Wireless Application Protocol Forum (WAP Forum) is an industry forum focusing on providing the specifications for developing advanced services for handheld mobile devices (e.g., mobile phones, communicators, PDAs, pagers and other devices). The WAP Forum was founded in June, 1997 and has since published several versions of WAP specifications, the latest version being WAP 2.0. The goal of these specifications is to specify the technologies that can be used by the whole industry (e.g., mobile device manufacturers, carriers, content providers, application developers) to offer interoperable mobile services on any available device and network technology. WAP Forum specifications are based on current Internet and Web technologies and as such they are aligned with the World Wide Web architecture with extensions specific to the wireless environment. WAP Forum membership is open to all industry participants.

Mission

The mission of the WAP Forum is:

- o   Developing Internet based specifications that are tailored for the mobile environment.
- o   Ensuring conformance and interoperability of mobile services.
- o   Collaborating with other standardization bodies.
- o   Disseminating information and collaborating with industry leaders.

In order to accomplish these goals, the WAP Forum has developed WAP specifications according to the following design principles:

- o   Use of existing standards: The specifications published by the WAP Forum are based on existing industry standards. The WAP Forum currently has several different relationships with other standardization bodies.
- o   Device independence: WAP specifications are designed to interwork with any device.
- o   Bearer independence: WAP protocol specifications are designed to work seamlessly over all air interfaces.
- o   Industry participation: the WAP Forum is an open forum for any industry participant.

Organization

WAP Forum organization consists of the following entities:

o   Charters: Authorized by the Board of Directors, charters are the main method for the definition of work items. Charters specify the responsibilities, deliverables and scope of the work that the group intends to accomplish.

o   Specification Working Groups: These groups are chartered to define architectures and write technical specifications.

o   Expert Working Groups: These groups are chartered to investigate new areas of technology, address industry and market viewpoints and provide domain-specific knowledge not directly tied to a single specification effort.

The following figure contains the specification working groups and expert working groups before the OMA announcement.

Specification Working Groups					Expert Working Groups
Architecture	Wireless Application Group (WAG)	Wireless Protocols Group (WPG)	Wireless Security Group (WSG)	Wireless Interoperability Group	Multimedia Expert Group (MMEG)
Standardization Track	Billing Drafting Committee	Next Generation			Billing Expert Group (BEG)
Architecture Consistency Group	Wireless Application Environment (WAE)	Provisioning			Wireless Developers Expert Group (WDEG)
Roadmap	Push Drafting Committee				Smart Card Expert Group (SCEG)
Privacy SIG	Multimedia Drafting Committee				Telematics Expert Group (TEG)
Immediate Messaging SIG	Location Drafting Committee				Service Provider Carrier Expert Group (SPCEG)
	Content Download Drafting Committee				E-Commerce Expert Group (ECOMEG)

WAP Forum Working Groups

Conformance and certification (WAP Forum Certification Program) activity is an integral part of the WAP Forum. The WAP Forum requires that any service or device wishing to use the WAP Forum Logos or trademarks must have WAP Forum Certification.

WAP Specification

The first set of WAP Forum specifications, WAP 1.0, was released in 1998. WAP 2.0 is the latest set of specifications and is tailored to take into account the latest development in mobile networks and mobile devices. It utilizes the most recent internet standards and protocols, but also provides backwards compatibility with the previous versions of WAP specifications.

The following items represent the major architectural components of WAP 2.0:

o   **Protocol Stack** - In addition to the WAP Stack introduced in WAP 1.0, WAP 2.0 supports both the WAP 1 WAP Stack and the common Internet stack services (e.g., Transmission Control Protocol (TCP), Transport Layer Security (TLS) and Hypertext Transfer Protocol (HTTP)).

o   **WAP Application Environment** - The WAP 2.0 Application Environment (WAE) is aligning its technologies with the standards being developed by World Wide Web Consortium (W3C). The definition of the XHTML Mobile Profile is based on XHTML modularization and XHTML Basic. In addition to the features provided by XHTML Basic, WAE also supports style sheets based on the Mobile profile of Cascading Style Sheets (CSS).

The WAP Programming Model, similar to the Web Programming Model, is based on the pull model. In addition, WAP has extended this model by adding support for push and telephony related features with Wireless Telephony Applications (WTA).

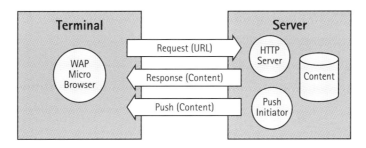

The WAP 2.0 Programming Model

Some of the additional features of the WAP 2.0 release are:

o   **WAP Push:** This service allows the applications in the server to send content to the client by using a Push Proxy.

o   **User Agent Profile (UAProf):** UAProf is used to help application developers tailor their applications to fit the capabilities and the preferences of mobile devices and consumers. UAProf is based on the W3C work on Composite Capability/Preference Profiles (CC/PP).

o   **Wireless Telephony Application:** WTA provides features that seamlessly integrate data services with wireless telephony functionality (e.g., making calls, answering calls, placing calls on hold, and redirecting calls).

o   **External Functionality Interface (EFI):** This framework provides access to applications and external devices (e.g., smart cards, Global Positioning System (GPS) devices, heart rate monitors, and digital cameras) that are not part of WAE but need to be accessed by the components of WAE.

- o **Persistent Storage Interface:** This provides an interface to storage services for organizing, accessing, storing and retrieving data on the mobile device or other connected memory device.

- o **Data Synchronization:** WAP 2.0 has adopted the SyncML language as its choice for the data synchronization solution.

- o **Multimedia Messaging Service (MMS):** MMS is the next generation mobile messaging solution providing a framework for a wide variety of message content types between mobile terminals and servers.

- o **Provisioning:** Mobile devices must be easy to set up to access a multitude of different WAP services. Provisioning specifies a standard framework for providing WAP clients with all the information needed to access mobile services.

- o **Pictogram:** This specification defines the framework for using tiny images (e.g., 6 - hourglass) in a consistent fashion.

Reference

http://www.wapforum.org

Third Generation Partnership Project

Global Initiative

The technical and political debate over Third Generation Mobile Systems was at the boiling point at the end of the 1990s. The International Telecommunication Union (ITU) dream of only one globally common third generation standard, "IMT 2000" (the ITU name for third generation), could not be realized because of the political situation, and the new aim was to produce a family of standards sharing the same architectural principles. The commercial advantages of global markets would not compensate for the losses in technical investments for the three biggest investors, the American, the European, and the Japanese, if they had to give up their technology. Each had developed their technology mostly independently of each other.

However, some regional standardization organizations did have similar technical views, and could possibly agree on a common standard. For example, Wideband Code Division Multiple Access (WCDMA) radio technology was being developed both in Europe and Japan. Discussions on creating a common standardization partnership were started early among the supporters of WCDMA. Finally, the 3rd Generation Partnership Project (3GPP) was formed in December 1998, to produce a full set of Universal Mobile Telecommunication System (UMTS) specifications.

The 3GPP was to a large extent formed around a compromise between Japanese and European telecom companies. The key enabling agreement was that the Japanese accepted the European-born Global System for Mobile Communications (GSM) based Core Network while Europeans accepted the WCDMA radio, which was viewed primarily as Japanese technology.

The European and Japanese companies also managed to lobby part of the American and Korean standardization community to join the original partnership. Later, China joined, bringing its flavor to the technology.

The 3GPP is a well-defined, agreed-upon and committed joint effort among its partners. The scope is to produce specifications for 3G, based on the evolved GSM Core Network and Universal Terrestrial Radio Access (UTRA). These specifications, with minor or no regional modifications, are in general terms respected as standards in the regional partner organizations. Moreover, the regional standardization organizations have submitted the 3GPP specification to ITU as an "IMT 2000" family member.

Organization

The 3GPP is composed of partners, members and observers working jointly in a functional organization where the technical specifications are developed. The following figure illustrates the structure and relations in the organization.

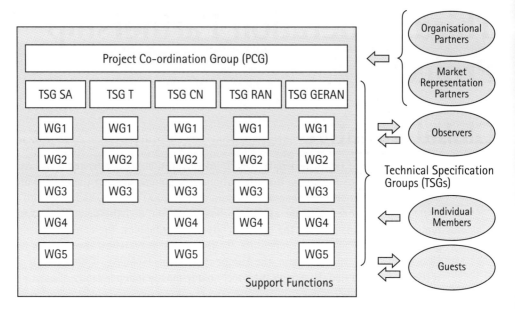

3GPP Organization

The functional organization of the 3GPP consists of a Project Co-ordination Group (PCG) which administers the work of Technical Specification Groups (TSGs), which in turn control the Working Groups (WGs). The partners and members participate in the work according to their defined roles, partners being responsible for the overall co-ordination, and individual members contributing to the technical work. A special support function is provided to help manage and produce standard documents. Observers and guests may follow the work and may in the future become partners or members respectively. The following sections provide further detail on the function of each organizational block.

Organizational Partners

The Organizational Partners (OPs) are the partnering regional standardization organizations, which have the authority to develop standards in their region. They participate in the work mainly at the PCG level, and also meet independently if needed. Their responsibility is to define the future of the 3GPP, in the sense of general policy and strategy. The Third Generation Partnership Project Agreement documents the scope of the 3GPP and is maintained accordingly. The OPs are also responsible for human and financial resource questions, as well as the creation or termination of TSGs. The OPs can also process procedural appeals directed to them.

In cooperation with the Market Representation Partners (MRPs), the OPs are also responsible for processing new partnership applications, and ultimately for dissolving the whole 3GPP organization.

The following regional standardization organizations are the 3GPP partners (use the provided Internet links for more information):

o   **Association of Radio Industries and Businesses (ARIB)** (http://www.arib.or.jp/arib/english/). The ARIB is responsible for radio system research development and standardization in Japan. It is accredited by the Minister of Public Management, Home Affairs, Posts and Telecommunications.

o   **China Wireless Telecommunication Standard Group (CWTS)** (http://www.cwts.org/cwts/index_eng.html). The CWTS is the national standardization body responsible for defining, producing and maintaining wireless telecommunication standards in China.

o   **European Telecommunications Standards Institute (ETSI)** (http://www.cwts.org/cwts/index_eng.html). The ETSI mission is to produce telecommunication standards in Europe and elsewhere. The ETSI is the main European telecommunication standardization organization, and its standards are widely used within and outside of Europe.

o   **T1** (http://www.t1.org/). Committee T1 is an American National Standards Institute (ANSI) accredited organization defining standards for interconnection and interoperability of the carrier network and consumers at various interfaces. T1 is one of several ANSI accredited organizations working in this area in North America.

o   **Telecommunications Technology Association (TTA)** (http://www.tta.or.kr/english/e_index.htm). The TTA is responsible for a broad range of telecommunication standardization activities including multimedia, mobile communication and conventional telephony. TTA is accredited by law in Korea.

o   **Telecommunication Technology Committee (TTC)** (http://www.ttc.or.jp/e/index.html). The TTC is responsible for research and standardization of protocols for interconnection of telecommunication systems and consumers in Japan.

Market Representatives

The Market Representation Partners are industry forums and consortiums, which have been invited to join the 3GPP, and have agreed to commit to the 3GPP scope. Their role is to provide guidance to the process so that the standards meet market requirements. They mainly work together with the OPs at the PSG level to manage the overall direction of the 3GPP work. Currently, eight market representatives have signed the 3GPP agreement (Internet links are provided for finding more information on these organizations):

o   **3G.IP** (http://www.3gip.org)

o   **Global Mobile Suppliers Association (GSA)** (http://www.gsacom.com)

o   **GSM Association** (http://www.gsmworld.com)

o   **IPV6 Forum** (http://www.ipv6forum.com)

o   **Mobile Wireless Internet Forum (MWIF)** (http://www.mwif.org)

o   **UMTS Forum** (http://www.umts-forum.org)

o   **Universal Wireless Communications Consortium (UWCC)** (http://www.uwcc.org)

o   **Wireless Multimedia Forum (WMF)** (http://www.uwcc.org)

Observers and Guests

The 3GPP goal is to produce globally applicable telecommunication standards. Since one element of this is the ability to expand into new geographical areas, mechanisms facilitating the introduction of new partners and members have been defined.

Prospective new partners possessing the qualifications to become an OP may become observers to familiarize themselves with the work. The participation rights of observers will be decided on a case-by-case basis by the existing OPs.

While following the work, observers are encouraged to contribute, in particular by bringing up regulatory issues which might lead to a situation where a 3GPP specification is not applicable in the area governed by that observer.

The following three standardization bodies have become observers (Please note the web links for further information):

- **Australian Communications Industry Forum (ACIF)** (www.acif.org.au)
- **Telecommunications Industry Association (TIA)** (www.tiaonline.org)
- **Telecommunications Standards Advisory Council of Canada (TSACC)** (www.tsacc.ic.gc.ca)

Similarly, a mechanism is available for prospective Individual Members. The OPs may assign a company the right to follow and contribute to some part of the technical work as a guest for a limited period of time.

Individual Members

3GPP membership is open to members of Organizational Partners who are willing to support 3GPP, contribute to its work, and utilize its results. The Individual Members are typically carriers, vendors and consultancy companies planning to become players in the third generation market place. Many member companies are large global enterprises, and therefore members via many OPs.

Individual Members are the carrying force for the technical work, which is driven by contributions from the member companies. The members also make their experts available to fill the chairperson positions in the TSGs.

Support Functions

The Support Team provides support functions. It is based at the ETSI premises in Sophia Antipolis, France. The team is responsible for managing technical meetings and documents. The support team prepares minutes from the regular 3GPP technical meetings, and executes the final editing of the output documents.

Project Coordination Group

The Project Co-ordination Group (PCG) oversees that the technical work is done according to the rules and principles of the partnership. The PCG ensures that the timeline of specification production is fulfilled. The PCG appoints the TSG chairmen based on elections in the TSGs and also allocates other resources for the TSGs. The PSG may also act as a body of appeals from Individual Members on technical or procedural matters.

Technical Specification Groups

Structure of the Technical Work

The technical work in the 3GPP is carried out in the Technical Specification Groups (TSGs) that are responsible for creating, approving and maintaining the Technical Specifications (TSs) and Technical Reports (TRs) in their responsibility area. The work is organized into two hierarchical levels, where each TSG controls several Working Groups (WGs). The detailed technical work takes place in the WGs, whereas the TSGs carry the main responsibility for approval of documents, discussion on new Work Items, and work co-ordination within the TSG.

TSG Services and System Aspects

The Technical Specification Group System Aspects (TSG SA) is responsible for all system wide aspects of the 3GPP system. This includes defining the system architecture, assigning functionality to sub-systems, and defining the main information flows. The TSG SA is also responsible for all aspects of services and bearer capabilities. Furthermore, security, network management, speech coding and billing topics are studied and specified by the TSG SA.

In addition to its very central technical work, the TSG SA is responsible for system wide co-ordination of all technical work in the 3GPP. All other TSGs report to TSG SA, and it is often referred to as the de facto plenary of the 3GPP. This arrangement makes TSG SA the most important technical forum of the 3GPP, since it already produces the technical materials prepared for the PCG.

The detailed work is carried out in the following five working groups:

o **TSG SA WG1, Services:** The stage 1 service work is carried out in this WG. This includes defining the framework and requirements for services, service capabilities and features. The aim is to be as independent of protocols as possible. The results are used by the other TSGs and WGs as the basis of their protocol work.

o **TSG SA WG2, Architecture:** This is one of the most influential WGs in the entire 3GPP, since it is responsible for defining the system architecture, including the high level capabilities of the whole system. Its decisions often affect many other WGs, and other WGs often turn to TSG SA WG2 requesting guidance on technical questions which seem to have system wide aspects.

- o **TSG SA WG3, Security:** This WG is responsible for all security aspects of the 3GPP system, including analyzing threats, developing security mechanisms and algorithms, and setting security requirements for protocols used in various interfaces. The work is often done in cooperation with WGs defining protocols and architectures.

- o **TSG SA WG4, Codec:** The codec WG is responsible for defining the codecs for speech, audio, video and multimedia, and setting the corresponding Quality of Service (QoS) and channel coding requirements for other WGs. The work takes end-to-end efficiency and interoperability into account.

- o **TSG SA WG5, Telecom Management:** This WG is responsible for the Telecommunication Management Network (TMN) of the 3GPP system and for all other technical and co-ordination aspects of management.

TSG Terminals

The Technical Specification Group Terminals (TSG T) is responsible for all aspects of Terminal Equipment (TE) interfaces. This includes TE performance specifications, and specifying the Universal Subscriber Identity Module (USIM) and its interfaces. TSG T is also responsible for end-to-end service interworking and service capability protocols, including messaging.

TSG T has the following three WGs:

- o **TSG T WG1, Mobile Terminal Conformance testing:** This WG is responsible for creating specifications on how to test that a mobile device conforms to 3GPP specifications. The work is further divided into studying radio frequency measurements, Electromagnetic Compatibility (EMC) and signaling.

- o **TSG T WG2, Mobile Terminal Services & Capabilities:** This WG specifies the Services and Service capabilities the mobile device will deliver in accordance with 3GPP specifications. This includes applications, features and interfaces in the device. Messaging is one important application being looked at in this WG.

- o **TSG T WG3, Universal Subscriber Identity Module:** This WG is responsible for developing USIM internal and USIM-to-Mobile device interface specifications. The work also includes SIM card roaming between different systems.

TSG Core Network

The Technical Specification Group Core Network (TSG CN) is responsible for the GSM/UMTS Core Network (CN), which is an evolution of the GSM and General Packet Radio Service (GPRS) CNs. This organization was inherited and moved from the ETSI to the 3GPP. It is the European component of the 3GPP compromise.

TSG CN is responsible for the layer three protocol between User Equipment (UE) and CN. It includes Call Control, Session Management and Mobility Management. Also, the interfaces with the CN and the interconnection to external networks are within the scope of TSG CN.

The work in the TSG CN is carried out in five WGs:

o **TSG CN WG1, MM/CC/SM (Iu):** This WG specifies the Call Control (CC), Session Management (SM) and Mobility Management (MM) protocols between the UE and the CN, over the Iu and radio interfaces. Also, the Short Message Service (SMS) protocol, as well as Session Initiation Protocol (SIP), are included.

o **TSG CN WG2, CAMEL:** WG2 is responsible for specifying the stage two and stage three protocols for Customized Applications for Mobile Network Enhanced Logic (CAMEL).

o **TSG CN WG3: Interworking with external networks.** This group specifies the interworking function needed for both the device in the 3GPP system and the device at the far end in external networks. The work also covers specification of the bearer capabilities both for circuit and packet switched services, especially data services.

o **TSG CN WG4:** Mobile Application Part (MAP) /GPRS Tunneling Protocol (GTP) / Basic Call Handling (BCH) / Supplementary Services (SS). This WG is responsible for stage two and three specification of protocols within the CN. These protocols relate to Supplementary Services, Basic Call Processing, Mobility Management within the Core Network, and Bearer Independent Architecture. Part of this work is applying existing standard protocols for use within CN.

o **TSG CN WG5, OSA:** The UMTS system uses Open Service Architecture (OSA) to facilitate service implementations. The WG is responsible for specifying the Application Programming Interfaces (APIs) for OSA.

TSG Radio Access Network

The Technical Specification Group Radio Access Network (TSG RAN) specifies the newly-defined UMTS Terrestrial Radio Access Network (UTRAN). When the work of TSG RAN was started, considerable information from ETSI and ARIB were merged to form the first set of draft specifications.

The UTRAN interfaces with the WCDMA radio interface for the devices and connects to GSM/UMTS CN via the Iu interface.

The TSG RAN includes four WGs:

o **TSG RAN WG1, Radio Layer 1 specification:** WG1 is responsible for all physical layer aspects (e.g., physical channel structures, mapping of transport channels to physical channels, physical layer multiplexing, channel coding, error detection, spreading, modulation, physical layer procedures, measurement provision and UE physical channel capabilities).

o **TSG RAN WG2, Radio Layer 2 specification and Radio Layer 3 Radio Resource (RR) specification:** This WG is responsible for all aspects of radio interface protocols in the radio interface of L2 and the radio part of L3. Part of WG2 work is applicable also to the GSM/Enhanced Data rates for GSM Evolution (EDGE) Radio Access Network (GERAN).

- **TSG RAN WG3, Iub specification, Iur specification, Iu specification and UTRAN Operations and Maintenance (O&M) requirements:** The WG is responsible for all protocol work in the UTRAN terrestrial interfaces Iu, Iur and Iub. In addition, the WG is responsible for the overall UTRAN architecture and the transport of O&M information between the Node B (denotes the UTRAN Base Station) and the Operations and Maintenance Center (OMC).

- **TSG RAN WG4, Radio performance and protocol aspects from a system point of view - Radio Frequency (RF) parameters and Base Station (BS) conformance:** This WG produces requirement specifications for Radio Link, System performance, and Radio Resource Management (RRM). In addition, BS conformance tests are specified in this WG.

TSG GSM/EDGE Radio Access Network

The 3GPP system is based on GSM CN technology, and the corresponding work has been transferred completely from the ETSI to the 3GPP. The fact that GSM work was being done in two different places created a lot of confusion, so the partners later agreed that GSM specifications work would be moved to the 3GPP. The remaining part includes the radio interface aspects that are now specified in the Technical Specification Group GSM/EDGE Radio Access Network (TSG GERAN).

This organization also emphasizes that GERAN is a prospective radio access technology for third generation services, so bringing it into the 3GPP strengthens the evolution of current GSM operators in that direction.

TSG GERAN is responsible for specifying all aspects of GSM/Enhanced Data rates for GSM Evolution (EDGE) Radio Access Network (GERAN) and its internal and external interfaces, including management and conformance testing. The TSG GERAN has the following five WGs:

- **TSG GERAN WG1, Radio Aspects:** This WG is responsible for EDGE radio interface L1 and RF, and the Ater interface (between Channel Codec Unit (CCU) and the Transcoder and Rate Adaption Unit (TRAU)).

- **TSG GERAN WG2, Protocol Aspects:** The WG is responsible for defining the various protocols in GERAN including radio interface L2 and L3 RR, A, Gb and Abis interfaces.

- **TSG GERAN WG3, Base Station Testing and O&M:** The WG has responsibility over Base Station conformance testing and GERAN-specific O&M matters.

- **TSG GERAN WG4, Terminal Testing - Radio Aspects:** This WG is responsible for creating specifications for GERAN terminal conformance testing in the area of radio interface L1 and Radio Link Control (RLC) / Medium Access Control (MAC).

- **TSG GERAN WG5, Terminal Testing - Protocol Aspects:** The responsibility of this WG is to create specifications for GERAN terminal conformance testing for all protocols above RLC/MAC, as well as for higher layer services.

Working Procedures and Methods

The 3GPP was established to produce documentation very quickly, by maximizing the usage of modern working methods like electronic document handling, and by allowing fast decision making at the appropriately low level of a flat organization. In fact, 3GPP has been successful in this respect. During its first year, it produced the first release of its specifications, Release 99, containing over 300 Technical Specifications (TSs) and Technical Reports (TRs). Together they contain well over 10,000 pages of high quality technical specifications. This is quite an achievement, although much of the documentation was based on existing documentation (e.g., the adopted GSM CN specification series).

3GPP work, like any tightly-scheduled standardization work, is done in committees, where often more than 100 people of over 20 nationalities meet to discuss technical, and sometimes not-so-technical, matters. Written contributions are presented in Temporary Documents (Tdocs), which are often submitted by the hundreds for each meeting. Most meeting days are well over ten hours long, making the meeting weeks rather intense.

Needless to say, in these circumstances a well-defined process is needed to allow anything sensible to be produced in a reasonable timeframe. The 3GPP Working Procedures and the 3GPP Working Methods are fundamental documents for the operation of 3GPP, defining how the 3GPP works.

The immediate responsibility for directing the work in each group lies with the chairperson. Chairs are elected for two-year terms. Candidates are volunteers from Individual Member companies. The chair tries to have all committee decisions made by consensus, though in extremely rare situations voting can be used to settle a dispute. Every group seems to have its own way of handling difficult situations. Some groups, indeed, have made it a goal not to vote, which shows their excellent working spirit.

The WGs meet four to eight times a year, normally for five working days each time. The WGs produce new versions of TSs and TRs for TSG approval. Communication between the groups is essential during the process for coherent specifications. To facilitate information exchange, the TSGs meet in the same location four times a year. PCG meetings are held biannually.

In addition to the scheduled meeting cycle, ad hoc meetings are often held to facilitate discussion, especially at the WG level. Each group also has its own e-mail list, where discussions are held and Tdocs are distributed before the face-to-face meetings. Also, the 3GPP Internet pages (http://www.3gpp.org) are used efficiently to share information, e.g., meeting invitations and all meeting documents. In fact, the 3GPP UMTS specifications are openly available on these pages.

In practical terms, specifications are developed based on written contributions from meeting participants. Initially, when a new specification is produced, an editor is selected to advance the work based on agreements made at the meeting. The contributions at this phase are in free written form (e.g., proposing a principle to be adopted), or text to be added or removed. When the specification reaches a certain state of maturity, the text will be frozen. Corrections and additions to frozen specifications are applied through a formal Change Request (CR) process that on one hand forces any change to be well premeditated, and on the other hand allows people to follow the extent and nature of necessary changes more easily.

New work areas are controlled by the TSGs and the PCG via the specified Work Item (WI) handling structure. WI is a generic name for new work areas, and WIs are categorized hierarchically into features, building blocks and work tasks. A feature represents the system level expression of the work, whereas a building block represents the corresponding functionality in a physical or logical entity, and a work task identifies the technical work needed in one of the 3GPP WGs.

Since the first release, Release 99, which was finalized in March 2000, the 3GPP has already produced another, Release 4 in June 2001, and is working towards the completion of Release 5 in 2002. Note that the release numbering was changed from production year to the document version number. The main scope of Release 99 was to introduce the new radio access network, UTRAN with a WCDMA radio interface. Release 4 includes major enhancements to the circuit switched side of the CN, and Release 5 adds a totally new subsystem, the IP Multimedia Subsystem (IMS) that handles connections to the internet, making the UMTS an integral part of MITA.

References

- Third Generation Partnership Project (3GPP) Partnership Project Description, 04/12/1998, a slide set of 47 slides.
- Third Generation Partnership Project Agreement, 04/12/1998, 12 pages.
- Third Generation Partnership Project 3GPP Working Procedures, 17/07/2000, 25 pages
- Third generation TR 21.900 V3.3.0 (2000-06), Technical Report, 3rd Generation Partnership Project; Technical Specification Group; Working Methods, (third generation TR 21.900 Release 1999), 28 pages.
- http://www.3gpp.org

The Internet Engineering Task Force

The Internet Engineering Task Force (IETF) is the home of the TCP/IP suite of protocols. The IETF does not define any protocols for link layers, and has only limited coverage of the presentation formats used over the Internet. However, more or less everything falling in between is in the scope of IETF standardization (e.g., the Internet Protocol itself (IPv4 and IPv6), IP Security (IPSec), Domain Name System (DNS), Transmission Control Protocol (TCP), Stream Control Transmission Protocol (SCTP), Session Initiation Protocol (SIP), Real-time Transport Protocol (RTP)). The programming interfaces for the protocols are usually not defined by the IETF, but the IP socket interface usually provides the basis for the programming model, which many use with IP protocols. An excellent introduction to the IETF is *The Tao of IETF - A Novice's Guide to the Internet Engineering Task Force* [TAO].

Organization of the IETF

The IETF work is conducted in Working Groups operating within one of the defined Areas. Each Working Group has a Charter and a set of Work Items. When a Working Group completes its Work Items and fulfills its charter, it can either attempt to recharter or to become dormant. Becoming dormant is a sign of success - the working group has solved the problem it was formed to solve. The Working Groups are overseen by the Area Directors.

Currently, there are eight different areas in the IETF:

- o Applications Area
- o General Area
- o Internet Area
- o Operations and Management Area
- o Routing Area
- o Security Area
- o Sub-IP Area
- o Transport Area
- o The User Services Area has closed recently

When a group of people are interested in starting a new working group, they can form a "Birds of a Feather" (BOF) session (taken from the saying, "Birds of a feather flock together"). If the BOF gathers sufficient interest and is thought to be working on a useful and solvable problem or problem area, a working group may be formed from the BOF. It is important to note that working groups are chartered to solve a particular problem in a reasonable amount of time.

The IETF and its Working Groups are open for anybody to participate. Participation is done on an individual basis, not on a per company basis. Contributions to the IETF are documented in Internet Drafts and the work takes place on working group e-mail lists. Face-to-face meetings scheduled triannually are reserved for reporting on the status of the Working Groups and solving any outstanding issues that could not be solved on the mailing lists. All the IETF contributions are governed by the Intellectual Property guidelines specified in RFC 2026 section 10.

Working Process

The working of the IETF is defined in *The Internet Standards Process* [BCP9]. In summary, Internet Drafts (IDs) are submitted to the IETF for temporary online distribution. The lifetime of an ID is six months, after which the draft is removed from the online repository if it has not been revised during that period. The numeric suffix before the file type extension identifies the ID revision, starting from *00*. When all technical issues in an ID have been solved, the Working Group chairs can call for a Working Group Last Call, for comments on the draft from the working group. After the comments have been satisfactorily addressed, the ID is sent to the Internet Engineering Steering Group (IESG) for the IESG last call. The IESG last call is the final call for comments on the ID from anyone interested. After all open issues are solved, the draft is sent to the RFC Editor for publication as *Request For Comments* (RFC).

RFCs are permanent IETF publications. Usually, the RFC text has been available as an Internet Draft before publication as an RFC. The RFCs are never modified once they have been published. New revisions of the RFCs are published with a new RFC number, which makes the older one obsolete. The old one will be available indefinitely.

Both IDs and RFCs are available online through many sources, for example, at the IETF web site (http://www.ietf.org). All IETF documents are published in ASCII format. The same text can also be made available in the PostScript format.

There are four classes of RFCs: Standards Track, Informational, Experimental and Historical. Standards Track RFCs specify the IETF protocol standards. Informational RFCs provide useful information about issues of interest to the IETF community, while Experimental RFCs specify protocols and practices for experimental use only. A third type, Best Current Practices (BCP) describes the current best thinking on the use of protocols. Protocols not recommended for further use can be retired from the Standards Track to Historical status.

The Standards Track includes three maturity levels. Each protocol starts with Proposed Standard status when approved by the IESG. When independent, interoperable implementations exist and operational experience is available, the standard can be revised and elevated to Draft Standard status. Elevation to Draft Standard is a strong recommendation of the usefulness and value of the protocol. It should be noted that if implementing the protocol requires the use of any licensed technologies, the requirement for independent implementations also includes independent exercises of the licensing processes. Internet Standard (or just Standard) is the final status on the Standards Track, requiring significant implementation and successful operational experience. Standards get a number on the Internet Standards series, in addition to an RFC number. For example, the Domain Name System specifications have the STD label STD0013.

The list of current Internet Standards Track documents can be found in STD1, currently RFC 2900 (http://www.ietf.org/rfc/rfc2900.txt). It should be noted that while there are thousands of RFCs, there are less than a hundred standards on the STD series.

Related Organizations

The Area Directors collectively form the Internet Engineering Steering Group. The IESG provides final review of all documents before they become RFCs. They are involved in approving new Working Groups and guide the standardization process with the IETF.

The Internet Architecture Board (IAB) is responsible for defining the overall architecture of the Internet, and provides guidance to the IESG. It provides technical advice and helps on a number of critical activities of the IETF.

IESG and IAB members are elected to serve a two year term through the Nomcom procedure [BCP10].

The IETF has a sister organization focusing on research activities, the Internet Research Task Force (IRTF). The IRTF is separate from, but often meets alongside, the IETF. Currently, the IRTF consists of the following research areas:

- Authentication Authorization Accounting Architecture
- End-to-End
- Group Security
- Internet Digital Rights Management
- Interplanetary Internet
- Network Management
- NameSpace
- Reliable Multicast
- Routing
- Services Management

The Internet Assigned Numbers Authority (IANA, http://www.iana.org) administers protocol parameter numbers and values for all IETF protocols. IANA, IAB, and IESG are chartered by the Internet Society (ISOC, http://www.isoc.org). The ISOC also funds the RFC editor function of the IETF.

Mobile Internet in the IETF

Many people believe that the future Internet will be dominated by mobile devices. This will certainly be true if all the existing mobile devices sold become Internet addressable devices, because there are already a lot more mobile devices being sold, and each year hundreds of millions more are added.

From that point of view, it may seem odd that so few of the working groups in the IETF are concerned with mobile networking. For quite a while, almost all activity was concentrated within the *mobile-ip* working group. Then there was the *roamops* working group, which met to investigate ways to handle roaming between different carrier domains by dial-up nodes. This is more properly considered *portable* computing, and there was not necessarily any provision for handovers. More recently, there have been results from the *pilc* (*Performance Implications of Layer-2 Characteristics*) working group that shed light on the effects of high-error rates or slow media on traditional Internet protocols such as IP and TCP. Most people in the pilc working group were interested mainly because wireless media are known to be slower and more error-prone than traditional wired media, such as Ethernet and fiber channel.

Lately, there has been a great deal of interest in the *seamoby* (*Seamless Mobility*) working group. *Seamoby* was chartered to investigate possibilities for micromobility, context transfer, and IP paging. Over the last year, those goals have evolved, so that by now micromobility and IP paging have been removed from the group charter. It has also been suggested that context transfer should proceed in conjunction with finding a way to identify viable target access routers that can satisfy the context features needed by the mobile nodes. This Candidate Access Router (CAR) discovery work has been added to the working group charter.

The *manet* (*Mobile Ad Hoc Networking*) working group has for quite a while now been investigating protocols for use with ad-hoc networks. This amounts to a substantially separate line of development for mobile networks, which by definition does not depend on Internet infrastructure for its operation. Since there is no dependence on the Internet, the *manet* work logically deserves its own chapter. For the purposes of this description of Mobile IPv6 and wireless networking in the Internet, *manet* has little relevance even though it is a very active and successful topic in its own domain.

Quality of Service (QoS) for mobile networking is widely expected to be investigated within the new *nsis* (*New Steps in Signaling*) working group. QoS is a huge missing piece that needs a lot of attention for mobile devices. Part of the solution may come from the *seamoby* working group and the context transfer protocol development. If a mobile node already has QoS established, transferring the QoS context to a new access router solves the mobility problem in some sense. However, it does not answer questions about how the QoS context is to be established in the first place by the mobile node.

One recent success story in the IETF has to do with header compression. This is very interesting for wireless voice communications, because headers usually are larger than the payload for voice applications. Especially for IPv6, this makes Voice over IP economically unfeasible, given the cost of spectrum licenses. In the *rohc* (*Robust Header Compression*) working group, a solution has been standardized that compresses IPv4 and RTP headers down to about three bytes, which is far more acceptable. Work is proceeding on solutions for TCP and SIP compression. The IPv6 solution is not yet underway, but IPv6 headers would be even more effectively compressible than IPv4 headers, and the RTP part will not change.

Lastly, it must be mentioned that a great deal of current effort related to mobile networking has been going on within the *aaa* (*Authentication, Authorization, and Accounting*) working group. Diameter has been chosen as the base AAA protocol, being preferred over RADIUS extension or Simple Network Management Protocol (SNMP) extension or Common Open Policy Service (COPS). In fact Diameter can best be understood as an extension to RADIUS, albeit developed outside of the RADIUS working group. An AAAv6 draft has been written and prototyped, but it seems that AAAv6 will have to wait a little while until Mobile IPv6 and Diameter are solid. Specification for the Mobile IPv4 Application for Diameter has been just about finished, and there are extensions for Mobile IPv4 that are being standardized to enable better operation with Diameter. These extensions include a Foreign Agent Challenge, a Mobile Node NAI, and several AAA Key Request and Reply extensions.

References

[BCP 9]   "The Internet Standards Process."

[BCP10]   BCP 10. "IAB and IESG Selection, Confirmation, and Recall Process: Operation of the Nominating and Recall Committees."

[TAO]   S. Harris, "The Tao of IETF - A Novice's Guide to the Internet Engineering Task Force." RFC 3160, August 2001.

International Telecommunications Union

According to the Constitution of the International Telecommunication Union (ITU) [coll1999], the ITU maintains and extends international cooperation among all its Member States for the improvement and rational use of telecommunications of all kinds, promotes and offers technical assistance to developing countries in the field of telecommunications, promotes the mobilization of the material, human and financial resources needed for its implementation, as well as provides access to information. In addition, the Union affects allocation of bands of the radio-frequency spectrum, allotment of radio frequencies and registration of radio-frequency assignments. Finally, in terms of space services, the ITU regulates any associated orbital position in a geostationary-satellite's orbit and any associated characteristics of satellites in other orbits, in order to avoid harmful interference between the radio stations of different countries.

History

The International Telecommunication Union (ITU) was founded in 1865, with the name International Telegraph Union, when 20 European states signed the first International Telegraph Convention [www.itu.int]. The first telegraph regulations dealt with common rules to standardize equipment, operating instructions and common international tariff and accounting rules. During the first 20 years the ITU concentrated only on telegraph issues, but in 1885 also matters concerning telephony were taken under discussion. A long leap was taken forward in 1903 when the ITU started wireless telegraphy studies and the first International Radiotelegraph Convention was signed in 1906.

The International Telephone Consultative Committee (CCIF) and the International Telegraph Consultative Committee (CCIT) were created in 1924 and 1925, respectively. The International Radio Consultative Committee (CCIR) was established in 1927. In 1932, the International Telegraph Union decided to combine the above-mentioned Telegraph and Radiotelegraph Conventions to form the International Telecommunication Convention. It was also decided that the name of the Union would be changed to the International Telecommunication Union.

After the United Nations was established, the ITU became a United Nations specialized agency for telecommunications in 1947. The following year, the headquarters of the ITU were transferred from Berne to Geneva. In 1956, the CCIF and the CCIT were merged into the International Telegraph and Telephone Consultative Committee (CCITT).

A Plenipotentiary Conference held in 1989 set up the Telecommunication Development Bureau (BDT) in order to give technical assistance to the developing countries. At the same Conference it was decided that a High Level Committee would be created to carry out an in-depth review of the structure and functioning of the Union. The recommendations of the High Level Committee were presented at the next Plenipotentiary Conference in 1992. The creation of three sectors was decided: the Radiocommunication Sector (ITU-R), the Telecommunication Standardization

Sector (ITU-T) and the Telecommunication Development Sector (ITU-D). Functions previously carried out by the International Frequency Registration Board (IFRB), CCIR, CCITT and BDT were integrated into the new sectors.

Most of the work in the ITU is currently done in the three Sectors established in 1992. Further changes in the management, functioning and structure of the ITU may happen in the 2002 Plenipotentiary Conference. These changes will be based on proposals made by the Working Group on ITU Reform (WGR) [c2001/25]. The WGR was established by the Council in 1999 under a mandate from the Plenipotentiary Conference in 1998. The function of the WGR was to review the management, functioning and structure of the Union and the rights and obligations of Member States and Sector Members.

The ITU has established the World Telecommunication Policy Forum (WTPF) in order to provide a forum where ITU Member States and Sector Members can discuss and exchange views and information on telecommunication policy and regulatory matters, especially on global and cross-sectoral issues. Although the WTPF neither processes regulatory outcomes or outputs with binding force, it does prepare reports and opinions for consideration by the members and by ITU working groups. The three WTPFs held so far are the WTPF '96 Global Mobile Personal Communications by Satellite (GMPCS), the WTPF '98 Trade in Telecommunication Services and the WTPF2001 Internet Protocol (IP) Telephony [www.itu.int].

The ITU has the leading role in preparing the World Summit on the Information Society (WSIS) [www.itu.int]. The Summit was proposed by the ITU Plenipotentiary Conference in 1998 and was endorsed by the United Nations General Assembly in December 2001. The Summit will be held in two phases: the first phase in Geneva, December 10-12, 2003 and the second phase in Tunisia in 2005. The anticipated outcome of the Summit is to develop and foster a clear statement of political will and a concrete action plan for facilitating the effective growth of the Information Society and to help bridge the digital divide.

One of the many responsibilities of the ITU is to organize once every four years ITU Telecom World, the world's largest telecommunication exhibition and forum. The next event will take place October 12-18, 2003 in Geneva [www.itu.int].

Structure of the ITU

ITU is an intergovernmental organization, where governments are represented through their telecommunication Administrations, which are called Member States in the ITU. Other entities are called Sector Members (e.g., Recognized Operating Agencies (ROA), Scientific or Industrial Organizations (SIO), regional and other international organizations). At the end of 2001 the ITU membership included 189 Member States, 653 Sector Members and 36 Associates [c02/35].

The structure of the ITU is given in the Constitution of the International Telecommunication Union [coll1999] and can be seen in the following figure. The Constitution together with the Convention form the basic instruments of the ITU. The provisions of both the Constitution and the Convention are further complemented by those of the Administrative Regulations, which consist of International Telecommunication Regulations and Radio Regulations. These regulate the use of telecommunications and shall be binding on all Member States.

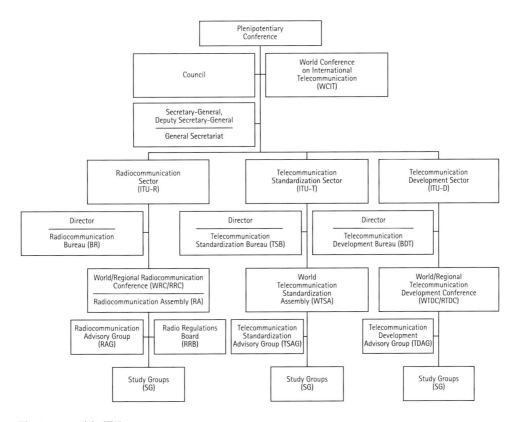

The structure of the ITU

The modification and approval of the Constitution and the Convention documents are the responsibility of the Plenipotentiary Conference. It convenes every four years and is composed of delegations representing the Member States. It is the highest decision making body in the ITU and determines the policy, structure and activities of the ITU and establishes the basis for the budget. It elects the Secretary-General, the Deputy Secretary-General and the Directors of the Bureaux of the Sectors as elected officials of the Union. It also elects the Member States which are to serve on the Council and the members of the Radio Regulations Board. The next Plenipotentiary Conference will be held in Marrakesh, Marocco, September 23-October 18, 2002 [www.itu.int].

During the four year period between Plenipotentiary Conferences, the governing body of the ITU is the Council which is composed of Member States elected by the Plenipotentiary Conference. The current Council consists of 46 Members. The seats in the Council are distributed among the five world regions: Americas (8 seats), Western Europe (8 seats), Eastern Europe (5 seats), Africa (13 seats), and Asia and Australasia (12 seats) [www.itu.int].

The duties of the Council are stated in the Constitution and the Convention of the ITU [col1999]. It is stated among other things that "The Council shall consider broad telecommunication policy issues in keeping with the guidelines given by the Plenipotentiary Conference in order to ensure that the Union's policies and strategy fully respond to the constantly changing telecommunication environment and shall prepare a report on the policy and strategic planning recommended for the Union, together with their financial implications." The Council shall review and approve the Union's biennial budget and consider the budget forecast for the two-year period following that budget. Each year the Council shall consider the report prepared by the Secretary-General on implementation of the strategic plan adopted by the Plenipotentiary Conference and shall take appropriate action. The Council shall submit to the Plenipotentiary Conference a report on the activities of the Union since the previous Plenipotentiary Conference and any appropriate recommendations. In addition, the Council shall perform any duties assigned to it by the Plenipotentiary Conference. One of these duties was the invitation for the Council to establish the Working Group on ITU Reform.

The Sectors of the ITU

According to the Constitution, the ITU has three Sectors: the Radiocommunication Sector, the Telecommunication Standardization Sector and the Telecommunication Development Sector. Although the structures of the Sectors are quite similar, the scope and activities of the three Sectors are different.

Radiocommunication Sector

The mission of the Radiocommunication Sector is, among other things, to ensure rational, equitable, efficient and economical use of the radiofrequency spectrum by all radiocommunication services, including those using satellite orbit, and to carry out studies and create recommendations on radiocommunication matters [coll1999].

The World Radiocommunication Conference (WRC) and the Radiocommunication Assembly (RA) normally convene every two to three years and the meetings may be associated with each other in order to improve the Sector's efficiency and effectiveness. WRC may revise the Radio Regulations, address any radiocommunication matter of worldwide character, and instruct the Radio Regulations Board (RRB) and the Radiocommunication Bureau (BR). In addition it may determine questions for study by the Radiocommunication Assembly and its Study Groups (SGs). The Radio Regulation Board considers reports of unresolved interference investigations carried out by the Bureau and formulates Recommendations. It also provides advice to Radiocommunication Conferences and Radiocommunication Assemblies.

The Radiocommunication Assembly is responsible for the structure, program and approval of radiocommunication studies. The RA approves and issues ITU-R Recommendations and Questions developed by the Study Groups. It is also responsible for approving the programme of technical work, which is carried out in Study Groups.

The Radiocommunication Advisory Group (RAG) reviews priorities, programmes, operations, financial matters and strategies related to the Radiocommunication Assemblies, Study Groups and the preparation of Radiocommunication Conferences. It also provides guidelines for the work of Study Groups.

The technical work needed to draft ITU-R Recommendations is carried out by Study Groups. Currently, there are seven Study Groups [www.itu.int].

ITU-R Study Groups and their scopes:

- **SG 1 - Spectrum Management**

  Principles and techniques for effective spectrum management, sharing criteria and methods, techniques for spectrum monitoring and long-term strategies for spectrum utilization and economic approaches to national spectrum management. Also, in association with the appropriate bodies of the ITU, facilitates the collection and dissemination of information concerning computer programs prepared for the implementation of relevant Recommendations.

  Assistance in matters within its competence to developing countries in cooperation with the Telecommunication Development Sector.

- **SG 3 - Radiowave Propagation**

  Propagation of radio waves in ionized and non-ionized media and the characteristics of radio noise, for the purpose of improving radiocommunication systems.

- **SG 4 - Fixed-Satellite Service**

  Systems and networks for the fixed-satellite service and inter-satellite links in the fixed-satellite service, including associated tracking, telemetry and telecommand functions.

- **SG 6 - Broadcasting Services (terrestrial and satellite)**

  Radiocommunication broadcasting (terrestrial and satellite), including vision, sound, multimedia and data services principally intended for delivery to the general public.

- **SG 7 - Science Services**

  Systems for space operation, space research, earth exploration and meteorology, including the related use of links in the inter-satellite service.

  Radio astronomy and radar astronomy.

  Dissemination, reception and coordination of standard-frequency and time-signal services, including the application of satellite techniques, on a worldwide basis.

- **SG 8 - Mobile, Radiodetermination, Amateur and Related Satellite Services**

  Systems and networks for the mobile, radiodetermination and amateur services, including related satellite services.

- **SG 9 - Fixed Services**

  Systems and networks of fixed services operating via terrestrial stations.

Work in Study Groups is driven by contributions from its membership. Contributions are discussed at meetings, with assessments made by Rapporteurs and the management team of the groups. There are several levels of meetings (e.g., by Study Groups, Working Parties, Rapporteurs).

The Working Party 8F (WP8F) in ITU-R is responsible for the overall system aspects of International Mobile Telecommunications 2000 (IMT-2000). WP8F has developed the vision of IMT-2000 and beyond, and is in close collaboration with the ITU-T Special Study Group (SSG) on *IMT-2000 and Beyond* in order to develop a common ITU vision of systems beyond IMT-2000. It has also been agreed that increased collaboration with external organizations is needed in the development of systems beyond IMT-2000.

Telecommunication Standardization Sector

The main functions of the ITU-T are to study technical, operational and tariff questions and to issue recommendations on them with a view to standardizing telecommunications on a worldwide basis [coll1999]. Technical or operating questions specifically related to radiocommunication are the responsibility of the Radiocommunication Sector. Currently, in ITU-T, high priority topics are IP-related issues, IMT-2000, and tariff and accounting issues. Other key topics include optical fiber transmission technologies, access networks, multimedia, security, Telecommunication Management Network (TMN), signaling, numbering and addressing, and global interconnectivity and interoperability.

At the end of December 2001, the Membership of ITU-T included 189 Member States, 452 Sector Members (179 Recognized Operating Agencies, 234 Scientific or Industrial Organizations including manufacturers, 39 Others (e.g., Regional and International organizations)) and 30 Associates [c02/35].

The World Telecommunication Standardization Assembly (WTSA) meets every four years to consider specific matters related to telecommunication standardization. It defines the general policy for the Sector, establishes the Study Groups and approves their work programs for each study period of four years and appoints the Study Group Chair and Vice-Chairs. In addition, the WTSA considers the reports of the Study Groups, approves, modifies or rejects draft Recommendations and considers the reports of the Telecommunication Standardization Advisory Group (TSAG). The WTSA may assign specific matters within its competence to the TSAG.

The TSAG acts between WTSAs reviewing priorities, programs, operations, financial matters and strategies for the Sector. It also checks to see that work is done and provides guidelines for the Study Groups. The most recent WTSA, held in October 2000 in Montreal, Canada, authorized TSAG to take action in some new areas. For instance, TSAG may create groups with a short lifetime to address items that require rapid reaction. It may also restructure and establish ITU-T Study Groups and assign Chairs and Vice-Chairs in response to changes in the telecommunications market.

Standardization work in ITU-T is carried out by 13 Study Groups, where Recommendations are developed on the basis of Questions. Work in the Study Groups is driven by contributions from its membership. Contributions are discussed at meetings (e.g., Study Group, Working Party, Rapporteur, Joint meeting), with assessments made by Rapporteurs and the management team of the Groups.

ITU-T Study Groups and their general areas of study as stated in Resolution 2 of WTSA-2000 and modified by TSAG [wtsa-res2up]:

- SG 2 - Operational Aspects of Service Provision, Networks, and Performance

    Responsible for studies relating to:

    - principles of service provision, definition and operational requirements of service emulation;
    - numbering, naming, addressing requirements and resource assignment including criteria and procedures for reservation and assignment;
    - routing and interworking requirements;
    - human factors;

- o operational aspects of networks and associated performance requirements including network traffic management, quality of service (traffic engineering, operational performance and service measurements);
- o operational aspects of interworking between traditional telecommunication networks and evolving networks;
- o evaluation of feedback from carriers, manufacturing companies and users on different aspects of network operation.

- o **SG 3 - Tariff and Accounting Principles Including Related Telecommunication Economic and Policy Issues**

  Responsible for studies relating to tariff and accounting principles for international telecommunication services and the study of related telecommunication economic and policy issues. To this end, SG 3 shall in particular foster collaboration among its Members with a view of establishing the lowest possible rates consistent with efficient service and taking into account the necessity to maintain the independent financial administration of telecommunications.

- o **SG 4 - Telecommunication Management, Including TMN**

  Responsible for studies regarding the management of telecommunication services, networks, and equipment using the telecommunication management network framework. Additionally, responsible for other telecommunication management studies relating to designations, transport-related operations procedures, and test and measurement techniques and instrumentation.

- o **SG 5 - Protection Against Electromagnetic Environment Effects**

  Responsible for studies relating to the protection of telecommunication networks and equipment from interference and lightning.

  Also responsible for studies related to electromagnetic compatibility (EMC), to safety and health effects connected with electromagnetic fields produced by telecommunication installations and devices, including mobile devices.

- o **SG 6 - Outside Plant**

  Responsible for studies relating to outside plants (e.g., construction, installation, jointing, terminating, protection from corrosion and other forms of damage from environment impact, except electromagnetic processes, of all types of cables for public telecommunications and associated structures).

- o **SG 9 - Integrated Broadband Cable Networks and Television and Sound Transmission**

  Responsible for studies relating to the use of cable and hybrid networks, primarily designed for television and sound program delivery to the home, as integrated broadband networks also carrying voice or other time-critical services, video on demand, and interactive services.

  Responsible also for studies relating to the use of telecommunication systems for contribution, primary distribution and secondary distribution of television, sound programs and similar data services.

- o SG 11 - Signaling Requirements and Protocols

    Responsible for studies relating to signaling requirements and protocols for Internet Protocol (IP) related functions, some mobility related functions, multimedia functions and enhancements to existing Recommendations on access and internetwork signaling protocols of Asynchronous Transfer Mode (ATM), Narrowband Integrated Services Digital Network (N-ISDN) and Public Switched Telephone Network (PSTN).

- o SG 12 - End-to-end Transmission Performance of Networks and Terminals

    Responsible for guidance on the end-to-end transmission performance of networks, terminals and their interactions, in relation to the perceived quality and acceptance by users of text, speech and image applications. This work includes the related transmission implications of all networks (e.g., those based on Plesiochronous Digital Hierarchy (PDH), Synchronous Digital Hierarchy (SDH), ATM and IP) and all telecommunications terminals (e.g., handset, hands-free, headset, mobile, audiovisual, and interactive voice response).

- o SG 13 - Multi-Protocol and IP-based Networks and Their Internetworking

    Responsible for studies relating to the internetworking of heterogeneous networks encompassing multiple domains, multiple protocols and innovative technologies with a goal to deliver high-quality, reliable networking. Specific aspects are architecture, interworking and adaptation, end-to-end considerations, routing and requirements for transport.

- o SG 15 - Optical and Other Transport Networks

    The SG is the focal point in ITU-T for studies on optical and other transport networks, systems and equipment. This encompasses the development of transmission layer related standards for the access, metropolitan and long haul sections of communication networks.

- o SG 16 - Multimedia Services, Systems and Terminals

    Responsible for studies relating to multimedia service definition and multimedia systems, including the associated terminals, modems, protocols and signal processing.

- o SG 17 - Data Networks and Telecommunication Software

    Responsible for studies relating to data communication networks, for studies relating to the application of open system communications including networking, directory and security, and for technical languages, the methods of using them, and other issues related to the software aspects of telecommunication systems.

- o SSG - Special Study Group on "IMT-2000 and Beyond"

    Responsible for studies relating to the network aspects of International Mobile Telecommunications 2000 (IMT-2000) and beyond, including wireless Internet, convergence of mobile and fixed networks, mobility management, mobile multimedia functions, internetworking, interoperability and enhancements to existing ITU-T Recommendations on IMT-2000.

During the operation of the CCITT (1956-1988), all of the Recommendations which had been completed during the four year period between Plenary Assembly meetings were approved by the Plenary Assembly. The approved Recommendations were published as books named after the color of their covers. The Recommendations, approved by the last Plenary Assembly in 1988, were issued in "Blue Books." The approval time of 4 years and the publication time of 2-4 years for Recommendations was much too long, however, so the procedures were changed so that the Study Group which was responsible for drafting a Recommendation could also approve it. When Recommendations were approved they were then published as separate documents. At the end of 2001 there were about 2,800 Recommendations in force.

In ITU-T, all Recommendations are numbered. The number of each Recommendation has a letter prefix referring to the series as well as a number identifying the particular subject in that series.

The series of ITU-T Recommendations [www.itu.int]:

- **A series** - Organization of the work of ITU-T
- **B series** - Means of expression: definitions, symbols, classification
- **C series** - General telecommunication statistics
- **D series** - General tariff principles
- **E series** - Overall network operation, telephone service, service operation and human factors
- **F series** - Non-telephone telecommunication services
- **G series** - Transmission systems and media, digital systems and networks
- **H series** - Audiovisual and multimedia systems
- **I series** - Integrated services digital network
- **J series** - Transmission of television, sound program and other multimedia signals
- **K series** - Protection against interference
- **L series** - Construction, installation and protection of cables and other elements in outside plants
- **M series** - TMN and network maintenance: international transmission systems, telephone circuits, telegraphy, facsimile and leased circuits
- **N series** - Maintenance: international sound program and television transmission circuits
- **O series** - Specifications of measuring equipment
- **P series** - Telephone transmission quality, telephone installations, local line networks
- **Q series** - Switching and signaling
- **R series** - Telegraph transmission
- **S series** - Telegraph services terminal equipment
- **T series** - Terminals for telematic services
- **U series** - Telegraph switching

- o **V series** - Data communication over the telephone network
- o **X series** - Data networks and open system communications
- o **Y series** - Global information infrastructure and internet protocol aspects
- o **Z series** - Languages and general software aspects of telecommunication systems

In October 2000, the World Telecommunication Standardization Assembly made some additional improvements to the approval procedure by adopting an Alternative Approval Process (AAP) [rec-a8]. The AAP can be used for technical Recommendations. Recommendations having policy or regulatory implications (e.g., numbering/addressing or tariff/billing/accounting) will be approved using the Traditional Approval Process (TAP), where a formal adoption by Member States is required.

The Alternative Approval Process allows, if a draft Recommendation is sufficiently mature, the Study Group chairman to request the director of the Telecommunication Standardization Bureau (TSB) to initiate the last call. The last call encompasses a four-week time period and procedures beginning with the director's announcement of the intention to apply the Alternative Approval Process. If no comments are received by the end of the last call period, the Recommendation is considered approved. The approved Recommendation will be available as a pre-published version on the ITU-T website in a couple of weeks. In 2001, about 190 new or revised technical Recommendations were approved by AAP. More than 60% of these were approved in less than two months after the texts were identified as mature [zhao2002].

ITU-T Recommendations, agreed by consensus in Study Groups, are non-binding standards. Nevertheless, Recommendations are generally complied with because they guarantee the interconnectivity of networks and enable services to be provided on a worldwide scale.

During the reorganization of the ITU-T Study Groups at WTSA-2000, a Special Study Group on "IMT-2000 and Beyond" was established. The SSG is responsible for studies relating to the network aspects of International Mobile Telecommunications 2000 (IMT-2000) and beyond, including wireless Internet, convergence of mobile and fixed networks, mobility management, mobile multimedia functions, internetworking, interoperability and enhancements to existing ITU-T Recommendations on IMT-2000.

As ITU-R Study Group 8 had earlier established Working Party 8F, which has (on the ITU-R side) the responsibility of developing IMT-2000 and beyond, it was decided that coordination between ITU-T and ITU-R is needed. The SSG collaborates with ITU-R Working Party 8F on the radio aspects of terrestrial elements and with ITU-R Working Party 8D for satellite elements. It was also decided that the SSG shall maintain strong cooperative relations with external Standards Development Organizations (SDOs) and especially with The Third Generation Partnership Project (3GPP) in order to avoid duplication of effort or overlapping projects.

The SSG and ITU-R WP 8F are currently collaboratively developing a common ITU vision of systems beyond IMT-2000. Cooperation has also been made with ITU-T SG 11, 3GPP, Third Generation Partnership Project 2 (3GPP2), European Telecommunications Standards Institute (ETSI), Standards Committee T1 - Telecommunications (T1), Telecommunications Technology Association (TTA), Telecommunication Technology Committee (TTC) and Association of Radio Industries and Businesses (ARIB).

In addition to the SSG's cooperation with other SDOs, other ITU-T Study Groups also have valuable cooperation with organizations like International Organization for Standardization (ISO), International Electrotechnical Commission (IEC), Internet Engineering Task Force (IETF), ETSI TIPHON, ATM Forum and MPLS Forum. Procedures for this cooperation are given in ITU-T Recommendations A.4 [rec-a4], A.5 [rec-a5], A.6 [rec-a6] and in WTSA-2000 Resolution 7 [wtsa-res7].

ITU-T is organizing a series of Workshops and Seminars in 2002. The aim of these events is to increase awareness of ITU-T studies and projects, coordinate the activities of the ITU-T Study Groups and benefit from the progress achieved in the ITU-R Study Groups and in other SDOs in order to avoid duplication of efforts. The subjects of the Workshops and Seminars are multimedia convergence, IPv6, security, IMT-2000 and systems beyond and IP/optical [www.itu.int].

Telecommunication Development Sector

The Constitution of the ITU determines the specific functions of the Telecommunication Development Sector. According to the Constitution [coll1999] the ITU-D shall promote the development, expansion and operation of telecommunication networks and services, particularly in developing countries by reinforcing capabilities for human resources development, planning, management, resource mobilization and research and development. It shall also mobilize resources to provide assistance to developing countries in the field of telecommunications and promote and coordinate programs to accelerate the transfer of appropriate technologies to developing countries.

Every four years ITU organizes the World Telecommunication Development Conference (WTDC), which serves as a forum for free discussion by all concerned with the Development Sector. The WTDC establishes work programs and guidelines for defining telecommunication development questions and priorities and provides direction and guidance for the work program of the Telecommunication Development Sector. The most recent World Telecommunication Development Conference (WTDC-02) was held March 18-27, 2002 in Istanbul, Turkey.

The Telecommunication Development Advisory Group (TDAG) reviews priorities, programs, operations, financial matters and strategies for activities in the Telecommunication Development Sector. It also provides guidelines for the work of Study Groups. The TDAG recommends measures, among other things, to foster cooperation and coordination with the Radiocommunication Sector, the Telecommunication Standardization Sector and the General Secretariat, as well as with other relevant development and financial institutions.

The Study Groups in ITU-D work with specific telecommunication Questions of general interest to developing countries and prepare draft Recommendations. Important tasks include the preparation of guidelines, handbooks, manuals and reports within the areas of competence of each Study Group; these pay particular attention to the needs and concerns of the least developed countries in furthering the work. There are currently two Study Groups in ITU-D. Study Group 1 deals with telecommunication development, strategies and policies while Study Group 2 deals with development, harmonization and management of telecommunication networks and services. In its final report to the Council [c2001/25], the Working Group on ITU Reform recommended that Study Groups in ITU-D should be replaced by project management groups.

ITU Reform

The Plenipotentiary Conference, held in 1998 in Minneapolis, USA, mandated that the Council establish the Working Group on ITU Reform (WGR), which would be an open workgroup of Member States and Sector Members. WGR was established in 1999, charged with reviewing the management, functioning and structure of the Union as well as the rights and obligations of its Member States and Sector Members. In addition, WGR should review the contribution of the Sector Members towards defraying the expenses of the Union.

The Strategic policies and financial matters of the ITU are discussed and decided upon at the Plenipotentiary Conference and in the Council. Only Member States are represented in these bodies, which is one reason why Sector Members complain that their role is not sufficiently recognized in the ITU and especially in ITU-T.

The decision-making process and possibilities to act are totally different in the ever-increasing number of forums and consortiums from those in the ITU. Forums and consortiums are usually driven by private industry, with governments having almost no role at all. In addition, forums and consortiums have a very flexible framework for the development of new standards.

The success of forums and consortiums puts more pressure on changing the ITU. The ITU is unique among international organizations in that it was founded on the principle of cooperation between governments and the private sector. For over 100 years, the Convention, and later the Constitution, have formed the rules for the work. However, due to the binding nature of these basic texts, they restrict the possibilities to change the working procedures between Plenipotentiary Conferences.

The Working Group on ITU Reform submitted its final report to the Council at the June 2001 meeting [c2001/25]. WGR recommended to the Council that it request the Director of the Telecommunication Standardization Bureau (TSB) to provide the Plenipotentiary Conference 2002 (PP-02) with a report of the activities of the Telecommunication Standardization Advisory Group (TSAG) in relation to a Pilot Forum. WGR also recommended that the Council ask PP-02 to consider whether a Forum should be established. WGR recommended a two-stage approval process for ITU-T technical Recommendations in which both Sector Members and Member States would participate.

At the June 2001 meeting, the Council did not make any big decisions concerning Reform of the ITU and for instance the much-debated proposal to establish a Pilot Forum for global standardization under the umbrella of the ITU was forwarded to the Plenipotentiary Conference 2002 for further discussion. In order to make some progress, the Council agreed that work on the Pilot Forum should continue within TSAG. At its subsequent meeting, TSAG considered, but did not support, the establishment of the Pilot Forum.

WGR made some recommendations which, if they were implemented, would require amendments or modifications to the ITU Constitution and Convention. The Council decided to set up a Group of Experts to prepare the required draft texts for amending the Constitution and Convention. The report of the Group of Experts was discussed in the meeting of the Council. The Council will transmit the report with its own comments to the Plenipotentiary Conference 2002. In addition to this, there are proposals from Member States on changes to the structure of the ITU Constitution and the Convention, in order to give more rights to the Sectors to decide matters relating to them.

Currently, it is impossible for a single standardization organization to be the focal point for all matters related to the rapidly changing telecommunication environment. Therefore, the ITU has significantly increased cooperation with other SDOs, forums and consortia. In order to facilitate the development of cooperative relationships with these organizations ITU-T has published three Recommendations in the A-series: A.4 "Communication process between ITU-T and forums and consortia" [rec-a4], A.5 "Generic procedures for including references to documents of other organizations in ITU-T Recommendations" [rec-a5] and A.6 "Cooperation and exchange of information between ITU-T and national and regional standards development organizations" [rec-a6]. According to the procedures described in these Recommendations ITU-T has published the following lists of qualified Organizations [www.itu.int].

Forums/Consortia Approved for Communication Process under Recommendation A.4 - "Communication process between ITU-T and forums and consortia" [rec-a4]:

- ATM Forum (ATM-F)
- DSL Forum
- E- and Telecommunication Information Services (ETIS)
- Frame Relay Forum (FRF)
- International Multimedia Telecommunications Consortium (IMTC)
- Internet Protocol Detail Record Organization (IPDR.org)
- IPv6 Forum
- Multi-Protocol Label Switching (MPLS) Forum
- Multiservice Switching Forum (MSF)
- Optical Internetworking Forum (OIF)
- Object Management Group (OMG)
- SDL Forum Society
- TeleManagement Forum (TM Forum)

Organizations Qualified for Including References in ITU-T Recommendations under Recommendation A.5 - "Generic procedures for including references to documents of other organizations in ITU-T Recommendations" [rec-a5] Procedures:

- Association of Radio Industries and Businesses (ARIB)
- ATM Forum
- Committee T1 (sponsored by ATIS, Alliance for Telecommunications Industry Solutions)
- China Wireless Telecommunication Standard Group (CWTS)
- DSL Forum
- ECMA Standardizing Information and Communication Systems
- European Telecommunications Standards Institute (ETSI)

- o   Institute of Electrical and Electronics Engineers (IEEE)
- o   Internet Society/Internet Engineering Task Force (ISOC/IETF)
- o   Japan Cable Television Engineering Association (JCTEA)
- o   Multi-Protocol Label Switching (MPLS) Forum
- o   National Institute of Standards and Technology (NIST)
- o   Optical Internetworking Forum (OIF)
- o   Object Management Group (OMG)
- o   Society of Cable Telecommunications Engineers (SCTE)
- o   Telecommunications Industry Association (TIA)
- o   TeleManagement Forum (TM Forum)
- o   Telecommunications Technology Association (TTA)
- o   Telecommunication Technology Committee (TTC)

National and Regional Standards Development Organizations approved for Cooperation and Exchange of Information under Recommendation A.6 - "Cooperation and exchange of information between ITU-T and national and regional standards development organizations " [rec-a6]:

- o   Association of Radio Industries and Businesses (ARIB)
- o   Committee T1 (sponsored by ATIS, Alliance for Telecommunications Industry Solutions)
- o   China Wireless Telecommunication Standard Group (CWTS)
- o   ECMA Standardizing Information and Communication Systems
- o   European Telecommunications Standards Institute (ETSI)
- o   Institute of Electrical and Electronics Engineers (IEEE)
- o   Japan Cable Television Engineering Association (JCTEA)
- o   National Institute of Standards and Technology (NIST)
- o   Society of Cable Telecommunications Engineers (SCTE)
- o   Telecommunications Industry Association (TIA)
- o   Telecommunications Technology Association (TTA)
- o   Telecommunication Technology Committee (TTC)

References

[c2001/25]	Final report to the Council by the Working Group on ITU Reform (WGR), Document C2001/25 ITU Council, 18-29 June 2001, Geneva.
[c02/35]	Report on the activities of the Union for 2001, Document C02/35 ITU Council, 25 April - 3 May 2002, Geneva.
[coll1999]	Collection of the basic texts of the International Telecommunication Union adopted by the Plenipotentiary Conference, 1999.
[rec-a4]	ITU-T Recommendation A.4 (10/2000) - Communication process between ITU-T and forums and consortia.
[rec-a5]	ITU-T Recommendation A.5 (11/2001) - Generic procedures for including references to documents of other organizations in ITU-T Recommendations.
[rec-a6]	ITU-T Recommendation A.6 (10/2000) - Cooperation and exchange of information between ITU-T and national and regional standards development organizations.
[rec-a8]	ITU-T Recommendation A.8 (10/2000) - Alternative approval process for new and revised Recommendations.
[wtsa-res2up]	Update of information to Resolution 2 of WTSA-2000, 2001.
[wtsa-res7]	Resolution 7 - Collaboration with the International Organization for Standardization (ISO) and the International Electrotechnical Commisson (IEC), World Telecommunication Standardization Assembly, 27 September - 6 October 2000, Montreal.
[www.itu.int]	International Telecommunication Union web page
[zhao2002]	Houlin Zhao: Opening address for the Workshop on Multimedia Convergence (IP Cablecom/Mediacom 2004/Interactivity in multimedia), 12-15 March 2002, Geneva.

European Telecommunications Standards Institute

The European Telecommunications Standards Institute (ETSI) is a non-profit organization whose mission is to produce telecommunications standards. It is an open forum uniting nearly 900 Members from five continents and 54 countries, consisting of manufacturers, carriers and service providers, administrations, research bodies and customers.

ETSI membership is open to interested organizations world-wide. A variety of working modes are available, from typical committee to global partnership projects (e.g., the Third Generation Partnership Project (3GPP)), including Forum Hosting and interoperability test event (e.g. Plugtest) facilities. ETSI was also among the first to make its deliverables freely available on the Web.

According to its Mission, ETSI shall, through international collaboration, pursue the objective of developing globally applicable deliverables, meeting the needs of the telecommunication and electronic communication community, while still fulfilling its duty to support European Union (EU) and European Free Trade Association (EFTA) regulations and initiatives.

The objective of ETSI is to produce and maintain widely implemented technical standards and other deliverables as required by its members. As a recognized European standards organization, the objective should be reached in a way so as to support and enhance competition in a unified European market for telecommunications and related areas. On the international level, ETSI contributes to world-wide standardization.

ETSI was founded in 1988. The creation of ETSI was based on changes in European telecom structures when the ministry, regulatory authority, carrier and service provider roles were separated in many cases. Efficiency, openness, quality and global competiveness were also considered. An important goal was to harmonize the European Telecoms market based on common technical standards. In 1987, the European Commission published its Green Paper on Telecommunications, which presented the idea of ETSI. In the same year, the Conference of European Postal and Telecommunications Administrations (CEPT), which at the time represented adminstrations and carriers, had in principle taken the steps and made the decision to establish ETSI in partnership with European manufacturing industry and customers. The original concepts - independence, openness and unity - lay firm ground for the future evolution of the ETSI organization.

Today, ETSI is a market-driven organization supporting global needs. Its members directly represent all aspects of industry, decide on the work program and allocate resources accordingly. Since its creation in 1988, ETSI's approach has been to provide comprehensive project and technical support to its Technical Bodies. ETSI has also been willing to evolve and develop as an organization in order to meet the requirements of its members.

ETSI Membership at the end of 2001:

- o 653 Full Members from 35 European countries
    - 6.9 % Administrations
    - 52.3 % Manufacturers
    - 14.9 % Carriers
    - 22.9 % Service Providers and Others
    - 3.0 % Users
- o 173 Associate Members from 19 non-European countries
- o 47 Observers from 18 countries

ETSI Structure

As the highest decision making authority in ETSI, the General Assembly (GA) is responsible for determining policy, agreeing on budgets, dealing with membership issues, appointing Board members, appointing the Director-General (D-G) and Deputy D-G, endorsing external agreements, and approving the Statutes and Rules of Procedure for the Institute. The GA normally meets twice a year.

The Board is a body which acts on behalf of the General Assembly between GA meetings, exercising the powers and functions delegated to it by the General Assembly.

The Technical organization consists of Technical Committees (TC), ETSI Projects (EP), Partnership Projects and Special Committees. One of the Special Committees, the Operational Coordination Group (OCG), will act as a focal point and a forum for coordinating the Technical Bodies (TCs/EPs), and between the Technical Bodies and the Secretariat. It shall resolve, as far as possible, any duplication of effort or conflict of technical views between the Technical Bodies to reinforce cooperation within the Technical Organization. Technical Body Chairpersons ensure that their Technical Body is properly represented in OCG meetings.

Expert teams, called Specialist Task Forces (STF), can be set up to support the Technical Organization to speed up production of urgent deliverables and deliverables requiring specific concentrated expertise, such as test specifications. STFs experts work together at the ETSI premises in Sophia Antipolis. Funding is provided to compensate companies for providing experts.

The ETSI Secretariat supports the activities of the Institute and is also located in Sophia Antipolis, the headquarters of the Institute. It is headed by the Director-General. From the beginning ETSI has had strong expertise within the Secretariat and is prepared to provide a variety of support functions to members participating in the standards and specifications development process.

Technical Organization

The following list of ETSI Technical Bodies describes the situation at the end of 2001. Some bodies, whose activities are related to the Mobile Internet, will be briefly described later.

Technical Committees

AT	Access and Terminals
ECMA TC32	Communication, Networks & Systems Interconnection
EE	Environmental Engineering
ERM	EMC and Radio Spectrum Matters
HF	Human Factors
JTC Broadcast	EBU/CENELEC/ETSI Joint Technical Committee
MSG	Mobile Standards Group
MTS	Methods for Testing & Specification
Safety	Telecommunications Equipment Safety
SEC	Security
SES	Satellite Earth Stations & Systems
SPAN	Services and Protocols for Advanced Networks
STQ	Speech processing, Transmission & Quality aspects
TM	Transmission and Multiplexing
TMN	Telecommunications Management Network

ETSI Projects

BRAN	Broadband Radio Access Networks
DECT	Digital Enhanced Cordless Telecommunications
M-COMM	Mobile Commerce
PLT	PowerLine Telecommunications
RT	Railway Telecommunications
SCP	Smart Card Platform
TETRA	TErrestrial Trunked RAdio
TIPHON	Telecommunications and Internet Protocol Harmonization Over Networks

ETSI Partnership Projects

3GPP	Third Generation Partnership Project
MESA	Public Safety Partnership Project

Special Committees

FC	Finance Committee
OCG	Operational Coordination Group
SAGE	Security Algorithms Group of Experts
USER	Special Committee User Group

Highlights of ETSI Technical Activities

Methods for Testing and Specification

Methods for Testing & Specification (MTS) is responsible for identifying and defining advanced specification and conformance testing methods, which take advantage of formal approaches and innovative techniques to improve the efficiency and economics of both the standard description and associated conformance testing processes. A major achievement was to finalize the first edition of version 3 of Tree and Tabular Combined Notation (TTCN-3), a flexible and powerful language capable of specifying many types of system tests over a variety of communication interfaces. TTCN-3 has also been approved by the International Telecommunication Union (ITU) as recommendation Z.140 and will be supported by development tools. Work on graphical presentation format for TTCN-3 has been initiated as a joint activity with the Object Management Group (OMG).

TTCN-3 can be used for protocol testing, supplementary service testing, module testing, the testing of platforms based on the Common Object Request Broker Architecture (CORBA), the testing of Application Programming Interfaces (APIs) and many other applications. The language is not restricted to conformance testing, and can be used for interoperability, robustness, regression, system and integration testing.

A significant achievement in 2001 was the development of the ASN.1 Encoding Control Notation (ECN) for the Third Generation Partnership Project (3GPP) in close collaboration with mobile telecommunications experts. ECN is much more flexible than previous ASN.1 encoding processes and is therefore particularly suitable for the 3GPP system.

As carriers acquire 3G equipment, they need to ensure interoperability with existing equipment. TC MTS has begun work on a document that will specify reference points within the architecture to help the carrier ensure that all of its equipment can be successfully integrated.

Broadband Radio Access Networks

The ETSI Project Broadband Radio Access Networks (EP BRAN) is responsible for standardizing Broadband Radio Access Networks (BRAN). EP BRAN produces specifications for three types of networks:

o **High Performance Radio Local Area Network type 2 (HIPERLAN2)** is intended for private use as a Wireless Local Area Network (WLAN)-type system as well as a complementary access mechanism in hot spot areas for public mobile network systems.

o **High Performance Radio Access (HIPERACCESS)** is intended for broadband multimedia fixed wireless access and back-haul for the Universal Mobile Telecommunications System (UMTS) as a flexible and competitive alternative to wired access networks. The standardization focuses on solutions optimized for frequency bands above 11 GHz, in particular the 31.8-33.4 GHz and 40.5-43.5 GHz bands, which have been identified by the World Radiocommunications Conference (WRC) 2000 as primary bands for fixed services.

o **High Performance Radio Metropolitan Area Network (HIPERMAN)** is aiming principally for the same usage as HIPERACCESS, but is targeted at different market segments and uses a different part of the spectrum; HIPERMAN standardization focuses on solutions optimized for frequency bands below 11 GHz.

Among the main achievements in 2001 in the HIPERLAN2 area was the completion of the Technical Report on the requirements for interworking with UMTS and other Third Generation (3G) networks. In addition, EP BRAN began work on producing the technical specifications for the architectures and protocols of such an interworking. By co-operating closely with the Multimedia Mobile Access Communications Promotion Council (MMAC) in Japan and the Institute of Electrical and Electronic Engineers (IEEE) in the USA, the EP BRAN aims to produce "generic" WLAN-3G interworking solutions independent of the access techniques used in the WLAN standard.

In 2001, work also progressed on the HIPERLAN2 Harmonized Standard and the project finalized a number of other technical specifications, including the IEEE 1394 Service Specific Convergence Sub-layer for restricted bridging and the home profile.

In HIPERACCESS, the two basic technical specifications (for the Physical layer (PHY) and the Data Link Control layer (DLC)) are close to completion. Only minor details need to be sorted out in the PHY specification, and the basic protocol has been defined for the DLC specification. The Convergence Layer (CL) technical specifications are scheduled for completion in 2002, together with the conformance test specifications (for PHY, DLC and CL).

The Technical Report on functional requirements for HIPERMAN has been completed, and significant progress was made on the technical specifications for PHY and DLC. In support of a single world-wide standard for fixed wireless access systems operating below 11 GHz, EP BRAN decided to use the IEEE 802.16a Orthogonal Frequency Division Multiplexing (OFDM) PHY and Medium Access Control (MAC) as a baseline for PHY and DLC respectively. Modifications are being made to the baseline documents, which will be aligned with the IEEE 802.16 Working Group to achieve a single global standard. The PHY and DLC technical specifications are expected to be final and approved in December 2002.

Telecommunications and Internet Protocol Harmonization Over Networks

Telecommunications and Internet Protocol Harmonization Over Networks (TIPHON) is responsible for the interoperability of multimedia communication services in the Next Generation Networks (NGN) environment. TIPHON started with a more limited scope of Internet Protocol (IP) Telephony including communication between IP-based networks and the Public Switched Telephone Network (PSTN), Integrated Services Digital Network (ISDN), and Global System for Mobile Communications (GSM). The ETSI Project TIPHON (EP TIPHON) has completed the specifications defining the requirements, architecture, and protocol profiles for interoperable IP telephony services. Phase 2 covered communication from the Switched Circuit Network (SCN) to IP telephony devices and SCN to SCN, using the IP network as a trunk connection. Release 3 provided a set of standards for operating basic telephony services and covers architecture, services, protocols, protocol profiles, numbering and naming, security profiles, and design guidelines for Quality of Service. Releases 4 and 5 will extend the basic call service to include multiple media applications and APIs.

The EP TIPHON methodology uses a new functional architecture, which separates the roles of application service provider and transport network operator. By using this approach, the service provider can offer unique application services composed of a number of standardized building blocks or service capabilities. A further innovation is the introduction of a technology-independent protocol framework known as the TIPHON meta-protocol. This is used to generate profiles for widely deployed industry protocols (e.g., H.323, SIP, H.248) enabling service interoperability in mixed protocol environments.

Third Generation Partnership Project

The Third Generation Partnership Project (3GPP) was established by ETSI and regional standards development organizations from Asia and the USA, with participation from industry groups and over 400 individual companies, to develop a complete set of globally applicable Technical Specifications for a 3rd Generation (3G) mobile system. The specifications are based on the evolved GSM core network and an innovative radio interface known as the UMTS Terrestrial Radio Access (UTRA), using Wideband Code Division Multiple Access (WCDMA) technology. GSM radio access was later included in the Partnership Project.

From the point of view of the ETSI partnership, 3GPP is treated as one of the ETSI Technical Bodies, so 3GPP-approved specifications also become ETSI Technical Specifications. ETSI hosts support functions for the 3GPP, called the Mobile Competence Center (MCC), with 27 international team members. ETSI also hosts the 3GPP Web site.

Other ETSI Activities

European Broadcasting Union (EBU) / European Committee for Electrotechnical Standardization (CENELEC)/European Telecommunications Standards Institute (ETSI) Joint Technical Committee (JTC) Broadcast is responsible for standardizing broadcast systems for television, radio, data and other new services via satellite, cable, Satellite Master Antenna TeleVision (SMATV) and terrestrial transmitters, and also for the transmission of programs. JTC Broadcast is mostly associated with Digital Video Broadcasting (DVB) and Digital Audio Broadcasting (DAB). DVB Satellite solutions are adopted world-wide. The basic transmission systems for both DVB and DAB were standardized by ETSI, and attention has now been concentrated on additional features (e.g., data broadcasting, interactivity, and transmission of Internet Protocol packets over broadcast channels). Of special interest is the over 1,000 page DVB Multimedia Home Platform (DVB-MHP) specification, which describes an Application Programming Interface (API) for interactive broadcast services.

The Smart Card Platform (SCP) Project is responsible for developing and maintaining a common Integrated Circuit (IC) Card platform for all mobile telecommunication systems, applying independent specifications for the interface with terminal equipment and for maintaining IC Card standards for general telecommunications and high security applications. Smart card activity relevant to GSM and 3GPP first passed to the 3GPP Partnership Project, but later ETSI created EP SCP to continue generic work on smart card specifications. Now EP SCP will become a Partnership Project.

Global Cooperation from the ETSI Perspective

To ensure the widest possible application of its deliverables, ETSI works in partnerships with other organizations around the world. Currently, ETSI has over 50 cooperation and Memoranda of Understanding (MoU) agreements.

As one of the three officially recognized European Standards Organizations (ESO), ETSI is participating and supporting initiatives within collaborative frameworks (e.g., the Information and Communications Technologies Standards Board (ICTSB), Global Standards Collaboration (GSC), and Global Radio Standardization (RAST)).

ETSI highlights the need for bridging the gap between different dimensions of traditional telecommunications, Internet Protocol (IP) and broadcasting environments by following market trends and creating key synergies.

ETSI has a number of cooperation agreements and MoUs with external bodies ranging from the European Commission, European Standards Organizations, European Free Trade Association (EFTA) and other European regulatory bodies, to National/Regional and International Standards Development Organizations (SDOs), to Fora, R&D bodies and Specification Providers. The key to all these co-operation frameworks is that the organizations concerned have a direct interest in telecommunications standardization or in research and development activities that might eventually lead to an input into the standards making process.

Support Functions

The ETSI Secretariat provides a wide range of support functions. Many of these have been essential for the rapid achievement of good quality standards. These support functions have also been a requirement for the success of 3GPP.

The Secretariat supports the Standards Making Process and the elaboration of timely high-quality ETSI deliverables by providing Technical Officers who facilitate the operations of various Technical Bodies. In addition, editing and approval support services are provided. In 2001, a total of 2,172 deliverables were published, containing 207,400 pages, which corresponds to more than one published standard per working hour, or one page published every 30 seconds.

With the help of its Members, the ETSI Secretariat has made efforts to increase market acceptance of ETSI standards and deliverables throughout the world by promotion, membership recruitment and distribution of deliverables.

In specific cases, the set-up of Specialist Task Forces (STFs) is decided by the members and managed by the Secretariat. This possibility to temporarily hire experts to do certain tasks or prepare parts of specifications has enhanced the quality of the results and made them available faster.

The Secretariat provides support to the GA, Board and OCG, provides legal advice and interpretation, and manages the external relations agreements and MoUs with other bodies. In addition, the ETSI Secretariat hosted 321 meetings in Sophia Antipolis in 2001, providing support to the technical organization.

FORAwatch is an activity and Web service that provides a good overview, with links to telecommunications-related fora and consortia.

Operating and developing the IT infrastructure, including the extensive Web pages and applications, is also one of its major support functions.

Competence and Service Centers in ETSI

The ETSI is applying and developing the model of competence and service centers, which address the needs of the organization in a horizontal manner by supporting several technical bodies within the ETSI. This complements the (partnership) project model for vertical technological issues.

The Mobile Competence Center was set up to support the 3GPP and other mobile standardization work. Hosted by the ETSI, it is now an integrated team of 27 persons, made up of the ETSI Secretariat Officers, assistants and contracted experts, drawn from more than a dozen countries across four continents and supported by both voluntary and funded resources. The MCC has been prolific. In 2000, for example, the Center managed 1,327 active specifications (461 on 3G and 1,156 on GSM) and processed a total of 5,716 change requests for GSM and 3G specifications.

The Protocol and Testing Competence Center (PTCC) provides support and services to ETSI Technical Bodies on the application of modern techniques for specifying protocols and test suites. The PTCC is also responsible for managing the Specialist Task Forces which develop test specifications for ETSI standards. In 2000, the PTCC managed/supported 16 STFs, three tasks for the MCC and assisted with the development of test specifications for 3G terminals. The Center also played a major role in the specification of the new version of TTCN-3.

The ETSI PLUGTESTS Service is a professional service specializing in the organization of interoperability testing events for any Telecommunication, Internet or Information Technology standard. Its customers are the ETSI Technical Bodies, Internet Engineering Task Force working groups, ITU Study Groups, or any forum or interest group which is developing a standard. An interoperability testing event is an opportunity for developers to get together to validate a draft standard and its implementations. The benefits are numerous and overall these events save companies and the industry considerable time and money. In 2001, PLUGTEST organized six events in Europe plus one co-organized in Japan. Up to twelve events are planned for 2002.

The ETSI's Fora Hosting Service is prepared to provide day-to-day administrative services for interested Fora. Currently the hosting service is provided to the Location Interoperability Forum (LIF). The ETSI is now looking to increase the number of hosted Fora.

Information Technology in Specifications Creation

ETSI has an information technology strategy where the use of electronic and telecom tools in specifications creation is maximized, including document availability.

The definition, development and maintenance of electronic working tools continues, both for use internally within the Secretariat for tasks (e.g., the automation of editing and document management), and for external use, in the interfaces between the Technical Organization and the Secretariat. The introduction of "paperless meetings" has brought a considerable saving in costs, and the use of Internet servers and services is steadily increasing. Practical solutions include the use of LANs and WLANs, distribution of meeting documents via a Web server and e-mail and the Web casting of some meetings.

Electronic working is now employed throughout the entire standardization process, and provides a speedier, streamlined process where members from all around the world can participate. The ETSI Collaborative Portal has been opened, which makes it easier to access information related to Technical Body activities. One of the goals is to allow organizations and experts to participate and be aware of what is going on without physically attending meetings.

In the approvals, electronic Web-based voting is used in the ETSI. Approved deliverables are available on the Web and on CD-ROM and DVD.

References

ETSI Annual Report & Activity Report 2001

ETSI Annual Report & Activity Report 2000

The Institute of Electrical and Electronics Engineers

The Institute Of Electrical and Electronics Engineers (IEEE) mainly focuses on technical publishing and organizing conferences. With more than 350,000 individual members in 150 countries, its products enjoy a very wide audience and visibility. It is no surprise that approximately 30% of the books related to the IEEE's expertise are published by it, or that thousands of conferences are annually organized under the name of IEEE. Nevertheless, the IEEE is a non-profit, technical professional association. Through the extensive member group, the IEEE is a leading authority in technical areas ranging from computer engineering, biomedical technology and telecommunications, to electric power, aerospace and consumer electronics, among others.

Technical publishing and organizing conferences is not, however, all the IEEE does. When speaking of the IEEE's main activities, we must not forget the third key area in its operation, standardization. In Europe it might not be as widely known as European Telecommunications Standards Institute (ETSI), for example, but especially in the USA it has an established role through its very extensive standardization activities. For example Ethernet is a standard created under the IEEE standardization process.

Working Groups

All standardization efforts within IEEE have been gathered under the IEEE Standards Association (IEEE-SA). IEEE-SA develops and publishes standards that include but are not limited to definitions and terminology, methods of measurement and tests; systems; products; technology; ratings structures; temperature limits and application guides; recommended practices; and safety.

There are two types of members in the SA: individuals and corporate. Individuals, including IEEE members of any grade, or non-IEEE members who are actively involved in standards development, are eligible for individual membership. Corporate membership can be acquired by (e.g., societies, corporate, universities and user groups) willing to actively participate in standards development. The most visible task of members in the standardization process is voting in the sponsor's ballot, the final actual phase of the lengthy and well-defined standardization process. For this standard development process there are working groups each of which is responsible for a certain standard.

Traditionally, IEEE working groups have been divided into the following ten main categories:

- o   Aerospace Electronics
- o   Broadcast Technology
- o   Communications
- o   Electromagnetics
- o   Information Technology

- Instrumentation and Measurement
- Marine Industry
- Medical Device Communications
- National Electrical Safety Code
- Power Electronics
- Power & Energy
- Quantities, Units and Letter Symbols
- Reliability
- Transportation Technology

From the perspective of the Mobile Internet, undoubtedly the most important and relevant working groups are within the Local Area Network (LAN) / Metropolitan Area Network (MAN) standards committee, IEEE 802, which is in the Information Technology category.

IEEE 802

The IEEE 802 LAN/MAN Standards Committee (LMSC) develops Local Area Network standards and Metropolitan Area Network standards. The most widely used standards are for the Ethernet family, Token Ring, Wireless LAN, Bridging and Virtual Bridged LANs. An individual Working Group provides the focus for each area.

The first meeting of the IEEE Computer Society "Local Network Standards Committee," Project 802, was held in February,1980. There was going to be one LAN standard, with speeds from 1 to 20 Mbps. It was divided into media or Physical layer (PHY), Media Access Control (MAC), and Higher Level Interface (HILI). The access method was similar to that for Ethernet, as well as the bus topology. By the end of 1980, a token access method was added, and a year later there were three MACs: Carrier Sense Multiple Access (CSMA) / Collision Detection (CD), Token Bus and Token Ring. Since then, other MAC and PHY groups have been added, and one for LAN security as well. The unifying theme has been a common upper interface to the Logical Link Control (LLC) sublayer, common data framing elements, and some commonality in the media interface. The scope of the work has grown to include MANs. Furthermore, higher data rates have been added. Nowadays there are tens of active standardization task groups creating either entirely new standards or amending and enhancing the existing ones in the IEEE 802 project. More than 800 individuals attend the plenary meetings regularly held three times a year.

Organization

LMSC is organized in to a number of Working Groups (WGs) and Technical Advisory Groups (TAGs) as well as a Sponsor Executive Committee (SEC):

802.0	SEC
802.1	High Level Interface (HILI) Working Group
802.2	Logical Link Control (LLC) Working Group (not active)

802.3	CSMA/CD Working Group
802.4	Token Bus Working Group (not active)
802.5	Token Ring Working Group (not active)
802.6	Metropolitan Area Network (MAN) Working Group (not active)
802.7	BroadBand Technical Advisory Group (BBTAG) (not active)
802.8	Fiber Optics Technical Advisory Group (FOTAG)
802.9	Integrated Services LAN (ISLAN) Working Group (not active)
802.10	Standard for Interoperable LAN Security (SILS) Working Group (not active)
802.11	Wireless LAN (WLAN) Working Group
802.12	Demand Priority Working Group (not active)
802.14	Cable-TV Based Broadband Communication Network Working Group (disbanded)
802.15	Wireless Personal Area Network (WPAN) Working Group
802.16	Broadband Wireless Access (BWA) Working Group
802.17	Resilient Packet Ring Working Group
802.18	Radio Regulatory TAG

The actual standardization runs in projects within working groups. Each project approved within an existing group is assigned a letter, for example 802.1d for MAC Bridges in the High Level Interface WG. A Study Group (SG) is formed when a new area is first investigated for standardization. The SG can be within an existing WG or TAG, or it can be independent of the WGs. A Task Force develops a new project in an existing group, while a new independent project creates a new WG.

No organization can exist without active members. Membership in LMSC is by WG/TAG, with voting rights granted if an individual has attended two of the last four meetings. Attendance at the interim meetings traditionally organized between the plenary meetings can to some extent be counted also. Attendance means going to at least 75 % of the 1/2 day meetings of the WG/TAG during the week. Credit is given for attendance at only one group per plenary meeting.

From an Idea to a Standard

Each standard starts as a group of people with an interest in developing the standard. In the first phase, a Project Authorization Request (PAR) specifying, e.g., the scope and the purpose of the project is created. Traditionally a Study Group which carries on from one plenary meeting to the next writes a PAR. The draft PAR is voted on by the SEC, and then goes to the IEEE Standards Board, New Standards Committee (NesCom) which recommends its approval as an official IEEE Standards project. The PAR also identifies which outside standards groups there will be liaisons with, for instance, International Telecommunication Union (ITU) for some international standards.

The real standardization work can start only after the PAR has been approved. Primarily the process relies on proposals submitted by individual working group members. Proposals are evaluated by the WG, and a draft standard is written and voted on the by the WG. The work progresses from technical to editorial/procedural as the draft matures. When the WG reaches enough consensus on the draft standard, a WG Letter Ballot is done to release it from the WG. It is next approved by the SEC and then goes to the Sponsor Letter Ballot. After the Sponsor Letter Ballot has passed and "No" votes are answered, the draft Standard is sent to the IEEE Standards Board Standards Review Committee (RevCom). Once recommended by RevCom and approved by the Standards Board, it can be published as an IEEE standard. The process from start to finish can take several years for new standards, though less for revisions or addenda.

Wireless 802 Working Groups

802.11 Wireless Local Area Network

802.11 started its work almost ten years ago in the early 1990s. The goal was to create a standard (PHY and MAC) for wireless LANs. It took several years before the group was ready to agree on all the details and the first 802 WLAN standard was published in 1997. Since then it has become the most widely-accepted and used WLAN standard deployed internationally, used on every continent. The original IEEE 802.11 standard itself consists of a single MAC but three different PHY layers. The MAC is based on a kind of listen-before-talk access scheme called Carrier Sense Multiple Access/Collision Avoidance (CSMA/CA). Two of the three PHYs in the original standard were specified for the 2.4 GHz Industrial, Scientific and Medical (ISM) band:

o   Direct Sequence Spread Spectrum (DSSS)
o   Frequency Hopping (FH)

Both of them provided 1 and 2 Mbps bit-rates. The majority of the products were based on the DSSS. The third PHY was based on Infrared (IR) communication.

Relatively soon after the original 802.11 standard was published, the WG began new projects to enhance data rates at the 2.4 GHz ISM band and to create a new PHY for the 5 GHz bands. The former resulted in a supplement IEEE 802.11b which was published in 1999. It enhanced the data-rates from the original 1 and 2 Mbps to 5.5 and 11 Mbps. Nowadays, practically all the 802.11 WLAN products are 802.11b products.

For 5 GHz PHY standardization, the IEEE 802.11 WG created a new task group TGa. In parallel with the TGb formulating the 802.11b amendment, the TGa created an OFDM-based PHY layer for the 5 GHz license-exempt U-NII bands. The effort was equally successful for the TGb, and the IEEE 802.11a supplement was published in 1999. With the new PHY, the 802.11 was now capable of providing bit-rates up to 54 Mbps.

Since then, the 802.11 WG has been actively defining new enhancements in various still ongoing task groups. Just recently, 802.11d was approved and published. It defines extensions to operate in additional regulatory domains and thus further enhances the international status of the 802.11 standard family. Today, there are five task groups within the WG:

o   TGe specifying MAC enhancements for Quality of Service (QoS)
o   TGf creating a recommended practice for Inter Access Point Protocol (IAPP)
o   TGg defining enhanced bit-rates for the 2.4 GHz ISM band

o   TGh specifying spectrum management functionalities for 802.11a for the European markets
o   TGi enhancing security capabilities of the 802.11 WLANs

Additionally, there is a new standing committee called WLAN Next Generation (WNG) to coordinate both the further activities within the WG and cooperation with other related standardization organizations such as European Telecommunications Standards Institute (ETSI) /Broadband Radio Access Networks (BRAN) in Europe and Multimedia Mobile Access Communications Promotion Council (MMAC) in Japan. Lately, the group has been discussing issues such as bit rate extensions for 5 GHz 802.11a, radio measurements and interworking with 3G cellular systems. One of the primary tasks of the standing committee is to make recommendations to the chairmanship of 802.11 and 802.15 about how to deal with new inititatives within the working group.

Enhancements to the Basic 802.11

The increased popularity of 802.11 has also meant that people have invented new usage scenarios, and devices are used for applications which were not thought of when the original standard was developed. Furthermore, new applications require higher data rates with the capacity of fixed LAN and fixed access networks is increasing. Therefore, the IEEE 802.11 standard has been and is being updated for higher link speeds, enhanced security, and better support for Quality of Service (QoS).

IEEE 802.11b, the dominant WLAN standard on the market, increases the data rates to 5.5 and 11 Mbit/s. It uses Complementary Code Keying (CCK) modulation which is based on DSSS technology. The multirate control of the 802.11 MAC protocol allows a terminal to switch from the 11 Mbit/s link speed down to 5.5 and further to 2 and 1 Mbit/s, if the radio channel quality cannot accommodate higher data rates.

A new 802.11 task group g (TGg), founded in March 2000, aims at further increasing data rates, but still utilizing the same 2.4 GHz band. The purpose of TGg is to investigate 802.11b interoperable technologies, which can support higher than 20 Mbit/s data rates on the 2.4 GHz band and other performance improvements to the existing 802.11b standard.

IEEE 802.11 has also developed a physical layer for the 5 GHz band. This 802.11a standard is based on the Orthogonal Frequency Division Multiplexing (OFDM) modulation. The physical layer achieves data rates from 6 Mbit/s up to 54 Mbit/s and the first products are already on the market.

The 802.11 Task Group i (TGi) addresses enhanced security features. The scope of the work includes 1) Upper Layer Authentication based on a new authentication framework based on the IEEE 802.1X standard, 2) Key distribution, and 3) Enhanced packet security.

The 802.11 Task Group e (TGe) addresses MAC modifications for enhanced Quality of Service. The objective is to be able to support voice and video type data services. The identified means to achieve this include: 1) prioritizing traffic to allow critical data to bypass queues, 2) control mechanisms to allow AP to allocate available bandwidth, and 3) more efficient methods for sending *lossy* data such as video (e.g., different retransmit behavior).

The 802.11 Task Group h (TGh) addresses proposals to modify the 802.11a PHY layer to meet European requirements for 5GHz. Europe requires that 5 GHz WLANs must detect when there is another system using the same channel and be able to change frequency automatically.

802.16 Broadband Wireless Access

The IEEE 802.16 Working Group on Broadband Wireless Access Standards develops standards and recommended practices to support the development and deployment of fixed broadband wireless access systems. Currently, the WG has three active task groups working according to their PARs.

The basis for all the work was set by a former Task Group 1, which developed an air interface for Fixed Broadband Wireless Access (FBWA) systems for 10-66 GHz. This work resulted in the MAC specification for the whole 802.16 and a PHY specification for the related licensed bands. Task Group 2 continues its work on the coexistence of FBWA systems. Currently, it is amending the recommended practice it created, which was published in the summer of 2001. Even though the basic coexistence standard deals with generic coexistence matters, it primarily addressed licensed bands 10-66 GHz. The amendment now being drafted will take into account the characteristics of the lower frequencies.

Earlier, the WG had Task Groups 3 and 4, which were developing amendments to the basic 802.16 standard created by TG 1. The focus of the TG3 was on licensed bands 2-11 GHz while the TG4 was developing an amendment covering MAC modifications and an additional PHY layer for license-exempt bands. Today, these TGs are working under a common PAR and continue their work as a single task group, TGa. The work has continued as planned and the draft developed by the task group is about to be included on a Sponsor Ballot.

802.15 Wireless Personal Area Network

The IEEE 802.15 Working Group develops Wireless Personal Area Network (WPAN™) standards for short distance wireless networks. These WPANs address the wireless networking of portable and mobile computing devices (e.g., PCs, Personal Digital Assistants (PDAs), peripherals, cell phones, pagers and consumer electronics), allowing these devices to communicate and interoperate with one another.

802.15 has four active task groups: Task Group 1 is deriving a standard based on the Bluetooth v1.x Foundation Specification; Task Group 2 has been working on coexistence issues, developing a recommended practice to facilitate the coexistence of WPAN™ and WLAN devices; Task Groups 3 and 4 are actively developing high rate and low rate standards for WPAN™. In addition, the WPAN WG has a study group, SG3a, which is defining a project to provide a higher speed PHY enhancement amendment to 802.15.3 for applications involving imaging and multimedia.

Reference

http://ieee802.org

Wireless Ethernet Compatibility Alliance

A few years ago, several WLAN standards and proprietary implementations were competing on the market. WLAN networking components from one vendor would only work with WLAN terminal equipment from the same manufacturer. In other words, consumers could not go and purchase a wireless access point from one vendor and use a wireless network adapter from another vendor on the same installation. This was widely recognized as being the main limiting factor in preventing wide-scale market acceptance of WLAN technology. The situation effectively limited WLAN usage to certain niche market segments, where multi-vendor capability was not of primary concern.

In 1999, the Wireless Ethernet Compatibility Alliance (WECA) was formed to tackle the interoperability problem by supporting the recently approved IEEE 802.11b standard and by testing and certifying WLAN multi-vendor interoperability. The founding members of the group include 3Com, Aironet, Intersil (formerly Harris Semiconductor), Lucent Technologies, Nokia and Symbol Technology. It is notable that today WECA has some 120 member companies.

In addition to certifying the interoperability of Wi-Fi™ (IEEE 802.11) products, WECA's mission includes promoting Wi-Fi™ as the global wireless LAN standard across all market segments.

Testing Process

Interoperability certification process in WECA can be divided into two main parts:

1) Definition of the test specifications, and
2) Testing products from various manufacturers against test specifications to make sure that they conform to the Wi-Fi™ standard.

The Agilent Technologies Interoperability Certification Lab (Agilent ICL) (former Silicon Valley Networking Labs) performs WECA testing. It operates as an independent test facility. However, only WECA members can submit products to the lab for Wi-Fi™ interoperability testing.

When a product meets the interoperability requirements as described in the WECA test matrix, the test laborator notifies WECA. WECA then grants a certification of interoperability, which allows the vendor to use the Wi-Fi™ logo on advertising and packaging for the certified product. The Wi-Fi™ logo is shown in the next figure.

Wi-Fi™ logo

The idea is that the Wi-Fi™ seal of approval assures the consumer of interoperability with other network cards and access points also bearing the Wi-Fi™ logo.

Organization

WECA activities are steered by a Board of Directors, which consists of a few representatives from the member companies. The WECA organization is presented in the next figure.

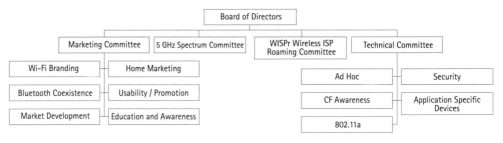

WECA organization

Reference

http://www.wi-fi.com

Bluetooth SIG

Early in 1998, five companies, Ericsson, IBM, Intel, Nokia and Toshiba, founded an industry consortium to develop a new wireless technology to connect different mobile devices. The newly created Special Interest Group (SIG) specified a system for a short-range radio communication suitable for small handheld devices with low current consumption. The system operates in an almost globally available, unlicensed 2.4 GHz Industrial, Scientific and Medical (ISM) frequency band. A fundamental objective of the five promoter companies was to produce an open specification, publicly available, and royalty free. This was achieved by a cross-license agreement of all intellectual properties within the specification of all companies outside the SIG. In order to participate, companies have to sign the SIG agreement. Today, the SIG comprises more than 2,800 companies. Then, in 1999, the promoter group was expanded by the addition of 3COM, Lucent, Microsoft and Motorola.

Bluetooth wireless technology is named after the Viking King of Denmark, Harald Blåtand (Bluetooth). Under the king's rule from 940 to 981, Harald Blåtand unified Denmark and Norway and christianized the two countries. Moreover, he promoted communication between these countries and is therefore a good symbol for Bluetooth wireless technology.

Based on early studies at Ericsson and Nokia, and based on many contributions from other members of the Bluetooth Special Interest Group, the 1.0 specification of the wireless technology was released on July 26th 1999. The specification is being improved successively with new releases, including new and improved functionality. The specifications consist of two documents, the core specification and the profile specification. The core specifies components (e.g., radio, baseband, link manager, and service discovery protocol), transport layer and interoperability with different communication protocols, while the profile document specifies the protocols and procedures required for different types of Bluetooth applications.

Interoperability is seen as one of the most critical success factors for technology. To foster interoperability the Bluetooth SIG has established a qualification program to test Bluetooth products for their compliance with the specification. An actual IPR license is tied to passing the qualification tests.

Organization

The Bluetooth SIG is a non-profit trade organization promoting and managing Bluetooth specifications. It owns the Bluetooth trademarks and the copyright license to the specifications. It also manages the qualification of products, including the granting of IPR licenses from the intellectual property owners to Bluetooth SIG members, as defined in the Bluetooth membership agreements. The SIG has defined three different membership options: promoter membership, associate membership and adopter membership. Each of the memberships has different rights and obligations involved.

To manage SIG processes, several different committees have been established. The SIG is managed by a nine-member Board of Directors, each appointed by the nine promoter members of SIG. General SIG policies and final specifications are ratified by the Board. The other committees are:

- Bluetooth Architecture and Review Board
- Bluetooth Qualification Review Board
- Bluetooth Test and Interoperability Committee
- Marketing Committee
- Regulatory Committee

Bluetooth Architecture and Review Board

The Bluetooth Architecture and Review Board (BARB) provides technical oversight and steers all policies of the technical working groups. BARB consists of nine appointed members from the promoter companies, four appointed members of the associate companies, and all the working group chairpersons. The Board decides on general Bluetooth architecture policies and approves draft versions of working group specifications before they are presented for ratification by the Board of Directors.

Working groups are established for specific problems, and groups work within a predefined charter on parts of the specifications. Working groups can consist of members from promoter and associate companies. Before a working group is finally established a study group prepares a charter which needs to be ratified by the Board of Directors. In addition, expert groups exist to provide industry inputs on certain problem areas, i.e., the security expert group.

Bluetooth Qualification Review Board

Since SIG manages the granting of IPR licenses to its members, it has established the Bluetooth Qualification Review Board (BQRB) to manage this process. This group consists of nine appointed members from the promoter companies. The main tool used to direct the license grant process is the Qualification Program, which is described in the Program Reference Document (PRD). The BQRB controls the activities of the Bluetooth Technical Advisory Board (BTAB), the Bluetooth Qualification Administrator (BQA) and the Bluetooth Qualification Bodies (BQBs).

Bluetooth Test and Interoperability Committee

The Bluetooth Test and Interoperability Committee (BTI) has several tasks, mainly in the area of testing and specification maintenance, focusing on achieving a high level of interoperability between Bluetooth devices. The BTI works together with the technical working groups to make sure that when specification is completed, there is also a test specification available, which is needed to run the Qualification Program. The BTI is also the SIG's interface with test equipment manufacturers and the main body in the development of formal test vectors. Additionally, the group manages existing specifications after they are formally ratified, by running the SIG's errata process.

Marketing Committee

The marketing committee is responsible for developing and executing activities that promote the Bluetooth specification, the Bluetooth Qualification program, membership in the Bluetooth SIG, and participation in the Qualification program. The marketing committee can establish sub-groups that may consist of members from both promoter and associate groups.

Regulatory Committee

The regulatory committee deals with regulatory bodies all around the world aiming to harmonize the regulatory conditions of the 2.4 GHz ISM band. The group also deals with aviation-related problems related to intentional radiators.

Bluetooth Qualification Program

The Bluetooth Qualification Program was established for two reasons: first, SIG has responsibility for intellectual property rights (IPRs) brought into the specifications by SIG members. These IPRs should only be licensed to compliant products. Through the qualification program, SIG determines which products are compliant with the specification and therefore licenses the IPRs to the applicant's product; second, as compliance with the specifications does not necessarily mean interoperability, the SIG fosters interoperability through special interoperability tests. Lower layer functionality is tested with blue units testing, while higher layers are tested with Designated Profile Interoperability Test platforms (DPITs) or profile interoperability tests. The SIG also organizes so-called unplug fests where manufacturers of different devices can test their implementations against each other in a confidential environment.

The entire qualification program is specified in the program reference document.

Authorities in the Qualification Program:

o **Bluetooth Qualification Administrator**

  The Bluetooth Qualification Administrator is responsible on behalf of the BQRB to administer the qualification program. The BQA is an individual person, appointed by the BQRB, who interfaces with SIG member companies and executes program policies.

o **Bluetooth Qualification Body**

  A Bluetooth Qualification Body is an individual person recognized by the BQRB responsible for checking declarations and documents against requirements, verifying the authenticity of product test reports, and listing products on the official Bluetooth Qualified Products List (QPL).

o **Bluetooth Qualification Test Facility**

  The Bluetooth Qualification Test Facility (BQTF) is a test facility that is recognized by the BQRB to test Bluetooth wireless products.

o **Bluetooth Technical Advisory Board**

The Bluetooth Technical Advisory Board (BTAB) is a forum consisting of all BQBs and BQTFs, responsible for advising the BQRB on technical matters relating to test requirements, test cases, test specifications and test equipment.

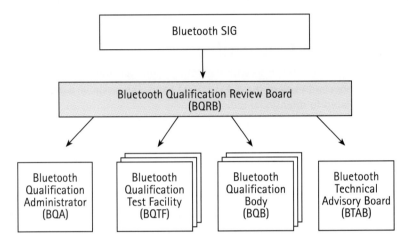

Reference

http://www.bluetooth.com

Open Source Development Lab and Carrier Grade Linux Work Group

The Open Source Development Lab (OSDL) is an independent, vendor-neutral, non-profit organization dedicated to enabling and guiding Linux® and Linux-based development for enterprise and carrier grade functionality worldwide. It was launched in August 2000 as a global consortium of 11 industry leading companies.

OSDL has grown rapidly and now occupies 14,000 square feet of lab space in Portland, Oregon, with another lab in Tokyo, Japan. OSDL intends to open a European site, but currently not enough interested European companies are available as sponsors to support such a site.

The lab has 25 full-time professional employees and is currently sponsored by 22 member companies, listed at the OSDL Web site [OSDL02]. It is apparent from this list of sponsors that the members are a heterogeneous lot, including computer hardware manufacturers, Linux distribution vendors and software development companies.

OSDL is governed by a board of directors, members of which are selected from sponsoring companies. The board also includes at-large members (currently two), who represent the Linux community in general. The lab director selects projects and manages the daily operation of the lab.

Several marketing and administrative people assist the lab director, and the lab facilities and computers require several engineers to keep everything running smoothly. In addition to these people, several software architects participate in the Open Source Software (OSS) community and projects.

The main idea behind OSDL is to create a non-profit and independent, but industry-backed, organization to assist in the development of Open Source Software which can be used in a Linux operating system installed in an enterprise-level environment. In practice, this means that OSDL offers high-end computer equipment for use without fee to OSS developers working on software aimed at business use. In this way, OSDL supports the highly complex projects that are needed to scale and harden Linux for real-life, mission-critical business applications.

Examples of projects supported by OSDL include scalability projects targeting 16-way and above systems and automated test platforms that validate the software on a number of systems maintained as a stable test environment. The key idea behind this is that OSS developers rarely have access to the high-end equipment used to run real-life business applications and thus cannot make the necessary improvements to their software which would allow it to scale from home use to enterprise level.

Except in very special circumstances, OSDL will not initiate Open Source projects itself; instead, OSS developers are free to propose their project to OSDL for approval. If OSDL agrees that the project fulfills the given requirements, then the use of OSDL resources is granted for the project:

1. The target OS for all development in the lab must be Linux.
2. The software must be licensed under an Open Source Initiative (OSI) approved Open Source license [OSI_02].
3. The source code must be publicly accessible during development.
4. The OSDL testing results must not be used for commercial purposes.
5. The project should interest more than a single company and the project maintainer should be identified.

Lists of proposed, active and completed projects can be browsed on the OSDL Web site [OSDL02]. At the moment, about two dozen projects are undergoing testing and validation within OSDL, about a dozen have been completed and several others are under evaluation.

Note that OSDL concentrates on projects which are approaching release quality since the available resources are mainly targeted for the testing and validation phase of the project. For example, OSDL has released the Scalable Test Platform (STP), which allows Linux developers to easily and automatically test their kernel upgrades and patches. Another key project is the Linux Scalability Effort, which attempts to measure and ensure the scalability of Linux.

The long-term goal of OSDL is to enhance positioning and image of Linux. A key element of this effort is to address business-critical topics of the Open Source software stack and to promote them to both the industry and the Open Source community. The OSDL mission statement is as follows:

"Providing Open Source developers with resources and guidance to build data center and Telco class enhancements into Linux and its Open Source software stack, enabling it to become the leading UNIX-like Operating System for e-Business deployment and development."

The word guidance is a recent addition to this statement and stems from the founding of the Carrier Grade Linux Work Group under OSDL. Basically, this addition means that OSDL will take a more active role in the standardization activities around Linux. Previously, OSDL had been trying to establish the Data Center Work Group, but the reception from Independent Software Vendors (ISVs) and Linux distribution vendors has been slow to emerge. Therefore, the advent of the Carrier Grade Linux Work Group gives OSDL added purpose and meaning.

OSDL does not attempt to address the entire spectrum of products in which Linux is used, but rather restricts its activities to the OSDL key focus areas: Carrier Grade and Data Center environments. Technically, these areas are not entirely separate, so cross-pollination of ideas between the two focus areas will likely occur. Additionally, many of the issues tackled in these areas will also benefit Linux installations in less demanding environments (e.g., desktop and embedded domains).

To address the Carrier Grade and Data Center focus areas, OSDL has established two work groups as separate organizations with their own hierarchies. OSDL provides management and resources for these groups, and the actual technical definition is done by a combination of OSDL employees and interested developers and companies from the Linux community and the industry.

The lab director is a member of the advisory board of both work groups, and a special Roadmap Coordinator (RC) is assigned to each work group. The RC is an OSDL employee who participates in the actual work done in work groups. In practice, this means that the RC is the chairperson of the technical group and acts as a project manager for the technical effort.

Carrier Grade Linux Work Group

OSDL Carrier Grade Linux Work Group (CGL-WG) was announced January 30, 2002 at the LinuxWorld Expo in New York. The initial members of the work group were Alcatel, Cisco, HP, IBM, Intel, MontaVista, Nokia, Red Hat and SuSE. Also currently involved is the Free Standards Group (FSG), whose goal is to drive compliance with the recently released version 1.1 of the Linux Standards Base (LSB) [LSB_02].

While Linux has proven to be a robust and relatively scalable operating system, it still lacks some qualities necessary for an operating system in carrier grade environments. For example, the Service Availability Forum [SAF_02] is defining a high-availability layer on top of operating systems that need to be supported from the perspective of the Linux OS. Therefore, CGL-WG has committed to being compatible with the SAForum specifications when they become available.

At the moment, Linux hardware drivers pose another technical difficulty. The normal Linux kernel reaction to the failing of a device is to generate a panic and shut down the system, resulting in unplanned downtime. Accordingly, all device drivers in carrier grade Linux need to be changed to handle efficiently the physical failure of a device. One technical solution to this problem is to trap and contain the failure, so the kernel can continue working without the device.

Overall, CGL-WG is an industry forum meant to guide and encourage development of commercial and open standard components on top of a Linux platform that has been enhanced to the level needed in carrier grade environments. The goals of the work group are to:

o   Enable the use of Linux and open standards based components, reducing development risks and costs by improving flexibility.

o   Facilitate the planning of product features that depend on Carrier Grade Linux through commonly agreed upon roadmaps and functionality.

o   Drive to create a critical mass for a Carrier Grade Linux ecosystem, enabling compatible platforms and components to be used in innovative products.

CGL-WG consists of an advisory committee, a marketing group and a technical board, the names of which are currently under revision although their functions will remain the same. The advisory committee steers the efforts of the whole work group, while the marketing group creates a marketing plan and handles public relations. The marketing group also works closely with the technical board to ensure that the Carrier Grade Linux specifications meet market requirements.

Actual development of specifications and support of development projects fall to a set of technical subgroups. There are currently four subgroups: Architecture, Requirements, Proof-of-Concept and Validation. The Technical Board oversees the work of the technical subgroups and coordinates their efforts into Carrier Grade roadmaps and specifications. The structure of CGL-WG is shown in following figure.

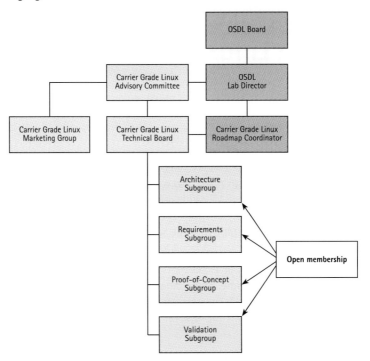

OSDL CGL-WG structure

Technical subgroups are open to anyone with a genuine interest in the work group's activity and a willingness to contribute time and expertise to the effort. In practice, this means that the subgroups are open to all OSDL member companies and to external entities who are willing to sign the OSDL IPR license paper, which means that the entity agrees not to make any IPR claims on its contributions.

The Roadmap Coordinator, an employee of OSDL, chairs the technical board. The Roadmap Coordinator chairs a weekly meeting of the Technical Board, which consists of active members of the technical group. The subgroup chairs are by definition active members, as each subgroup chair is chosen from the subgroup's active members. Note that active members are limited to one per entity, and in this limitation the subgroup chair counts as one.

Subgroup chairs have a free hand to arrange the working of their subgroups, but are responsible to the RC. Just as the RC can be thought of as the project manager, the subgroup chairs can be considered subproject managers. The RC still has overall responsibility and reports weekly to the advisory committee.

Scope of Carrier Grade Linux Work Group

The scope of the Carrier Grade Linux Work Group is limited to Linux operating system enhancements. Here, the definition of operating system includes the needed interfaces toward middleware and hardware. Also, some tools are included to address specific issues like kernel debuggers and kernel testing tools. The scope has been solidified as an architecture definition, shown in the following figure.

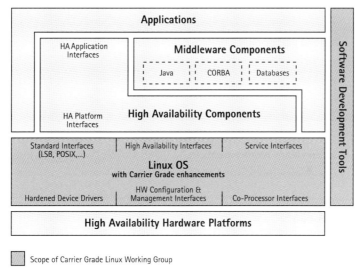

OSDL CGL-WG architecture

Existing standards like LSB and POSIX will be used whenever possible and, as mentioned previously, the specifications produced by the SAForum will also be embraced. The intent is to focus on needed new technology. So, in the first phase, the CGL-WG will map out all OSS projects whose output can be used in the creation of carrier grade Linux systems. In the second phase, gaps will be identified and addressed by the creation of new Open Source projects under the guidance of the Proof-of-Concept subgroup.

As the previous figure illustrates, in addition to the aforementioned device driver hardening, management of the carrier grade hardware is critical in ensuring that there is no single point of failure in the system. A third hardware-related area addresses co-processors (e.g., the network processor), needed in emerging packet switched networks to achieve needed performance levels. The Linux kernel needs to be checked and, if necessary, modified to allow these kinds of co-processors to be used for maximum benefit.

The following list shows some additional requirements for Carrier Grade Linux that will be addressed in the CGL-WG:

- o **Heartbeat and watchdog timers:** these are needed for monitoring the system and ensuring no deadlocks go undetected
- o **Remote software installation:** this enables maintenance of a system without physical access
- o **Ethernet aggregation:** physical lines are duplicated so connections will not be lost in case of a hardware failure
- o **Timeliness in the Linux kernel:** both pre-emption and low-latency scheduling will be considered to achieve a soft real-time solution; currently hard real-time solutions are not considered
- o **Kernel dump functionality:** the ability to analyze the reason for kernel failures
- o **Thread performance enhancement:** currently threading is expensive in Linux, so possible ways to enhance it will be investigated
- o **IPv6 and SCTP support:** there are already Open Source projects addressing these issues, and they will be supported
- o **Enhanced file-systems (RAID, journaling & clustering):** all possible measures will be investigated to secure availability of physical disks
- o **Middleware requirements (e.g., CORBA, J2SE™, databases):** reliability and performance issues with middleware support will be addressed

Conclusions

The OSDL Carrier Grade Linux Work Group is a remarkable attempt by several leading industry companies to create a unified operating system platform for the carrier grade environment, where an availability figure as high as six nines (i.e., 99.9999%) is required. The ambitious goal of the group is to have version 1.0 of the specifications available by the end of 2002 as can be seen from the following figure which shows the current roadmap for the CGL-WG.

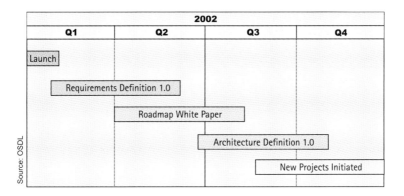

OSDL CGL-WG roadmap [OSDL02]

The initial specifications will not cover all the requirements for carrier grade Linux (e.g., SAForum requirements will be missing), but it will be a basis for first implementations and will continue to be enhanced on a tight schedule. Also, just as Linux itself will continue to evolve, so too will the carrier grade Linux systems evolve and specifications be updated.

References

[OSDL02] http://www.osdl.org

[OSI_02] http://opensource.org/licenses

[SAF_02] http://www.saforum.org

[LSB_02] http://www.linuxbase.org

Service Availability Forum

The origins of the Service Availability Forum (SAF) go back to early 2001, when a number of companies started to talk about the need for a standardized development and execution environment for highly available network elements. The motivation was to create an eco-system for open, compatible third-party software components, enabling equipment manufacturers (e.g., Nokia), to develop carrier class network elements in an open fashion, using software components from third parties when applicable.

Service availability is made up of two key criteria: high availability and service continuity. High availability refers to the devices and systems that make up the network. These must be up and running, available to provide services and applications to the consumers.

Service continuity means that the sessions are maintained, despite the failure of individual systems or components. For example, to provide uninterrupted service, the state of the application session for each user must be preserved during switchovers.

Open Standards for Service Availability

To make it easier to satisfy this need, leading industry players (i.e., Nokia, Motorola, Intel, IBM, HP, GoAhead Software, Radisys and Force Computers), came together in the Service Availability Forum. Later, Sun Microsystems, Compaq, Siemens, Huawei Technologies, MontaVista, WindRiver, Stonesoft and Fujitsu-Siemens, joined as well.

In an open system, service availability software is used to manage critical network resources in order to maintain the desired level of availability. The Service Availability Forum will develop standard programming interfaces to link this software with the resources that need to be managed.

Currently, there is no such thing as a standardized availability management interface. This means that the availability management software must be customized for each resource it needs to manage, increasing the cost of development and producing fragile systems that are not portable. With the way things stand now, an application must be re-written when the infrastructure underneath it changes. Clearly, standards allowing systems to be easily managed for availability and interoperability are essential.

The Service Availability Forum was formed to create an open, standard programming interface that will allow the rapid building of network elements that deliver highly dependable voice, data and multimedia services.

The two initial programming interface specifications are located between the applications and Service Availability middleware and Service Availability middleware and the operating system.

422

Java™ Community Process

The Java™ Community Process (JCP) is a framework under which the international Java community develops and maintains Java technology specifications for some aspects of Java technology (e.g., Java language, virtual machine, platform editions and different application programming interfaces).

Even if not a typical standardization forum, JCP forms a de facto standardization process for Java related specifications.

JCP was originally established by Sun Microsystems in 1998 to respond to the needs of the growing Java community. The process which is in effect today (The Java Community Process Program, version 2.0 - JCP2) was established by Sun in June 2000 partially based on input from leading companies active in Java development.

JCP Agreement and Process

Document

The formal JCP process is described in two founding documents:

1) Java Specification Participation Agreement (JSPA), which is a legal agreement (between Sun and a JCP member) that all parties wishing to work in the JCP expert groups need to sign and which defines the legal aspects of the process, including all IPR grants that members have to grant related to the developed specifications and other output of the process; and

2) the JCP2 Process description, which defines the operational details of the process.

Membership

To join the Java community you need to sign the JSPA and pay an annual fee. There is no fee for Sun licensees. Currently, there are approximately 500 Java Community members.

Process

The life of a specification and other JCP output moves in different phases, from initiation to the final output and maintenance phase. In each phase, certain scrutiny and controls are applied by the expert group, executive committee, community and public (e.g., reviews of the specification and approvals to go forward in the process to the next phase).

In the JCP, all work towards the output is done in expert groups, which are lead by a Specification Lead.

Each proposed and approved specification proposal Java Specification Request (JSR) and related expert group always produces three main and mandatory results (output):

1. **Specification**
2. **Reference Implementation (RI)** - a prototype or proof of the specification's concept implementation
3. **Technology Compatibility Kit (TCK)** - a suite of test tools and documentation that allows any implementer of a specification to test compliance with the specification

The Specification Lead is responsible for having all three results created, and as such has much power and responsibility in the process.

Guidance Body

Another important body is the Executive Committee (EC), which is a control/guidance body consisting of parties selected to this role by the Java Community. There are two Executive Committees, one for controlling the work on the Java™ 2 Standard and Enterprise Editions (J2SE™/J2EE™) related specifications, and one for Java 2 Micro Edition (J2ME™) related specifications.

EC members are elected by the community through a ballot for a 3-year term, with 5 seats rotating each year. Nominees to the ballot are either nominated by Sun (10 ratified seats) or by the community (5 elected seats), while Sun has a permanent seat.

Decision-making in the EC is based mostly on a simple majority of votes cast (yes/no/abstain). However, to approve new Java platform editions or propose changes to the Java language specification a two-thirds majority of votes is required.

Sun has dedicated a team to host and run the operations of the JCP. The Project Management Office (PMO) hosts the JCP Web sites, arranges ballots and takes care of the general flow of the specifications in the process.

Life Cycle of a Specification

Specifications flow in the process through the following stages:

1. Any JCP member can suggest a specification to be created in the JCP. This suggestion is provided in writing in the form of a Java Specification Request. The filing party often solicits support by some other parties already at this point, which will help JSR to go through the next step. This form then describes the proposal and indicates the Specification Lead and the initial members of the expert group.

2. The JSR is posted to the jcp.org Web site for a 14-day JSR review, during which anyone can comment on it. At the end of the JSR review, a JSR approval ballot will take place by the respective Executive Committee. If JSR does not pass, the JSR initiators can correct it and submit it again for ballot. If that fails the JSR will be closed.

3. After a JSR is approved, the Spec Lead will form an Expert Group for the JSR. The Spec Lead can decide the size and membership of the EG, and can also very freely decide on the group's working style, as long as it is compatible with JCP rules.

4. After the Expert Group has worked on the specification and decided that the first draft is good enough for a wider audience, the specification will be placed for Community Review by the Project Management Office. The goal of the Community Review is to get community comments and prepare the specification for Public Review by uncovering and fixing problems with the draft specification. The length of the Community Review is decided by the Spec Lead, and is usually between 30-90 days. The spec shall be frozen for the last 7 days of the Community Review for the EC to complete the Draft Specification Approval Ballot.

    The Draft Specification Approval Ballot will be done by the respective Executive Committee, who decides whether the specification can proceed to a Public Review. EC considers the quality of the specification, the comments received by the community and the original goals of the JSR.

    At this point, at the latest, the Executive Committee needs to take a position regarding the intended licensing terms of the Reference Implementation and the Technology Compatibility Kit, which the Spec Lead needs to inform them of before the start of the Community Review.

5. After it passes the Community Review ballot, the specification proceeds to Public Review, during which the specification can be seen and commented on by any party having Internet access. The Spec Lead needs to ensure that the comments and suggestions from the public are considered. The public review period is from 30 to 90 days, as decided by the Spec Lead.

    After the close of the Public Review, the Expert Group prepares the Proposed Final Draft, which is then used to complete the RI and TCK.

    The Final Draft Proposal is placed on the JCP public Web site for community members and the public to see and download. The Specification Lead is responsible for producing the RI and TCK. If the specification needs to be changed due to some findings in RI/TCK development, the EG will work to correct the specification accordingly and the revised specification with a list of changes will be placed on the JCP Web site again.

6. When the EG is satisfied with the RI and the TCK, as well as its coverage, and when the RI passes the TCK, the Spec Lead can submit the specification for a Final Approval Ballot by the EC. When submitting the Final Draft the Spec Lead will as well provide instructions on how EC members can obtain the RI/TCK for evaluation.

    The Final Approval Ballot is a 14-day-period when the EC approves/disapproves the Final Draft and associated RI and TCK.

    If the specification does not pass the Final Approval Ballot, the Spec Lead has 30 days to take corrective actions in response to the concerns of the EC. If no response is received, the JSR will be closed and the EG will disband. For a failed ballot one Final Reconsideration Ballot will be executed based on the Spec Lead's response.

7. After the Final Approval Ballot, an approved Specification is posted on the JCP Web site and an announcement is made to the community and public. The EG has then completed its task and can pursue other challenges.
8. The specification moves to Maintenance mode after Final Approval, and the Spec Lead assumes the post of Maintenance Lead.

JCP contains rules of how a specification is maintained and updated. This process can be fairly light, i.e., the *Minor revision process*, in which the Maintenance Lead may make the proposed changes by herself, unless the EC in the Exception Ballot chooses to defer some of the proposed changes to a new JSR. An alternative process for maintenance and updates is to set up a new JSR to revise a Spec.

JCP Process 2 contains, on top of the main flow, provisions for different kinds of exceptions (e.g., handling an implementer's TCK test case complaints, dealing with uncooperative expert group members, finding another party to continue the tasks of the Maintenance Lead who wishes to relinquish ownership).

Specification Lead Position

It should be noted that JCP vests much power in a Spec Lead. On the other hand, a Spec Lead has extensive responsibilities. A Spec Lead is responsible for creating the output (i.e., the Spec, the RI and the TCK). JCP rules do not mandate the way the Spec Lead does this; for instance, there are no provisions about how decisions in the EG are made, and in the end the Spec Lead may decide the procedures in EG.

Another aspect of the Spec Lead's power is the possibility to set the licensing terms for the RI and TCK, with the only qualification that the terms need to be non-discriminatory, fair and reasonable. The Executive Committee can also provide guidance on the overall acceptability of the intended licensing terms, but in practice the EC is not able to influence the terms too much.

It should be noted that a Spec Lead (and later the Maintenance Lead) makes a long term commitment to keep the specification updated and the respective RI and TCK in sync with the specification.

Intellectual Property

Intellectual property rules are defined in the JSPA.

At the present time, JCP's legal and process documents are being revised, so here we will only briefly summarize the existing IPR rules.

The general model is that first the Community members grant the Spec Lead and Sun wide, royalty-free licenses with respect to members' contributions (e.g., specs, materials, ideas) which are to be incorporated into the specification and associated RI and TCK, to copy/disclose/distribute the specification for Community and Public Review, and very wide rights with respect to the contribution as part of RI and TCK.

Sun and the Spec Lead will jointly own the copyright of the final specification.

Based on the inbound grants, as described above, Sun and/or the Spec Lead shall then further allow implementations of the Specs, provided they fulfill strict compatibility requirements (e.g., implementing the whole spec without subsetting/supersetting, pass the TCK).

The RI and TCK will be licensed by Sun or the Spec Lead, as appropriate, on non-discriminatory, fair and reasonable terms.

Essential patents in a contribution can be chosen either to be opted out or licensed under fair, reasonable and non-discriminatory terms.

At the time of this writing, a new JSPA is being prepared in the ongoing JSR, which if successful will change the existing IPR and compatibility requirement rules, among other things, towards a more balanced model between all JCP participants.

One aspect of the new rules would be that the JSPA would not prohibit independent compatible open source implementations of a Specification. Furthermore, at the discretion of the Specification Lead, an Expert Group could release its own Reference Implementation and TCK under open source License and in general TCKs would be available free of charge to not-for-profit organizations.

World Wide Web Consortium

The World Wide Web Consortium (W3C) is an industry organization founded to further the development of the World Wide Web by promoting interoperability and encouraging an open forum for discussion. In this role, W3C commits to leading the technical evolution of the Web.

W3C's long term goals for the Web are:

1. **Universal Access**: To make the Web accessible to all by promoting technologies which take into account the vast differences in culture, education, ability, material resources and physical limitations of users on all continents.
2. **Semantic Web**: To develop a software environment which permits each user to make the best use of the resources available on the Web.
3. **Web of Trust**: To guide the Web's development with careful consideration for the novel legal, commercial and social issues raised by this technology.

W3C is not an official standards body, but as with many other information technologies, in particular those owing their success to the rise of the Internet, the Web must evolve at a pace unrivaled in other industries. This makes it difficult for traditional standards organizations to develop standards which meet the needs of the developing industry-which is not to say that it is not difficult for W3C as well. But, since its inception in 1994, the consortium has created a large number of important technical specifications. These specifications have been adopted as de facto standards for the Web. These specifications include Hypertext Markup Language (HTML), Hypertext Transfer Protocol (HTTP), Cascading Style Sheets (CSS), eXtensible Markup Language (XML) and Resource Description Framework (RDF).

Operation and Organization

W3C is hosted by the MIT Laboratory for Computer Science in the USA, the National Institute for Research in Computer Science and Control (Institut national de recherche en informatique et en automatique - INRIA) in France and Keio University in Japan. Work at the consortium is organized around activities, each of which belongs to one of five domains:

- **Architecture Domain**: develops the underlying technologies of the Web.
- **Document Formats Domain**: works on formats and languages that will present information to users with accuracy, beauty and a higher level of control.
- **Interaction Domain**: seeks to improve user interaction with the Web and to facilitate single Web authoring to benefit users and content providers alike.
- **Technology and Society Domain**: seeks to develop a Web infrastructure to address social, legal and public policy concerns.
- **Web Accessibility Initiative (WAI)**: pursues accessibility of the Web through five primary areas of work: technology, guidelines, tools, education and outreach and R&D.

Each activity consists of one or more working groups, each typically with the goal of producing a new specification. Each specification goes through the W3C process, and the ones the consortium membership approves eventually become official recommendations of the organization.

The day-to-day activities of the consortium are run by the W3C management team, assisted by an advisory board (at the time of writing, the nine elected members of the board are from Adobe, Boeing, IBM, IPR Systems, Microsoft, Nokia, Opera Software, SAIC and Sun Microsystems). Ultimately, decisions are sanctioned by the advisory committee, where each member organization has one representative. The current W3C membership consists of over 500 companies and other organizations.

Current Activities of Interest

Several W3C activities, working groups or specifications are related to the Mobile Internet technologies.

Hypertext Markup Language

Hypertext Markup Language is the dominant content markup language on the World Wide Web. The purpose of the HTML Activity is to develop the next versions of HTML. Recent specifications include:

- **eXtensible Hypertext Markup Language (XHTML) 1.0** is a recast of the HTML 4.01 specification as an XML application.

- **XHTML Modularization** is a decomposition of XHTML into a collection of abstract modules that provide specific types of functionality. These modules can be combined and extended by document authors as well as application and product designers to make it feasible for content developers to deliver content on a broader spectrum of diverse platforms.

- **XHTML Basic** specifies the minimal set of modules required from an XHTML host language document type, plus some other features (e.g., images, forms, and basic tables) and object support. It is designed for Web clients that do not support the full set of XHTML features (e.g., mobile phones, PDAs, pagers and set-top boxes).

Cascading Style Sheets

Style Sheets describe how documents are rendered (e.g., presented on screen, in print, through audio); they allow the separation of document structure and content from how the document is to be presented to consumers. W3C has promoted the use of style sheets on the Web since the consortium's inception. The Style Activity has produced two W3C recommendations (CSS1 and CSS2).

Web Services

The consortium has formed two working groups to work on Web Services Architecture and Description, respectively. There is considerable pressure to form widely adopted standards in this area, particularly since the current (emerging) product solutions are based on specifications that are rapidly becoming de facto standards.

Additionally, the W3C Web Services Activity has a working group for XML Protocols. Its goal is to develop technologies that allow communication in a distributed environment, using XML as the encapsulation language. Solutions developed allow a layered architecture on top of an extensible and simple messaging format, providing robustness, simplicity, reusability and interoperability. Current specifications under development include the next version of the Simple Object Access Protocol (SOAP).

Semantic Web and Resource Description Framework

The Semantic Web Activity is a continuation of the Consortium's earlier Metadata Activity that created the RDF representation standard. The new activity will promote the development of technologies that transition the Web towards richer semantic content. Current work in this activity includes the RDF Core Working Group and the Web Ontology Working Group creating a standard language for ontology creation on the Web.

Graphics and Multimedia

Scalable Vector Graphics (SVG) is a language for describing two-dimensional graphics in XML. It allows for three types of graphic objects: vector graphic shapes, bitmap images and text. Graphical objects can be grouped, styled, transformed and composed into previously rendered objects.

The Synchronized Multimedia Integration Language (SMIL) enables simple authoring of interactive audiovisual presentations. It is typically used for rich media or multimedia presentations that integrate streaming audio and video with images, text or any other media type.

Device Independence

There is significant work underway to integrate Web technologies into various new kinds of devices (e.g., mobile devices and TV sets). In this process, we are faced with the possibility that services for those devices may not interoperate with each other or with the existing Web. Not only would that cause fragmentation of the Web space, but it would also make device-independent authoring impossible. Through its origins and heritage, W3C has a particular interest in device-independent Web access and authoring. The Device Independence Activity was created by merging the Consortium's earlier Mobile Access and TV and the Web activities.

Earlier work in this area led to the creation of the Composite Capability/Preference Profiles (CC/PP), a representation format which allows Web clients to describe their functional characteristics as well as user preferences.

Conclusions

W3C is striving to lead the Web to its full potential by developing common protocols that promote evolution and ensure interoperability. Nokia, as a long term active member of W3C, believes in - and strongly supports - the creation and adoption of open, interoperable standards for the Internet and for the World Wide Web, and is thus committed to supporting the core mission of the consortium.

Reference

http://www.w3.org

Web Services Fora

Standards organizations operating in the Web services space can be loosely organized into groups. The Web legacy group can be characterized as those organizations with a direct history of Web and Internet participation who are now reacting to the industry-generated inertia surrounding Web services. Groups focused on Web services can be characterized as a set of new organizations solely created to address Web services requirements. The last group can be characterized as those who provide some value for mobility-aware Web services and are typically telephony-oriented organizations.

For mainstream (non-mobile) aspects of Web services, the key standardization organizations are the World Wide Web Consortium (W3C) and the Web Services Interoperability Organization (WS-I). These organizations have assumed responsibility for producing a coherent vision for Web services: a plan for Web service technologies from low-level messaging to high-level service orchestration. The W3C is producing a consistent Web services architecture. Other organizations (e.g., The Internet Engineering Task Force (IETF), Java Community Process (JCP), Object Management Group (OMG), 3rd Generation Partnership Project (3GPP), WAP Forum, Liberty Alliance and Location Interoperability Forum (LIF)) have also standardization actions related to Web services.

OASIS & UN/CEFACT

OASIS, originally SGML Open (1993), is a non-profit, international consortium creating interoperable industry specifications based on public standards (e.g., eXtensible Markup Language (XML) and Standard Generalized Markup Language (SGML), as well as others that are related to structured information processing. UN/CEFACT is the United Nations body whose mandate covers worldwide policy and technical development in the area of trade facilitation and electronic business:

o **ebXML**. A suite of business and technical specifications which exists to transition the Electronic Data Interchange (EDI) community Business-to-Business (B2B) -based processing to be compatible with the Web and XML. ebXML defines business models and a method that enables implementation of a trading partner's business metadata that is inserted by a trader into an ebXML registry. This metadata package, the Collaboration Protocol Profile (CPP), contains a trader's supported business processes and business service interface requirements. During runtime, another trader queries the ebXML query looking for a service provider meeting its business criteria. After discovering a match, a Collaboration Protocol Agreement (CPA) is generated by the new trading partners, describing the messaging service and the agreed business process requirements. All ebXML documents use either Document Type Definition (DTD) or XMLSchema for document vocabularies and use Unified Modeling Language (UML) as the modeling language to create business processes. ebXML defines a Universal Description, Discovery and Integration (UDDI) registry compatibility component and describes both Simple Object Access Protocol (SOAP) and Multi-purpose Internet Mail Extensions (MIME) as messaging protocols.

- o **ebXML Registry**. Interoperable registries and repositories, with an interface that enables submission, query and retrieval. Part of the ebXML effort.
- o **Web Services Interactive Applications (WSIA)** - Extending Web Services to the user interface where the device capabilities and business logic of the service provider, service broker, and the consumer are all leveraged to provide the correct experience.
- o **SAML** - Web Services authentication using XML syntax. Security Assertion Markup Language (SAML) assertions are created and digitally signed by an authentication authority. SAML uses XML-DSIG and is targeted for single sign-on systems.
- o **ebXML Messaging Service** - Develops and recommends technologies for secure and reliable messaging exchanges in electronic business transactions. They have produced a Message Service Specification for riding ebXML content on top of SOAP with Attachments (SwA) messages. SwA allow MIME multipart payloads.

UDDI.org

UDDI.org is an initiative established by Microsoft, IBM and Ariba to develop a standard for an online registry for the publishing and dynamic discovery of Web services offered by businesses. UDDI allows a business to locate potential business partners and form business relationships on the basis of the services they provide. The primary target is integration and semi-automation of business transactions in B2B e-commerce. To achieve this, UDDI provides a registry for businesses and the services they offer, described using an XML schema defined by the UDDI specification.

Web Services Interoperability Organization

Web Services Interoperability Organization is a new industry consortium established in February 2002. Its principal aim is to promote interoperability between Web service technologies and tools. Its founders include the major Web service and IT tool vendors including Microsoft, IBM, BEA and Hewlett-Packard.

WS-I is a systems integration group whose charter is to take specifications from a variety of organizations and, through the use of test suites, certify their level of interoperability. An interoperable set of Web service specifications is grouped together to form a WS-I profile. WS-I promotes that WS-I profiles will be bundled with interoperability test suites, sample applications, monitoring and analysis tools, and best practices documentation.

The first WS-I profile has been delivered. It is labeled WS-I Basic and consists of XML Schema 1.0, SOAP 1.1, Web Services Description Language (WSDL) 1.1 and UDDI 2.0.

To attain critical mass in the marketplace, coherent packaging of the many Web services technologies is essential.

DAML-S

DARPA is funding a five-year research program entitled *DARPA Agent Markup Language* (DAML), conducting research into various Semantic Web technologies. As part of this program, the DAML-S Coalition (DAML for Services) is using the DAML+OIL ontology language based on W3C Resource Description Framework (RDF) to build an upper-level ontology for Web services. DAML-S -capable agents will be able to semantically describe their services in such a way that other agents will be able to discover, invoke, compose and monitor the execution of these services. Descriptions communicate not only what the inputs and outputs are but also how the service works and what the logical relationships between the inputs and the outputs are. DAML-S is complementary to WSDL and SOAP and will leverage these technologies on a lower-level layer.

RosettaNet

RosettaNet is an industry consortium which promotes standards for e-business vocabularies, processes and the messaging infrastructure. RosettaNet standards include the RosettaNet Business Dictionary, the RosettaNet Technical Dictionary, the RosettaNet Implementation Framework (RNIF) and RosettaNet Partner Interface Processes. RosettaNet uses XML across its specifications. Of special interest to the Web services community are the business, domain-specific vocabularies and the Implementation Framework. The Implementation Framework defines a message protocol using MIME multipart and digitally signed MIME (S/MIME). The messaging framework addresses security concerns of authentication, authorization and non-repudiation. RosettaNet is a B2B focused Web service that, besides XML, is adopting its own solutions for standardized vocabularies and messaging.

Object Management Group

The Object Management Group (OMG™) was founded in April 1989. In October 1989, it began independent operations as a non-profit-making corporation. Through the OMG's commitment to developing technically excellent, commercially viable and vendor independent specifications for the software industry, the consortium now includes about 800 members. The OMG is moving forward in establishing the Common Object Request Broker Architecture (CORBA®), as the *Middleware that's Everywhere* through its worldwide standard specifications: CORBA/Internet Inter-Object Request Protocol (IIOP), Object Services, Internet Facilities and Domain Interface specifications, Unified Modeling Language (UML) and other specifications supporting analysis and design.

The OMG was formed to create a component-based software marketplace by hastening the introduction of standardized object software. The organization's charter includes the establishment of industry guidelines and detailed object management specifications to provide a common framework for application development. Conformance to these specifications will make it possible to develop a heterogeneous computing environment across all major hardware platforms and operating systems. Implementations of OMG specifications can be found on many operating systems across the world today. OMG's series of specifications detail the necessary standard interfaces for Distributed Object Computing. Its widely-popular Internet protocol IIOP is being used as the infrastructure for technology companies (e.g., Netscape, Oracle, Sun, IBM). These specifications are used worldwide to develop and deploy distributed applications for vertical markets (e.g., Manufacturing, Finance, Telecoms, Electronic Commerce, Real-time systems and Health Care).

OMG defines object management as software development that models the real world through the representation of objects.

The OMG is structured into three major bodies, the Platform Technology Committee (PTC), the Domain Technology Committee (DTC) and the Architecture Board. The consistency and technical integrity of work produced in the PTC and DTC is managed by an overarching Architectural Board. Within the Technology Committees and Architectural Board are contained all the Task Forces, special interest groups (SIGs) and Working Groups which drive the technology adoption process of OMG.

OMG Modeling Specifications

UML is a graphical language that expresses program design in a standard way, allowing design tools to interchange models. It is an object-oriented language which standardizes an impressive number of diagram types including Class and Object diagrams, Structure diagrams, Use Case diagrams, and more.

The Meta-Object Facility (MOF™) is a standardized repository for meta-data - that is, the descriptions and definitions of the fundamental concepts that applications work with.

The XML Metadata Interchange (XMI™) is a stream format for the interchange of metadata including the UML models created during analysis and design activities.

The Common Warehouse Metamodel (CWM™) completes the OMG data modeling standards. It provides standard models for table-based, file-based and object-based datastores.

Recent and Forthcoming Enhancements

The Minimum CORBA specification defines a static subset of CORBA. The target is primarily an embedded device, ultimately an ORB printed on silicon. The forthcoming Smart Transducers proposals target even smaller devices (e.g., actuators and sensors). The focus is on management interfaces. In telecommunication end-user devices minimum CORBA may be appropriate but a *micro CORBA*, that is a Generic Inter-ORB Protocol (GIOP) engine and a light-weight activation mechanism, is worth examining.

The Real-Time CORBA specification addresses timing constraints on the ORB-level. The focus is on the allocation of resources and on the predictability of system execution. The interfaces and mechanisms provided by Real-Time CORBA facilitate a predictable combination of the ORB and the application. The application manages resources by using the Real-Time CORBA interfaces while the ORB mechanisms coordinate the activities that comprise the application. The Real-Time ORB relies on the underlying real-time operating system to schedule threads that represent activities being processed and to provide mutexes to handle resource contention. The forthcoming Extensible Transport Framework for Real-Time CORBA and the Real-Time Notification proposals will be an important infrastructure service for real-time applications.

The Fault-Tolerant CORBA specification aims to provide robust support for applications requiring a high level of reliability. The specification supports a range of fault tolerance strategies, including request retry, redirection to an alternative server, passive (primary/backup) replication and active replication which provides more rapid recovery from faults.

The Wireless Access and Terminal Mobility in the CORBA specification introduces relocatable object reference and terminal mobility transparency by using the bridging concept of the CORBA specification. The Wireless CORBA also supports vertical handoffs between different transport domains through introducing a generic GIOP tunnel.

The forthcoming proposals for GIOP SCTP/IPv6 protocol mapping is an important step towards the possibility to use CORBA as the control plane platform of UMTS. Two requests for proposals (RFPs) are out targeting middleware interoperability: CORBA/Simple Object Access Protocol (SOAP) interworking RFP and CORBA/Wireless Application Protocol (WAP) interworking RFP, which are important enablers of middleware interoperability.

The OMG has been the leading forum in specifying interoperability bridges between CORBA and other middleware platforms. The OMG has started a comprehensive development of a new architecture called the Model Driven Architecture™ (MDA™). The objective is to interrelate IDL specifications, UML modeling, Meta-Object Facility and XML Metadata Interchange (XMI). The forthcoming MDA might provide a useful starting point for tools supporting interoperability between parts of an application running on different middleware platforms.

Reference

http://www.omg.org

Liberty Alliance

The Liberty Alliance is a consortium which aims to create open, federated, single sign-on identity standards.

Membership and Organization

Membership in the consortium is open to all commercial and non-commercial organizations. It is composed of 17 founder members: American Express, AOL Time Warner, Bell Canada, Citigroup, France Telecom, General Motors, Global Crossing, Hewlett-Packard Company, Mastercard International, Nokia, NTT DoCoMo, Openwave Systems, RSA Security, Sony Corporation, Sun Microsystems, United Airlines and Vodafone. Several others are sponsor members.

As illustrated in the following figure, the Liberty Alliance is organized into three working groups:

o **Technology:** Responsibilities include specifying the Engineering Requirements Document, as well as specifying the architecture, messages and protocols for current as well as future Liberty-enabled clients/browsers.

o **Marketing:** Responsibilities include specifying the marketing value proposition and use cases, as well as specifying the Marketing Requirements Document.

o **Public Policy:** This group is concerned with legal implications as well as privacy issues.

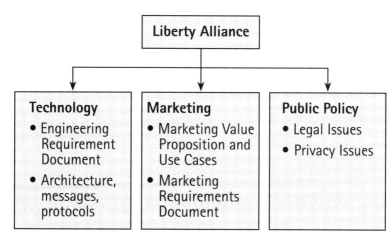

Working groups within the Liberty Alliance

Phased Approach

As illustrated in the next figure, the Liberty Alliance has decided on a phased approach:

- **Phase 1:** This phase deals with business-to-business sharing within a circle of trust, and the verification of identities across multiple devices within this circle of trust. A circle of trust could be a business alliance, where the involved businesses agree to specific terms and conditions to securely share user identity. It is anticipated that businesses maintain their customer databases and the relationship with the customer.

- **Phase 2:** This phase expands the scope of Phase 1 to include user profiles and preferences. User preferences (or attributes) are shared across businesses, with the user being in control of what attributes may be shared and with whom. This phase deals with personalized services centered around the user.

- **Other phases:** Other phases may be specified to expand the scope beyond phases 1 and 2.

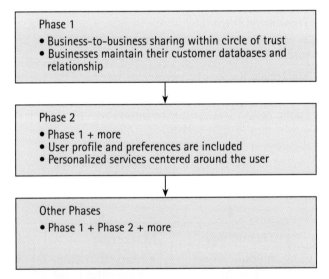

Phased approach adopted by the Liberty Alliance

Liberty Applicability and Architecture

Liberty specifies mechanisms for achieving identity federation, single sign-on, termination of identity federation as well as single logout.

The following figure illustrates a case where a user federates his identity at Service Provider 1 (SP1) with his identity at Service Provider 2 (SP2). Liberty specifies the federation protocol in Step 5.

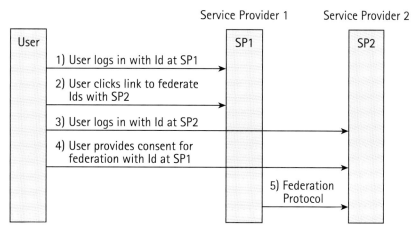

Scenario illustrating identity federation

When a Single Sign-On service is made available, a user who has already authenticated himself with SP1 need not authenticate himself with SP2. Instead, the sign-on at SP1 lends itself transparently to SP2. In other words, a single sign-on is achieved.

Consider the case where the user has authenticated herself with SP1, and with the single sign-on service is currently at SP2. After having utilized the services of SP2, the user wishes to logout from all SPs that she currently has an authenticated session with. With the aid of the single logout protocol, if the user indicates a single logout at SP2, this ensures that she is logged out of all concurrent, authenticated SP sessions.

A user who has federated his identity at SP1 with his identity at SP2, may decide to terminate such federation. Alternatively, for certain reasons, either SP1 or SP2 may decide to terminate such federation, which can be done by the Liberty federation termination protocol.

Liberty takes the approach of separating the message schemas from their associated profiles and bindings. Message schemas are proposed for the single sign-on and federation protocol, the federation termination notification protocol, as well as the single logout protocol. Associated with the single message schema, Liberty specifies several profiles and bindings. Depending on the particular scenario, the appropriate Liberty profile may be applied. Such scenarios include cases where the client is using either a Web browser, a WAP browser or a Liberty-enabled browser.

Conclusions

The Liberty Alliance has undertaken the task of developing open, federated identity standards which will be widely deployed. The timeline is aggressive, so that the standards developed meet the needs of the marketplace. In order to be effective, a phased approach has been adopted. While initial specifications deal with mechanisms for achieving identity federation, single sign-on, federation termination and single logout, subsequent phases will deal with attribute sharing.

Reference

http://www.projectliberty.net

Location Inter-Operability Forum

Nokia, Motorola and Ericsson established the Location Interoperability Forum (LIF) as a global forum to address the complexity and multiplicity of solutions and the market situation. The forum should define, develop and promote an inter-operable location services solution that is open, simple and secure. This solution allows user appliances and Internet-based applications to obtain location information from mobile networks independent of their air interfaces and positioning methods.

LIF's purpose is to define, develop and promote through the global standard bodies and specification organizations a common and ubiquitous location services solution.

Such a solution will:

1. Define a simple and secure access method that allows user appliances and Internet applications to access location information from mobile networks, irrespective of their underlying air interface technologies and positioning methods.
2. Promote a family of standards-based location determination methods and their supporting architectures, which are based on Cell-ID and Timing Advance, Enhanced Observed Time Difference (E-OTD), AFLT, and Mobile Station based Assisted Global Positioning System (A-GPS).
3. Establish a framework for influencing the global standard bodies and specification organizations to define common methods and procedures for testing and certifying the LIF-recommended access methods and positioning technologies.

LIF is made up of a mix of carriers, equipment manufacturers and service providers responsible for deploying equipment utilizing this solution. Its members will define this solution and submit it to the working standards groups. The members will then support the solution developed in LIF in the appropriate existing global standard bodies and specification organizations and in the deployment of their systems and services.

Reference

http://www.locationforum.org

Mobile Commerce Fora

The growing interest in mobile commerce, combined with understanding of its open nature, is resulting in a growing number of standardization activities. Some of these activities, directly relevant to the mobile aspect of commerce, are described below. They are here organized by the date they were announced.

Mobile Electronic Transaction Initiative

The Mobile Electronic Transactions (MeT) [metsite] Initiative was created in April 2000 by major mobile device manufacturers wishing to establish a framework for secure mobile transactions. Currently, there are two categories of members:

- o **Sponsors** (open to mobile device manufacturers): the current members are Ericsson, NEC, Nokia, Panasonic, Siemens and SonyEricsson.
- o **Associate members**: currently over 50 mobile commerce stakeholders (e.g., carriers, financial institutions, service and content providers).

The MeT initiative is global and open. Open briefings, workshops and Associate Member Summits have been held to inform others about new developments and share ideas on how to develop the framework in the right direction.

The MeT initiative attempts to influence and support existing open standards (e.g., the Wireless Application Protocol (WAP)) when appropriate, but it also specifies new ones when needed. A specific concept or functionality might be missing or may not be viable, for example, for implementation in the Mobile domain. MeT supports open standards and a fair and non-discriminatory implementation environment. MeT also takes a wide view in specifying viable concepts and technologies and how they will work together in a seamless solution.

MeT work is concentrated on the following areas:

- o **Architecture.** MeT defines three basic models: remote environment, local environment and private environment. The remote environment model addresses transactions carried over a remote connection (e.g., WAP), i.e., when the consumer is not physically present at the shop. The local environment discusses transactions conducted when the consumer is present at the point of sales, using some form of local connection (e.g., Bluetooth). Finally, the personal environment assumes that the consumer is performing the transaction through a certain device (e.g., a computer, set-top box) but the authentication and transaction approval is handled by the Personal Trusted Device (PTD).
- o **Core functions.** An important target of the initiative is to specify the architecture and the core functionality of the PTD. A PTD is a mobile device incorporating a comprehensive set of functionality supported by security elements and thus capable of performing secure mobile transactions. MeT makes sure that core functions can be reasonably implemented within the limited environment of the mobile device.

- **Consistent consumer experience.** The MeT objective of ensuring a consistent consumer experience independent of device, service and network has been welcomed by the market. A consistent consumer experience increases a consumer's trust in services, especially ones in which money is handled. A consistent consumer experience puts certain requirements on the design of user interaction (e.g., in the form of the secure environment indicator) while allowing significant flexibility in implementation.
- **Payment.** MeT is examining existing payment systems and protocols in order to identify their relevance to the PTD concept and implementation in mobile devices. MeT is working with several payment fora in order to ensure that MeT PTD can meet this requirement.
- **Security.** Security is seen as the major enabler of mobile transactions. MeT is leveraging security defined by other fora (e.g., WAP and Wireless Identity Module (WIM)), in order to achieve the set of features required for transactions. This includes using security features in different environments (e.g., the Personal Transaction Protocol (PTP)).
- **Ticketing.** MeT is examining non-monetary transactions in the form of tickets. MeT explores use cases and existing protocols in order to identify the core set of features that enable PTD to offer support for ticketing. Such features (e.g., ticket database) are then translated into work items for other groups or fora.

Release 1 of the MeT specifications was published in February, 2001 and release 2 is expected in the third quarter of 2002.

Mobey Forum

The Mobey Forum [mobey] is a financial industry-driven forum whose mission is to encourage the use of mobile technology in financial services. The formation of the Mobey Forum was publicly announced by the world-leading financial institutions and mobile manufacturers on the 10th of May 2000. Currently, there are three categories of Membership:

- Founder Members
- Membership (open to financial institutions and mobile device manufacturers)
- Associate Membership is open to all others.

Mobey Forum aims to achieve its mission by:

- Raising the awareness of mobile financial service implementations
- Facilitating the open provisioning of mobile financial services
- Identifying business considerations and working to obtain the interoperability of the technical and security requirements for the mobile finance industry, in order to promote competition, and
- Acting as an active liaison between various standardization forums in the mobile industry and the financial industry, so as to promote competition.

Mobey Forum works through its workgroups, as described below.

- o **Business Workgroup.** The objective of this workgroup is to investigate business models and to elaborate on business requirements for mobile commerce applications. The workgroup addresses its objectives by providing input to the Requirements and Technology workgroup, promoting pilot projects, coordinating activities with other mobile and electronic commerce bodies, elaborating on recommendations and best practices and by creating business requirements and models for the financial industry. The list of items which will be addressed include: payment systems and m-billing; payer's and payee's m-commerce infrastructure; trust infrastructure for mobile commerce; financial applications on the chip card in mobile devices.

- o **Requirements and Technology Workgroup.** The scope of this workgroup is to obtain interoperability of the financial industry's technical and security requirements for mobile services and technology, to exchange information on future mobile technologies and their impact on mobile financial transactions, to exchange information on the future requirements of financial services, and to create technical and security requirements of the financial industry for mobile services and technology.

- o **Rules and Regulations Workgroup.** This group is an active liaison between various standardization forums in the mobile and financial industries. This includes, but is not restricted to, the WAP Forum and the Global Platform. Furthermore, the group is encouraging members to participate in the relevant standardization bodies as well as to collect and study aspects of legislation and their consequences for the financial industry. This includes studying competition, legislation concerning electronic identification, electronic signatures and their archiving and patents and intellectual property rights. Also, this group is concerned with the work of the Mobey Forum and other related issues. The group can provide statements regarding internal disputes of the Mobey Forum if requested by the board.

- o **Marketing Workgroup.** This group is responsible for acquiring new members and building awareness of the forum and its goals within the financial services and mobile commerce industries. This will be done by planning and coordinating publicity operations, by planning, coordinating and taking care of public relations materials and events, producing and distributing information, by acquiring research on the issues involved and by developing and carrying out marketing operations.

Mobey Forum has recently released its Preferred Payment Architecture document which summarizes the Forum's views on the technology evolution in remote mobile payment.

References

metsite   http://www.mobiletransaction.org
mobey   http://www.mobeyforum.org

Wireless Village

The Wireless Village initiative is about building a community around new and innovative mobile Instant Messaging and Presence Services (IMPS). Instant Messaging (IM) and Presence is moving from the desktop to the Mobile domain. Ericsson, Motorola and Nokia recognize the need for an industry standard for mobile IMPS, so these companies have formed the Wireless Village initiative to ensure interoperability of wireless messaging services and IM in particular.

The specifications will be used for exchanging messages and presence information between mobile devices, mobile services and Internet-based instant messaging services, all fully interoperable and leveraging existing Web technologies. Through its supporters, the Wireless Village initiative aims to build a vibrant community of consumers and global business partners where Web and Mobile domains converge. The Institute of Electrical and Electronics Engineers' (IEEE) Industry Standards and Technology Organization (IEEE-ISTO) provides day-to-day administrative support to the Wireless Village initiative.

The goal of the Wireless Village initiative is to ensure interoperability of mobile instant messaging and presence services while building a community both around the initiative and through the deployment of innovative new IMPS services. The strategy of the Wireless Village initiative is to help the carrier succeed in attracting and retaining consumers, leveraging their investment in current 2G and 2.5G as well as emerging 3G networks and increasing profits, by providing a comprehensive solution that addresses both carrier requirements and consumers' needs. The Wireless Village solution enables the carrier to leverage the existing customer base, Short Message Service (SMS) usage patterns and business models. At the same time, the carrier will be attracting new consumers, enabling partnerships with existing IM providers, providing new value-add services, and building their own IMPS communities.

IMPS Solution

The Wireless Village Instant Messaging and Presence Service (IMPS) includes four primary features:

- o Presence
- o Instant Messaging
- o Groups
- o Shared Content

Presence is the key enabling technology for the Wireless Village initiative. In the Web domain, consumers have been able to announce their status to authorized recipients, facilitating instant messaging. Presence can be seen as a service of its own or as an enabler for other applications, (e.g., Instant Messaging), which has been the first application to utilize presence capabilities.

Instant Messaging is a familiar concept in both the Mobile and Web domains. Wireless Village enables interoperable mobile IM in concert with other innovative features to provide an enhanced consumer experience.

Discussion groups and chat communities are a fun and familiar concept on the Internet. The Wireless Village initiative enables both, carriers and consumers, to create and manage groups. Consumers can invite their friends and family to chat in group discussions, while carriers can build common interest groups where consumers can meet each other online.

Shared Content allows consumers and carriers to set up their own storage area where they can post pictures, music and other multimedia content to share with other individuals and groups in an IM or chat session.

These features, taken in part or as a whole, provide the basis for innovative new services that build upon a common interoperable framework. The Wireless Village initiative will use its community of supporters as a forum in which to test that framework.

Wireless Village Organization

The Wireless Village has three founding sponsor members (Ericsson, Motorola and Nokia) and over 150 supporting members. Any company can become a supporter of the Wireless Village with no extra participation fees. To become a Wireless Village supporter, interested parties should visit the Wireless Village homepage and complete the application form.

The Wireless Village has four working groups: the business committee (BizCom), the technical committee (TechCom), the marketing committee (MarCom) and the interoperability committee (IOPCom). BizCom acts as the highest governing body for the initiative, taking care of the strategic and day-to-day operational issues. TechCom takes care of all the technological development while MarCom steers the PR and marketing efforts of the initiative. IOPCom takes care of defining the interoperability tests and the process itself.

BizCom, TechCom and IOPCom are all open for supporter participation. There are mailing lists for each of these groups as well as periodic face-to-face meetings.

Technical Specification Groups

The Wireless Village Technical Committee provides the Wireless Village initiative with the end-to-end technical solution to achieve the overall goal and strategy. TechCom is responsible for investigating IMPS technology, identifying requirements, and working out the end-to-end solution in an overall IMPS interoperability framework.

TechCom operates under the supervision of BizCom. The TechCom Standing Committee organizes TechCom's activities and is responsible for its overall progress, while the TechCom Program Management handles document management, daily administration, logistics, and other administrative work. The TechCom Chairperson manages and coordinates the TechCom Standing Committee and reports the progress and escalation to BizCom.

TechCom membership is open to all sponsors and supporters of the Wireless Village initiative. A maximum of two delegates from each member company may participate in TechCom's plenary meetings and conference calls.

Working Procedures and Methods

TechCom operates through scheduled meetings, ad hoc meetings and ad hoc conference calls. The general issues, Change Requests (CR) and Other Input Documents (OID) are discussed in the meetings and conference calls.

The Change Request (CR) is the mechanism enabling members to provide input and contributions relating to the technical specifications. The CR template must be used in submitting a CR.

The Other Input Document (OID) is the mechanism enabling members to provide input and contributions relating to other miscellaneous topics (e.g., including, but not limited to, issues for discussion, the new input of specifications for discussion or approval). The OID template must be used in submitting an OID.

When needed, the TechCom Chairperson may call for an ad hoc meeting or conference call to address specific topics or to discuss CR's and/or OID's that have not been introduced or discussed conclusively in the scheduled meetings. The scope of the ad hoc meeting or conference call may be limited to a WG only.

Decision Making in TechCom

TechCom decisions are made by the TechCom Chairperson based on a general consensus of the members. TechCom is responsible for extending every effort to arrive at a general consensus.

If, however, the progress of the TechCom work becomes stalled on an issue where a general consensus cannot be reached, the TechCom Chairperson may request an informational vote and report both the underlying issue and the result of the informational vote to BizCom for further guidance and directives. All participants who are present can vote, with the limitation of only one vote per company.

Issues in TechCom

Currently, in 2002, the topics in TechCom include interoperability issues as well as future releases of specifications. The main emphasis in Wireless Village work in 2002 is on product implementation and interoperability.

VI EPILOGUE

o Epilogue

Epilogue

The Mobile Internet does not simply refer to accessing the conventional Internet from a mobile device. We will not spend our time browsing Internet pages for content as we do today, although this will still be possible. Instead, we will use services and applications to access content, make transactions, do business, link up with friends and family, play games, watch videos and listen to and download music. More importantly, we will use the Mobile Internet to help manage our lives and give us more time to do the things we choose to, regardless of our location.

The new environment is about enabling people to shape their own Mobile World through personalized communication technology. It is also a place where companies do business by matching consumer needs with their service portfolio or product offering. To ensure the success of services in the Mobile World, they must be highly user-friendly. The consumer should be able to ignore the underlying technologies and enjoy the richness of the services regardless of the access method.

Three success factors for enabling service consumption are:

o   Consumer acceptance of the services,
o   A healthy business system, and
o   Understanding technology life cycles and maturity levels.

Nokia has a good understanding of the mobile device consumer segments, the general attitudes and needs of these segments and the characteristics of technology adoption in each of them. This knowledge is applied to new services and, through product and user interface categorization, enables an optimal consumer experience and ease of use in each of the segments.

Understanding the profit-making logic of the Mobile and Web domains, and applying this knowledge to the innovation possibilities in the Mobile World will combine the best of these domains. It is important for all constituents of the Mobile World to unlock innovation potential through solid business models, so that the flow of new services and evolving consumer needs match. This consumption of new services benefits carriers through volume and through higher average revenue per user.

All technology choices should emphasize service enabling; technical architectures can be perceived as end-to-end service enablers. These technical architectures should lead technology evolution and the related selections, establish a framework for standardization efforts, and enable early system verifications with reference implementations.

Having a common framework for services and understanding this technical foundation will be important when companies interact with each other and consider introducing new services to consumers. To facilitate this, Nokia is active in open global standardization fora, such as Open Mobile Alliance, to develop a comprehensive application and service architecture for the Mobile Internet. The ultimate objective of Nokia is to create a user-friendly Mobile Internet experience for everyone.

Epilogue

The Mobile Internet Technical Architecture aims to provide seamless interoperability between all interaction modes, any network environment and with any type of access. Identifying the relevant communication modes, defining the required supporting key technologies and driving industry participation to develop a common Mobile Internet platform are steps towards the needed solutions. MITA is a common Mobile Internet platform for applications, a technical framework for Mobile services, and a technical foundation for an evolution path to the Mobile World.

As has been presented in the technologies section, the Mobile World will be built on a wide range of technologies. The target of this part of the book has been to provide understanding of the role and major features of these components. It is important to recognize that most of these technologies are being continuously improved and new technologies are being developed at universities, companies, research institutes and as a part of standardization processes. Thus, the technologies described in this book are only a snapshot of the continuous development process.

The future and significance of an individual technology is not solely dependent on the technology itself. There are also other influencing factors, such as legislation and standardization. However, the most important factor is always the consumers, who vote on the significance of a technology with their wallets.

The standardization section introduced various standardization forums and some basic challenges for all of them. Standardization has been and will be a very powerful tool for generating growth in the communications industry and for benefiting its consumers. It is the task of standardization forums to search for the right balance in their work so that high quality standards are available at the right time.

We are in middle of an IP Convergence-driven technology evolution from three network environments: mobile networks, the Internet and Intranets towards a unified Mobile Internet. The role of end-to-end solutions, new technologies, such as SIP and Web services, together with increasing programmability in mobile devices will lead us to a personalized Mobile World where the requirements of consumers drive the development of the Mobile Internet. To achieve this goal, mobile devices, networks and service solutions must work together as one and developers must have access to develop new services. An open solution benefits all (e.g., profitable business scenarios call for interoperability, short development cycles, large volumes and, most of all, global reach). Unless there is a commonly accepted architectural solution, markets will be fragmented and require separate parameters, and be much smaller than a single global market.

As part of the open architecture development, Nokia invites all developers, service providers, carriers and other industry players to participate in further clarifying and specifying the market requirements and technical solutions required to provide the Mobile Internet needed for superior end-to-end services.

VII APPENDIX

o Glossary

VII. APPENDIX

Appendix A: Glossary

2G	Second Generation	
3DES	Triple DES	
3G	Third generation	
3GPP	Third Generation Partnership Project	
3GPP2	Third Generation Partnership Project 2	
A/D	Analog-to-digital	
AAA	Authentication, Authorization, and Accounting	
AAC	Advanced Audio Coding	
AAP	Alternative Approval Process	
ABNF	Advanced Backus-Naur Form	
AC	Admission Control	
ACID	Atomicity, Consistency, Isolation and Durability	
ACIF	Australian Communications Industry Forum	
AD	Area Director	
ADI	Application Development Interface	
ADPCM	Adaptive Differential Pulse Code Modulation	
AES	Advanced Encryption Standard	
A-GPS	Assisted Global Positioning System	
AH	Authentication Header	
AIC	Access Independent Connectivity	
AII	Access Independent Interface	
AKA	Authentication and Key Agreement	
All-IP RAN	All-IP Radio Access Network	
ALF	Application Level Framing	
ALG	Application Level Gateway	
AMR	Adaptive Multi-Rate	
AMR-NB	AMR Narrowband	
AMR-WB	AMR Wideband	
AMS	Application Management Software	
ANSI	American National Standards Institute	
AO	Application Originated	
AOA	Angle Of Arrival	
AP	Access Point	
API	Application Programming Interface	

ARIB	Association of Radio Industries and Businesses
ARL	Authority Revocation List
ARP	Address Resolution Protocol
ARPU	Average Revenue Per User
ARQ	Automatic Repeat Request
AS	Announcement Server
AS	Application Server
AS	Authentication Server
ASCII	American Standard Code for Information Interchange
ASN.1	Abstract Syntax Notation One
ASP	Application Service Provider
AT	Access and Terminals
AT	Application Terminated
ATM	Asynchronous Transfer Mode
ATM	Automatic Teller Machine
ATM-F	ATM Forum
AtoM	Any Transport over MPLS
AVT	Audio/Video Transport
AWT	Abstract Windowing Toolkit
B2B	Business-to-Business
B2BUA	Back-to-Back User Agent
B2C	Business-to-Consumer
BARB	Bluetooth Architecture and Review Board
BBTAG	BroadBand Technical Advisory Group
BCH	Basic Call Handling
BCP	Best Current Practices
BDT	Telecommunication Development Bureau
BE	Best Effort
BEC	Backward Error Correction
BEP	Bit Error Probability
BG	Border Gateway
BGCF	Breakout Gateway Control Function
BGP	Border Gateway Protocol
BizCom	Business Committee
BLER	Block Error Ratio
BMP	Windows Bitmap
BNEP	Bluetooth Network Encapsulation Protocol

BOF	Birds of a Feather	
BOV	Business Operational View	
BPF	Berkeley-Packet-Filter	
BPSK	Binary Phase Shift Keying	
BQA	Bluetooth Qualification Administrator	
BQB	Bluetooth Qualification Bodies	
BQRB	Bluetooth Qualification Review Board	
BQTF	Bluetooth Qualification Test Facility	
BR	Radiocommunication Bureau	
BRAN	Broadband Radio Access Networks	
BS	Base Station	
BSA	Binding Security Association	
BSC	Base Station Controller	
BSD	Berkeley Software Distribution	
BSS	Base Station System	
BSS	Basic Service Set	
BSSGP	Base Station System GPRS Protocol	
BTAB	Bluetooth Technical Advisory Board	
BTI	Bluetooth Test and Interoperability Committee	
BTP	Business Transaction Protocol	
BTS	Base Station	
BWA	Broadband Wireless Access	
CA	Certification Authority	
CA	Collision Avoidance	
CAMEL	Customized Applications for Mobile Network Enhanced Logic	
CAR	Candidate Access Router	
CBA	Command Button Area	
CBR	Constant BitRate	
CC	Call Control	
CCK	Complementary Code Keying	
CC/PP	Composite Capability/Preference Profiles	
CCIF	International Telephone Consultative Committee	
CCIR	International Radio Consultative Committee	
CCIT	International Telegraph Consultative Committee	
CCITT	International Telegraph and Telephone Consultative Committee	
CCS	Cascading Style Sheets	
CCU	Channel Codec Unit	

CD	Collision Detection
CD	Compact Disc
CDC	Connected Device Configuration
CDMA	Code Division Multiple Access
CDR	Call Detail Record
CDR	Common Data Representation
CEK	Content Encryption Key
Cell-ID	Cell Identity
CENELEC	European Committee for Electrotechnical Standardization
CEPT	Conference of European Postal and Telecommunications Administrations
CGI	Common Gateway Interface
CGL-WG	OSDL Carrier Linux Work Group
cHTML	Compact HTML
CIDR	Classless Inter-Domain Routing
CIF	Common Intermediate Format
CIR	Carrier-to-Interference Ratio
CL	Convergence Layer
CLDC	Connected Limited Device Configuration
CLI	Command-Line Interface
CLIP	Classical IP
CLP	Common Line Protocol
CLR	Certificate Revocation List
CMT	Cellular Mobile Telephone
CN	Core Network
CoA	Care-of Address
COD	Content Object Descriptor
COFDM	Coded Orthogonal Frequency Division Multiplex
COPS	Common Open Policy Service
CORBA	Common Object Request Broker Architecture
CoT	Care-of Address Test
CoTI	Care-of Address Test Initiate
CPA	Collaboration Protocol Agreement
CPCS	Common Part Convergence Sublayer
CPIM	Common Profile for Instant Messaging
CPL	Call Processing Language
CPP	Collaboration Protocol Profile
CPS	Connection Processing Server

CPU	Central Processing Unit	
CR	Change Request	
CRC	Cyclic Redundancy Check	
CRL	Certificate Revocation List	
CRM	Customer Relationship Management	
CS	Circuit Switched	
CSCF	Call State Control Function	
CSD	Circuit Switched Data	
CSMA	Carrier Sense Multiple Access	
CSP	Client-Server Protocol	
CSS	Cascading Style Sheets	
CTP	Cordless Telephony Profile	
CTS	Clear To Send	
CWM	Common Warehouse Metamodel	
CVM	Java™ virtual machine supporting CDC	
CVOPS	C Virtual Operating System	
CWTS	China Wireless Telecommunication Standard Group	
D/A	Digital-to-Analog	
DAB	Digital Audio Broadcasting	
DAD	Duplicate Address Detection	
DAML	DARPA Agent Markup Language	
DCF	Distributed Coordination Function	
DCOM	Distributed Component Object Model	
DECT	Digital European Cordless Telecommunications	
DER	Distinguished Encoding Rules	
DES	Data Encryption Standard	
DFRD	Device Family Reference Design	
D-G	Director-General	
DHCP	Dynamic Host Configuration Protocol	
DiffServ	Differentiated Services	
DIFS	Distributed InterFrame Space	
DLC	Data Link Control	
DLPI	Data Link Provider Interface	
DM	Device Management	
DMCA	Digital Millennium Copyright Act	
DNS	Domain Name System	
DNS-Sec	DNS Security	

DoD	Department of Defense
DPIT	Designated Profile Interoperability Test platform
DPSCH	Dedicated Physical SubChannel
DRM	Digital Rights Management
DS	Distribution System
DSA	Digital Signature Algorithm
DSCP	Differentiated Services Code Point
DSL	Digital Subscriber Line
DSP	Digital Signal Processor
DSUI	Differentiated Services Urgency/Importance
DSS	Distribution System Services
DSSS	Direct Sequence Spread Spectrum
DTC	Domain Technology Committee
DTD	Document Type Definition
DTMF	Dial Tone Multi-Frequency
DUNP	Dial-Up Networking Profile
DVB	Digital Video Broadcasting
DVB-T	Digital Video Broadcasting Terrestrial
DVD	Digital Versatile Disc
E.164	International public telecommunication numbers for ISDN, ITU-T
E2E	Enterprise-to-Enterprise
EAIF	External Application Interface
EAP	Extensible Authentication Protocol
EBU	European Broadcasting Union
ebXML	Electronic Business XML
EC	Executive Committee
ECDSA	Elliptic Curve DSA
ECMA TC32	Communication, Networks and Systems Interconnection
ECML	Electronic Commerce Modeling Language
ECN	Encoding Control Notation
ECN	Explicit Congestion Notification
ECSD	Enhanced Circuit Switched Data
EDGE	Enhanced Data Rates for Global Evolution
EDI	Electronic Data Interchange
EE	Environmental Engineering
EEMA	European Forum for Electronic Business
EFI	External Functionality Interface

Glossary

EFR	Enhanced Full Rate
EFTA	European Free Trade Association
EGW	Multimedia E-mail Gateway
EGPRS	Enhanced GPRS
EIRP	Effective Isotropic Radiated Power
EIS	Enterprise Information System
EJB	Enterprise JavaBeans™
EMC	Electromagnetic Compatibility
EMS	Enhanced Messaging Service
EMV	EuropayMastercardVisa payment protocol
ENUM	E.164 Number Mapping
E-OTD	Enhanced Observed Time Difference
EP	ETSI Project
EP BRAN	ETSI Project Broadband Radio Access Networks
EP TIPHON	ETSI Project TIPHON
EPC	Enhanced Power Control
ERC	European Radiocommunications Committee
ERM	EMC and Radio Spectrum Matters
ESMTP	Extended Simple Mail Transfer Protocol
ESO	European Standards Organization
ESP	Encapsulating Security Payload
ESS	Extended Service Set
E-TCH	Enhanced Traffic Channel
ETIS	E- and Telecommunication Information Services
ETSI	European Telecommunications Standards Institute
EU	European Union
FACCH	Fast Associated Control Channel
FAX	Facsimile
FaxP	Fax Profile
FRWA	Fixed Broadband Wireless Access
FC	Finance Committee
FCC	Federal Communication Commission
FDD	Frequency Division Duplex
FH	Frequency Hopping
FHSS	Frequency Hopping Spread Spectrum
FOTAG	Fiber Optics Technical Advisory Group
FP	File Transfer Profile

FR	Full Rate
FRF	Frame Relay Forum
FSG	Free Standards Group
FSV	Functional Service View
FTP	File Transfer Protocol
GA	General Assembly
GAP	Generic Access Profile
GCC	GNU C Compiler
GERAN	GSM/EDGE Radio Access Network
GFSK	Gaussian Frequency Shift Keying
GGSN	Gateway GPRS Support Node
GIF	Graphics Interchange Format
GIOP	Generic Inter-Object Request Protocol
GMA	Gateway Mobility Agent
GMM	GPRS Mobility Management
GMPCS	Global Mobile Personal Communications by Satellite
GMSK	Gaussian Minimum Shift Keying
GMT	Greenwich Mean Time
GOEP	Generic Object Exchange Profile
GPL	General Public License
GPRS	General Packet Radio Service
GPS	Global Positioning System
GRA	GERAN Registration Area
GRX	GPRS Roaming Exchange
GSA	Global Mobile Suppliers Association
GSC	Global Standards Collaboration
GSM	Global System for Mobile Communications
GT	Generic Technology
GTD	Geometric Time Difference
GTP	GPRS Tunneling Protocol
GTP-U	GTP for the user plane
GUI	Graphical User Interface
GUP	Generic User Profile
GW	Gateway
HA	Home Agent
HAck	Handover Acknowledgement
HC	Handover Control

HF	Human Factors
HI	Handover Interface
HILI	High Level Interface
HIPERACCESS	High Performance Radio Access
HIPERLAN	High Performance Radio Local Area Network
HIPERLAN2	High Performance Radio Local Area Network type 2
HIPERMAN	High Performance Radio Metropolitan Area Network
HLR	Home Location Register
HoT	Home Address Test
HoTI	Home Address Test Initiate
HR	Half Rate
HSCSD	High Speed Circuit Switched Data
HSDPA	High Speed Downlink Packet Access
HSP	Handset Profile
HSS	Home Subscriber Server
HTML	HyperText Markup Language
HTTP	HyperText Transfer Protocol
IAB	Internet Architecture Board
IACC	In-Advance Credit Check
IANA	Internet Assigned Numbers Authority
IAP	Internet Access Provider
IAPP	Inter Access Point Protocol
IC	Integrated Circuit
ICL	Interoperability Certification Lab
ICMP	Internet Control Message Protocol
I-CSCF	Interrogating CSCF
ICTSB	Information and Communications Technologies Standards Board
ID	Internet Draft
ID	User Identification
IDE	Integrated Development Environment
IDL	Interactive Data Language
IDL	Interface Definition Language
IEC	International Electrotechnical Commission
IEEE	The Institute Of Electrical And Electronics Engineers
IEEE-ISTO	IEEE Industry Standards and Technology Organization
IEEE-SA	IEEE Standards Association
IETF	Internet Engineering Task Force

IESG	Internet Engineering Steering Group
IFH	Intelligent Frequency Hopping
IFRB	International Frequency Registration Board
IGMP	Internet Group Multicast Protocol
IIOP	Internet Inter-Object Request Broker Protocol
IKE	Internet Key Exchange
ILP	Integrated Layer Processing
IM	Instant Messaging
IM/P	Instant Messaging and Presence
IMAP	Internet Message Access Protocol
IMEI	International Mobile Station Equipment Identity
IMPACT	International Marketing and Promotional Activities
IMPP	Instant Messaging and Presence Protocol
IMPS	Instant Messaging and Presence
IMR	IP Multimedia Register
IMS	IP Multimedia Subsystem
IMSI	International Mobile Subscriber Identity
IMT-2000	International Mobile Telecommunications 2000
IMTC	International Multimedia Telecommunications Consortium
IN	Intelligent Network
IntP	Intercom Profile
IntServ	Integrated Services
IOPCom	Interoperability Committee
IOT	Interoperability Testing
IP	Internet Protocol
IPDC	IP Datacasting
IPDL	Idle Period Down Link
IPDR.org	Internet Protocol Detail Record Organization
IPNG	Internet Protocol Next Generation
IPR	Intellectual Property Right
IPSec	IP Security
IPv4	Internet Protocol, version 4
IPv6	Internet Protocol, version 6
IR	Incremental Redundancy
IR	Infrared
IRC	Internet Relay Chat
IrDA	Infrared Data Association

IRTF	Internet Research Task Force	
ISDN	Integrated Services Digital Network	
ISLAN	Integrated Services LAN	
ISM	Industrial, Scientific and Medical	
ISO	International Organization for Standardization	
ISOC	Internet Society	
ISP	Internet Service Provider	
ISV	Independent Software Vendor	
IT	Information Technology	
ITU	International Telecommunication Union	
ITU-D	ITU Telecommunication Development Sector	
ITU-R	ITU Radiocommunication Sector	
ITU-T	ITU Telecommunication Standardization Sector	
IWE	Interworking Element	
IWF	Interworking Function	
IVR	Interactive Voice Response	
J2EE™	Java™ 2 Platform, Enterprise Edition	
J2ME™	Java™ 2 Platform, Micro Edition	
J2SE™	Java™ 2 Platform, Standard Edition	
JAD	Java™ Application Descriptor	
JAIN	Java™ APIs for Integrated Networks	
JAM	Java™ Application Manager	
JAR	Java™ Archive file	
JAXM	Java™ API for XML Messaging	
JAXP	Java™ API for XML Processing	
JAXR	Java™ API for XML Registries	
JAX-RPC	Java™ API for XML-based RPC	
JCP	Java™ Community Process	
JCTEA	Japan Cable Television Engineering Association	
JDBC	Java™ Database Connectivity API	
JDK	Java™ Development Kit	
JMS	Java™ Messaging Service	
JNDI	Java™ Naming and Directory Interface	
JNI	Java™ Native Interface	
JPEG	Joint Photographic Experts Group	
JSPA	Java Specification Participation Agreement	
JSP™	JavaServer™ Pages	

JSR	Java Specification Request
JTC	Joint Technical Committee
JTC Broadcast	Joint Technical Committee on Broadcasting
JVM	Java™ Virtual Machine
KVM	Java™ Virtual Machine supporting CLDC
L2	Layer 2
L2CAP	Logical Link Control and Adaptation Protocol
L3	Layer 3
LA	Link Adaptation
LAN	Local Area Network
LAP	LAN Access Profile
LC	Load Control
LCD	Liquid Crystal Display
LCDUI	Limited Capability Device User Interface
LDAP	Lightweight Directory Access Protocol
LGPL	Library GNU Public License
LIF	Location Interoperability Forum
LLC	Logical Link Control
LMA	Leaf Mobility Agent
LMM	Local Mobility Management
LMP	Link Manager Protocol
LMSC	IEEE 802 LAN/MAN Standards Committee
LMU	Location Measurement Unit
LSB	Linux Standards Base
LSP	Label Switched Path
LSR	Label Switched Router
M2M	Machine-to-Machine
MAC	Media Access Control
MAC	Message Authentication Code
MAN	Metropolitan Area Network
MANET	Mobile Ad Hoc Networking
MAP	Mobile Application Part Protocol
MarCom	Marketing Committee
MBM	Multi Bitmap file
MCC	Mobile Competence Center
M-COMM	Mobile Commerce
MCS	Modulation Coding Scheme

Glossary

MCU	Multiparty Conferencing Unit
MCU	Multipoint Control Unit
MD5	Message Digest 5
MDA	Model Driven Architecture
MEO	Medium Earth Orbit
MESA	Public Safety Partnership Project
MeT	Mobile Electronic Transactions
MExE	Mobile Station Application Execution Environment
MFN	Multi Frequency Network
MG	Media Gateway
MGC	Media Gateway Controller
MGCF	Media Gateway Control Function
MGW	Media Gateway
MHP	Multimedia Home Platform
MIBA	Mobile Internet Business Architecture
MIDI	Musical Instrument Digital Interface
MIDP	Mobile Information Device Profile
MII	Mobile Internet Interfaces
MIME	Multi-purpose Internet Mail Extensions
MITA	Mobile Internet Technical Architecture
MLD	Multicast Listener Discovery for IPv6
MM	Mobility Management
MMA	Multimedia Message Adaptation
MMAC	Multimedia Mobile Access Communications Promotion Council
MMCTL	Mobility Management Controller
MMS	Multimedia Messaging Service
MMSC	Multimedia Messaging Service Center
MMS-IOP	MMS Interoperability Group
MMU	Multimedia Unit
MMUSIC	Multiparty Multimedia Session Control
MO	Mobile Originated
MOF	Meta-Object Facility
MOTO	Mail Order/Telephone Order
MoU	Memoranda of Understanding
MPE	Multi-Protocol Encapsulation
MPEG	Moving Pictures Experts Group
MPEG-4	MPEG version 4

MPI	Mobile Internet Protocol Interfaces
MPLS	Multi-Protocol Label Switching
MRFC	Media Resource Function Control
MRFP	Media Resource Function Processing
MRP	Market Representation Partner
MRV	Mobile Rights Voucher
MS	Mobile Station
MSC	Mobile Services Switching Center
MSDU	MAC Service Data Unit
MSF	Multiservice Switching Forum
MSG	Mobile Standards Group
MSI	Mobile Internet Software Interfaces
MSISDN	Mobile Station International ISDN Number
MSS	Media Subsystem
MSS	Mobile Satellite Services
MT	Mobile Terminated
MTP3B	Message Transfer Part 3 Broadband
MTS	Methods for Testing and Specification
MTU	Maximum Transmission Unit
MVC	Model-View-Controller
MWIF	Mobile Wireless Internet Forum
MX	Mail eXchange
N-ISDN	Narrowband Integrated Services Digital Network
NACC	Network Assisted Cell Change
NAMP	Nokia Artuse Messaging Platform
NAPTR	Naming Authority Pointer
NAR	New Access Router
NAS	Network Access Server
NAT	Network Address Translation
NATP-PT	Network Address/Port Translation - Protocol Translation
NAT-PT	NAT - Protocol Translation
NesCom	New Standards Committee
NGN	Next Generation Networks
NIST	National Institute of Standards and Technology
NLOS	Non Line Of Sight
NNA	Naming, Numbering and Addressing
NNI	Network-Node Interface

NNTP	Network News Transfer Protocol	
NOKOS	NOKia Open Source license	
NRT	Non-Real-Time	
NSIS	Next Steps In Signaling	
NTA	Sofia Transaction API	
NTP	Network Time Protocol	
NTR	Sofia Transaction API for RTSP	
NUA	Sofia User Agent API	
O&M	Operations and Maintenance	
OBEX	Object Exchange	
OCG	Operational Coordination Group	
OCSP	Online Certificate Status Protocol	
OFDM	Orthogonal Frequency Division Multiplexing	
OID	Other Input Document	
OIF	Optical Internetworking Forum	
OMA	Open Mobile Alliance	
OMC	Operations and Maintenance Center	
OMG	Object Management Group	
OP	Organizational Partner	
OPP	Object Push Profile	
OR	Octal Rate	
ORB	Object Request Broker	
OS	Operating System	
OSA	Open Service Architecture	
OSDL	Open Source Development Lab	
OSI	Open Source Initiative	
OSI	Open Systems Interconnection	
OSPF	Open Shortest Path First	
OSPFv6	Open Shortest Path First for IPv6	
OSS	Open Source Software	
OSS	Operation Support Systems	
OTA	Over The Air	
OTD	Observed Time Difference	
OTDOA	Observed Time Difference Of Arrival	
OWL	Ontology Web Language	
OWLAN	Operator Wireless Local Area Network	
PACCH	Packet Associated Control Channel	

PAI	Platform Adaptation Interface
PAN	Personal Area Networking
PAR	Previous Access Router
PAR	Project Authorization Request
PC	Power Control
PC	Personal Computer
PCF	Point Co-ordination Function
PCG	Project Co-ordination Group
PCM	Pulse Code Modulation
PCMCIA	PC Memory Card International Association
P-CSCF	Proxy-Call State Control Function
PDA	Personal Digital Assistant
PDC	Personal Digital Communication
PDCP	Packet Data Convergence Protocol
PDH	Plesiochronous Digital Hierarchy
PDP	Packet Data Protocol
PDTCH	Packet Data Traffic Channel
PDU	Protocol Data Unit
PFC	Packet Flow Context
PGP	Pretty Good Privacy
PHB	Per Hop Behavior
PHY	Physical
PHP	Hypertext Preprocessor
PID	Program Identifier
PII	Personally Identifiable Information
PIM	Personal Information Management
PIM	Protocol Independent Multicast
PIN	Personal Identification Number
PJAE	PersonalJava™ Application Environment
pJava	PJAE, PersonalJava™ Technology
PKC	Public Key Cryptography
PKI	Public Key Infrastructure
PLCP	Physical Layer Convergence Protocol
PLMN	Public Land Mobile Network
PLT	PowerLine Telecommunications
PMD	Physical Medium Dependent
PMO	Project Management Office

PNG	Portable Network Graphics	
POI	Proof Of Identity	
POP	Post Office Protocol	
POP3	Post Office Protocol 3	
POP	Proof Of Possession	
POS	Point-Of-Sale	
PPP	Point-to-Point Protocol	
PRD	Program Reference Document	
PS	Packet Scheduler	
PS	Packet Switched	
PS	Presence Server	
PSCH	Physical SubChannel	
PSK	Phase Shift Keying	
PSS	Packet Switched Streaming	
PSTN	Public Switched Telephone Network	
PTC	Platform Technology Committee	
PTCC	Protocol and Testing Competence Center	
PTD	Personal Trusted Device	
PTP	Personal Transaction Protocol	
QCIF	Quarter-CIF	
QM	Quality Manager	
QoS	Quality of Service	
QPL	Bluetooth Qualified Products List	
QPSK	Quadrature Phase Shift Keying	
QR	Quarter Rate	
R&D	Research and Development	
RA	Registration Authority	
RA	Radiocommunication Assembly	
RAB	Radio Access Bearer	
RADIUS	Remote Authentication Dial-In User Service	
RAG	Radiocommunication Advisory Group	
RAN	Radio Access Network	
RANAP	RAN Application Protocol	
RAST	Global Radio Standardization	
RB	Radio Bearer	
RC	Roadmap Coordinator	
RC5	Rivest Cipher 5	

RDF	Resource Description Framework
RED	Random Early Detection
REL	Rights Expression Language
RevCom	Review Committee
RF	Radio Frequency
RFC	Request For Comments
RFID	Radio Frequency ID
RFP	Request For Proposal
RI	Reference Implementation
RIP	Routing Information Protocol
RIPng	Routing Information Protocol for IPv6
RLC	Radio Link Control
RM	Resource Management
RMI	Remote Method Invocation
RMS	Record Management Store
RNC	Radio Network Controller
RNIF	RosettaNet Implementation Framework
RNSAP	Radio Network Subsystem Application Part
ROA	Recognized Operating Agencies
ROHC	Robust Header Compression
RPC	Remote Procedure Call
RR	Radio Resource
RRA	Requirement Requires Applying
RRB	Radio Regulations Board
RRC	Radio Resource Control
RRC	Regional Radiocommunication Conference
RRM	Radio Resource Management
RSA	Rivest Shamir Adleman
RSVP	Resource Reservation Protocol
RT	Railway Telecommunications
RT	Real-Time
RTCP	Real-time Transport Control Protocol
RTD	Real Time Difference
RTDC	Regional Telecommunication Development Conference
RTP	Real-time Transport Protocol
RTS	Request To Send
RTSP	Real-Time Streaming Protocol

RTT	Round Trip Time	
S	Signaling	
S/MIME	Secure/Multi-purpose Internet Mail Extensions	
SA	Security Association	
SA	Selective Availability	
SACCH	Slow Associated Control Channel	
SA Forum	Service Availability Forum	
Safety	Telecommunications Equipment Safety	
SAGE	Security Algorithms Group of Experts	
SAI	Service Area Identifier	
SAML	Security Assertion Markup Language	
SAP	Service Access Point	
SAP	Session Announcement Protocol	
SAR	Segmentation And Re-assembly	
SAR	Servlet Archive	
SCCP	Signaling Connection Control Part	
SCN	Switched Circuit Network	
SCP	Smart Card Platform	
S-CSCF	Serving CSCF	
SCTE	Society of Cable Telecommunications Engineers	
SCTP	Stream Control Transmission Protocol	
SDAP	Service Discovery Application Profile	
SDE	Service Discovery Engine	
SDH	Synchronous Digital Hierarchy	
SDK	Software Development Kit	
SDO	Standards Development Organization	
SDP	Session Description Protocol	
SDP	Service Discovery Protocol	
SDPng	SDP next generation	
SDS	Service Discovery Service	
SEC	Security	
SEC	Sponsor Executive Committee	
SES	Satellite Earth Stations and Systems	
SFN	Single Frequency Network	
SG	Signaling Gateway	
SG	Study Group	
SGML	Standard Generalized Markup Language	

SGSN	Serving GPRS Support Node
SHA-1	Secure Hash Algorithm 1
S-HTTP	Secure HyperText Transfer Protocol
SI	Service Indication
SIFS	Short InterFrame Space
SIG	Special Interest Group
SIGCOMP	Signaling compression
SIIT	Stateless IP/ICMP Translation Algorithm
SILS	Standard for Interoperable LAN Security
SIM	Subscriber Identity Module
SIMPLE	SIP for Instant Messaging and Presence Leveraging Extensions
SIO	Scientific or Industrial Organization
SIP	Session Initiation Protocol
SIP-CGI	Common Gateway Interface for SIP
SIR	Signal-to-Interference Ratio
SLF	Subscriber Locator Function
SLP	Service Location Protocol
SM	Session Management
SMATV	Satellite Master Antenna Television
SMIL	Synchronized Multimedia Integration Language
SMP	Symmetric Multi-Processing
SMS	Short Message Service
SMSC	Short Message Service Center
SMSGW	SMS Gateway
SMTP	Simple Mail Transfer Protocol
SNAP	Subnetwork Access Protocol
SNMP	Simple Network Management Protocol
SOAP	Simple Object Access Protocol
SOAP-DSIG	SOAP Digital Signature
SP	Service Provider
SP	Synchronization Profile
SPA	Self Provided Application
SPAN	Services and Protocols for Advanced Networks
SPD	Security Policy Database
SPI	Service Provisioning Infrastructure
SPP	Serial Port Profile
SPSCH	Shared Physical SubChannel

SRES	Signed RESponse	
SS	Station Services	
SS	Supplementary Services	
SS7	Signaling System No. 7	
SSA	Subsystem Architecture	
SSCF	Service Specific Co-operation Function	
SSCOP	Service Specific Connection Oriented Protocol	
SSG	Special Study Group	
SSI	System Software Interfaces	
SSL	Secure Sockets Layer	
SSO	Single Sign-On	
SSP	Server-to-Server Protocol	
SSPI	Security Service Application Programming Interface	
STA	Station	
STF	Specialist Task Force	
STP	Scalable Test Platform	
STQ	Speech processing, Transmission and Quality aspects	
SwA	SOAP with Attachments	
SWA	Systems Software Architecture	
SVG	Scalable Vector Graphics	
SVR4	System V Release 4.x	
SWT	Standardized Web Services Technologies	
SyncML	Synchronization Markup Language	
T1	Standards Committee T1 - Telecommunications	
TA	Timing Advance	
TAP	Traditional Approval Process	
TC	Technical Committee	
TC	Tunnel Client	
TCH	Traffic Channel	
TCK	Technology Compatibility Kit	
TCP	Transmission Control Protocol	
TCP/IP	Transmission Control Protocol/Internet Protocol	
TCS	Telephony Control System	
TDAG	Telecommunication Development Advisory Group	
TDD	Time Division Duplex	
TDMA	Time Division Multiple Access	
Tdoc	Temporary Document	

TE	Terminal Equipment
TechCom	Technical Committee
TELNET	Terminal Emulation Protocol
TETRA	Terrestrial Trunked Radio
TF	Transport Format
TGW	Terminal Gateway
TIA	Telecommunications Industry Association
TIPHON	Telecommunications and Internet Protocol Harmonization Over Networks
TLS	Transport Layer Security
TM	Transmission and Multiplexing
TM Forum	TeleManagement Forum
TMN	Telecommunication Management Network
TOA	Time Of Arrival
TR	Technical Reports
TRAU	Transcoder and Rate Adaptation Unit
TRX	Transceiver
TS	TimeSlot
TS	Technical Specification
TSACC	Telecommunications Standards Advisory Council of Canada
TSAG	Telecommunication Standardization Advisory Group
TSB	Telecommunication Standardization Bureau
TSG	Technical Specification Group
TSG CN	Technical Specification Group Core Network
TSG GERAN	Technical Specification Group GSM/EDGE Radio Access Network
TSG RAN	Technical Specification Group Radio Access Network
TSG SA	Technical Specification Group System Aspects
TSG T	Technical Specification Group Terminals
TTA	Telecommunications Technology Association
TTC	Telecommunication Technology Committee
TTCN	Tree and Tabular Combined Notation
TTI	Transmission Time Interval
TTP	Trusted Third Party
TU	Transaction User
TU3	Typical Urban 3 km/h
UA	User Agent
UAC	User Agent Client
UAProf	User Agent Profile

UAS	User Agent Server	
UDDI	Universal Description, Discovery and Integration	
UDP	User Datagram Protocol	
UDP	Unacknowledged Data Protocol	
UDVM	Universal Decompressor Virtual Machine	
UE	User Equipment	
UHF	Ultrahigh Frequency	
UI	User Interface	
UID	Unique Identification code	
UML	Unified Modeling Language	
UMTS	Universal Mobile Telecommunication System	
UNI	User-Network Interface	
U-NII	Unlicensed National Information Infrastructure	
URI	Uniform Resource Identifier	
URL	Uniform Resource Locator	
USB	Universal Serial Bus	
USER	Special Committee User Group	
USIM	UMTS Subscriber Identity Module	
USSD	Unstructured Supplementary Service Data	
UTC	Universal Time Co-ordinates	
UTF-8	Unicode Transformation Format 8	
UTRA	UMTS Terrestrial Radio Access	
UTRAN	UMTS Terrestrial Radio Access Network	
UWCC	Universal Wireless Communications Consortium	
VAS	Value Added Service	
VASP	Value Added Service Provider	
VC	Virtual Cursor	
VGA	Video Graphics Array	
VHF	Very High Frequency	
VLAN	Virtual Local Area Network	
VM	Virtual Machine	
VoiceXML	Voice Extensible Markup Language	
VoIP	Voice over IP	
VPN	Virtual Private Network	
W3C	World Wide Web Consortium	
WAE	Wireless Application Environment	
WAI	Web Accessibility Initiative	

WAP	Wireless Application Protocol
WARC	World Administrative Radio Conference
WAV	Windows Audio File
WBMP	Wireless BitMap
WCDMA	Wideband Code Division Multiple Access
WCIT	World Conference on International Telecommunication
WDCTL	WLAN Device Controller
WECA	Wireless Ethernet Compatibility Alliance
Web-Ont	W3C Web Ontology Working Group
WEP	Wired Equivalent Privacy
WG	Working Group
WGR	Working Group on ITU Reform
WI	Work Item
WIM	Wireless Identity Module
WIM	WAP Identity Module
WINS	Windows Internet Name Service
WIP	WLAN Interworking Protocol
WLAN	Wireless Local Area Network
WMF	Wireless Multimedia Forum
WML	Wireless Markup Language
WMLS	Wireless Markup Language Script
WPAN	Wireless Personal Area Network
WP8F	ITU-R Working Party 8F
WPKI	Wireless Public Key Infrastructure
WRC	World Radiocommunication Conference
WSCL	Web Services Conversation Language
WSDL	Web Services Description Language
WSFL	Web Services Flow Language
WS-I	Web Services Interoperability Organization
WSIA	Web Services Interactive Applications
WSIL	Web Services Inspection Language
WSIS	World Summit on the Information Society
WS-License	Web Service License Language
WSP	Wireless Session Protocol
WS-Security	Web Service Security Language
WTA	Wireless Telephony Applications
WTAI	Wireless Telephony Applications Interface

WTDC	World Telecommunication Development Conference
WTLS	Wireless Transport Layer Security
WTP	Wireless Transport Protocol
WTPF	World Telecommunication Policy Forum
WTSA	World Telecommunication Standardization Assembly
WV	Wireless Village
WWW	World Wide Web
XACML	eXtensible Access Control Markup Language
xDSL	any Digital Subscription Line
XHTML	eXtensible HyperText Markup Language
XHTML MP	XHTML Mobile Profile
XKMS	XML Key Management Specification
XLANG	Web Services for Business Process Design
XMI	XML Metadata Interchange
XML	eXtensible Markup Language
XML-DSIG	XML Digital Signature
XMLP	XML Protocol
XMLP-WG	XML Protocol Working Protocol
XSD	XML Schema Definition
XSL	eXtensible Stylesheet Language
XSLT	eXtensible Stylesheet Language Transformations
xSP	any Service Provider

Index

Symbols

2.4 GHz frequency band 186
3COM 141
5 GHz frequency band 186
802.11 Wireless Local Area Network 404
802.15 Wireless Personal Area Network 406
802.16 Broadband Wireless Access 406

A

Access control 300
Access Independent Connectivity 28
Address Autoconfiguration 212
Adleman 302
Andrew Tanenbaum 58
anycast 77
Apache 57
Application Framework 24
Application Programming Interfaces 38
Applications and the Technical Architecture 17
Architecture Frameworks 30
Architecture Specifications 23
ARM 59
Asymmetric Encryption 301
Authentication 32, 47, 300
Authentication Methods and Technologies 253
   Biometric Authentication 259
   Diameter 260
   Hardware Token Based Authentication 255
   Kerberos Authentication 256
   key distribution and management 253
   Onetime Passwords 254
   Password Authentication 254
   prime categories 253
   Public Key Cryptography 258
   Remote Authentication Dial-In User Service 259
   Single Sign-On 260
   Smart Card Authentication 255
   Subscriber Identity Module Authentication 256
   Symmetric Key Based Solutions 255
Authorization 32, 300
Availability 32

B

Berkeley Standard Distribution 58
Biometric Authentication 259
Bluetooth 195
   Bluetooth Lower Layers 196
   Bluetooth Middleware Protocols 197
   Connectivity scenarios 195
   Future Developments 199
   Profiles 199
   Technical overview 196
   Usage scenarios 195
Bluetooth Special Interest Group 195, 409
   Bluetooth Architecture and Review Board 410
   Bluetooth Qualification Program 411
   Bluetooth Qualification Review Board 410
   Bluetooth Test and Interoperability Committee 410
   committees 410
   Marketing Committee 411
   Organization 409
   Regulatory Committee 411
Broadband Radio Access Networks 394
Browsing 9, 24, 45, 265
   Extensible HyperText Markup Language 265
   WAP 2.0 265
   XHTML Basic 266
   XHTML Mobile Profile 266

C

care-of-address 211
Carrier Grade Linux Work Group 413, 415
   architecture 417
   goals 415
   Scope 417
   structure 416
Cascading Style Sheets 269, 430

Classless Inter Domain Routing 209
Code Division Multiple Access 2000 155
Concepts 327
Confidentiality 32, 300
Connected Device Configuration 237
Connected Limited Device Configuration 237
Content Adaptation 29
Content Formats 29
Correspondent Node 77
Crystal DFRD 146

D

DAML-S 435
DARPA Agent Markup Language 130
Data Synchronization 29, 50, 243
   SyncML 243
   SyncML Technology 244
Delivery 47
Development Cycle 64
Device Family Reference Designs 145
Device Management 29, 45
Diameter 260
Differentiated Services 81
Diffie-Hellman 302
Digital Alpha 59
Digital Certificates 306
   Certificate Details 308
   Certificate Revocation 308
   Certificate Revocation Lists 308
   Issuing a Certificate 307
   Online Certificate Status Protocol 308
   Transitive Trust 309
   Using a Certificate 307
Digital Equipment Corporation User Society 59
Digital Rights Management 49, 313
   Access Control 319
   Asymmetric Encryption 319
   Basic System Model 315
   Connection Protection 319
   Content Delivery 317
   Content Distribution Methods 321
   Digital Fingerprints 321
   Encryption 317
   logical parts 315
   Rights Assignment 317
   Security Levels 320
   Security Technologies 319
   Session Protection 319
   Standardization 323
   superdistribution 314
   Symmetric Encryption 319
   System Models 314
   Tamper Resistance 320
   Technologies Related to DRM 318
   Time value of content 316
   Variations System Model 316
   Voucher Creation 317
   Watermarking 321
Digital Signatures 304
Digital Video Broadcasting – Terrestrial Network 201
   Datacasting Profiles 203
   Enabling multicast 205
   Hybrid solution 205
   IP Datacasting 201
   IP Multiprotocol Encapsulation 204
   Mobility 202
   Radio Characteristics 201
   Radio Network Design Issues 202
   speeds 202
   Technical Overview 201
Direct Sequence Spread Spectrum 193
Directories 31
Distributed Component Object Model 112
Document Type Definition 268
Dynamic Host Configuration Protocol 71, 77

E

E.164 85
ebXML 433
Electronic Business XML 119
Electronic Document Interchange Based Web Services 119
Encryption 301
Enhanced Data Rates for Global Evolution 155
European Telecommunications Standards Institute 391
   Broadband Radio Access Networks 394
   Competence and Service Centers 398
   Global Cooperation 397
   Information Technology in Specifications Creation 399

Index

Methods for Testing & Specification 394
Partnership Projects 393
Projects 393
Structure 392
Support Functions 397
Technical Committees 393
Third Generation Partnership Project 396
eXtensible HyperText Markup Language 430
extension headers 76

F

File Transfer Protocol (FTP) 73
Free Software Foundation 58
Frequency Hopping Spread Spectrum 193

G

Gartner Group 59
Generic Technologies 143
Global Positioning System 336
GNU C Compiler 58
GNU Hurd 58
Group Management 51
GSM Evolution towards 3G/UMTS 169
GSM/EDGE Radio Access Network 169
    Architecture 170
    classes of traffic 173
    Control channel 177
    Control Plane QoS mechanisms 174
    GSM Evolution towards 3G/UMTS 169
    Incremental Redundancy 179
    Legacy Interfaces 170
    Link Adaptation 180
    Medium Access Control 177
    modulation 178
    New Interfaces 171
    Packet Data Control Protocol 174
    Performance 179
    Physical Layer 177
    Quality of Service 173
    Radio Link Control 176
    Radio Protocols 174
    Radio Resource Control 175
    Traffic channels 177
    User Plane QoS mechanisms 174

H

Hall, John 59
Hash Algorithms 302
High Speed Downlink Packet Access 164
HIPERACCES 395
HIPERLAN2 394
HIPERLAN 186
HIPERMAN 395
Home Address 77
home agent 211
Home Agent Discovery 220
Home Location Register 85
HTTPD Accelerator 61
Hypertext Markup Language 430
Hypertext Transfer Protocol 72, 81

I

ICMP 90
ICMPv6 215
ID 47
IDC 57
Identity 9
IEEE 802 402
IEEE 802.11 187
    access point 188
    Architecture 188
    Basic Channel Access 191
    Direct Sequence Spread Spectrum 193
    distribution system 188
    Enhancements 405
    family of standards 187
    Frequency Hopping Spread Spectrum 193
    Infrared 193
    Medium Access Control (MAC) Layer 191
    Physical (PHY) Layer 192
    portal 188
    Reference Model 187
    Services 189
    station 188
IGMP 90
IMT-2000 155
Industrial, Scientific and Medical (ISM) band 186
Infrared 193
Ingress Filtering 219

Instant Messaging 49, 287
   message flow 290
Instant Messaging and Presence Service 295
Institute of Electrical and Electronics Engineers
      401
   IEEE 802 402
   Wireless 802 404
   Working Groups 401
Integrated Services 94
Intel 58
Interaction Modes 8
   Browsing 9
   Messaging 9
   Rich Call 9
International Telecommunications Union 375
   History 375
   ITU Reform 386
   Radiocommunication Sector 378
   Sectors 378
   Series of Recommendations 383
   Structure 376
   Study Groups 380
   Telecommunication Development Sector 385
   Telecommunication Standardization Sector 380
Internet Application Protocols 71
Internet Control Message Protocol 89
Internet Engineering Task Force 75, 369
   Organization 369
   Related Organizations 371
   Working Process 370
Internet Group Multicast Protocol 89
Internet Inter-Object Request Broker Protocol 112
Internet Multicast 89
   ICMP 90
   Managing a Group 90
   multicast group 90
   Service Announcement Protocol 92
   Service Description Protocol 92
   Service Discovery 92
   Services 92
Internet Multimedia Architecture 81
Internet Protocol Version 4 75
   Header 75
Internet Protocol Version 6 75
   address prefixes 78

address space 75
Address Types 77
anycast 77
Correspondent Node 77
Dynamic Host Configuration Protocol 77
Enhancements 77
extension headers 76
Header 76
Header Handling 76
Home Address 77
Internet Engineering Task Force 75
IPSec 77
multicast 77
scoping of addresses 79
Stateless Address Autoconfiguration 77
unicast 77
Internet Protocols 25, 69, 74
   Common Internet Application Protocols 71
   Domain Name System (DNS) 71
   Dynamic Host Configuration Protocol (DHCP)
      71
   File Transfer Protocol (FTP) 73
   Hypertext Transfer Protocol (HTTP) 72
   Multipurpose Internet Mail Extensions (MIME)
      73
   Network News Transfer Protocol (NNTP) 73
   Related Protocols 71
   Simple Mail Transfer Protocol (SMTP) 73
   Stream Control Transmission Protocol (SCTP) 70
   Transmission Control Protocol (TCP) 69
   Transport Protocols 69
   User Datagram Protocol (UDP) 70
Internet Service Provider 89
IP Datacasting 201
IP Multicast 89
IP Multiprotocol Encapsulation 204
IPSec 77
IPv5 75
Issuing a Certificate 307

J

Java 235
   Connected Device Configuration 237
   Connected Limited Device Configuration 237
   Foundation Profile 237

Java 2 Platform, Micro Edition 236
   Midlet Networking 240
   Midlet Provisioning 239
   Mobile Information Device Profile 237
   Mobile Information Device Profile Version 1.0 237
   Mobile Information Device Profile Version 2.0 238
   Mobile Media 241
   Personal Profile 237
Java 2 Platform, Micro Edition 236
Java Community Process 423
   Agreement and Process 423
   Guidance Body 424
   Intellectual Property 426
   Life Cycle of a Specification 424
   Membership 423
   Process 423
   Specification Lead Position 426
Java™ Technology 44

K

Kerberos Authentication 256
Kernel HTTPD Accelerator 61
kernel modules 60
Key System Specifications 27

L

Layer 2 Broadcast 91
Layer 2 Multicast 91
Liberty Alliance 439
   Architecture 440
   identity federation 441
   Liberty Applicability 440
   Membership 439
   Organization 439
   Phased Approach 440
   Working groups 439
Library GNU Public License 64
Linus Torvalds 58
Linux 57
   Apache 57
   Architecture 59
   ARM 59
   Carrier-Grade Linux 149
   Development Cycle 64
   Distribution Packages 61
   Fault Resistance 152
   Features 59
   Firewall 64
   Free Software Foundation 58
   GNU C Compiler 58
   GNU Hurd 58
   Jon Hall 59
   Kernel HTTPD Accelerator 61
   Kernel modules 60, 65
   Library GNU Public License 64
   Media Terminal 57
   microkernel 60
   Minix 58
   Modules 60
   Monolithic 59
   Networking 63
   Origins 58
   PowerPC 59
   Real-Time Support 62, 151
   Routing 64
   scheduler 60
   Security 63
   Sparc 59
   Support for Replication 150
   University of Helsinki 59
   Unix 58
Location 28
Location Inter-Operability Forum 443
Location Technologies 331
   Angle of Arrival 334
   Carrier applications 331
   Cell Coverage 332
   Commercial applications 331
   Enhanced Observed Time Difference 337
   Global Positioning System 336
   Government applications 331
   Idle Period Down Link 339
   Methods 332
   Observed Time Difference of Arrival 339
   Received Signal Levels 333
   requirements 332
   Round Trip Time 335
   Time of Arrival 337
   Timing Advance 334

M

Message Authentication Codes 303
Messaging 9, 25, 273
   Multimedia Messaging Service 276
   Short Message Service 275
microkernel 60
Microsoft 141
Middleware 247
   Application Requirements 247
   CORBA 249
   DCOM 249
   Internet Protocols 248
   Java 249
   Programming Models 249
Minix 58
Mobey Forum 446
Mobile Commerce Fora 445
   Mobey Forum 446
   Mobile Electronic Transaction Initiative 445
Mobile Commerce Services 325
Mobile Digital Rights Management 49
Mobile Domain
   Authentication Methods and Technologies 253
   Bluetooth 195
   Browsing 265
   Characteristics 231
   Data Synchronization 243
   Digital Video Broadcasting – Terrestrial Network 201
   GSM/EDGE Radio Access Network 169
   Instant Messaging 287
   Java 235
   Location Technologies 331
   Messaging 273
   Middleware 247
   Mobile Payment 325
   Multimedia Messaging 275
   Multimedia Sessions 231
   operating systems 141
   Presence 287
   Standardization 343
   Symbian 141
   Wideband Code Division Multiple Access Technology 155
   Wireless Local Area Networks 185
   Wireless Village 295

Mobile Information Device Profile 237
Mobile Information Device Profile Version 2.0 238
Mobile Internet Business Architecture 18
Mobile Internet Interfaces 32
Mobile Internet Technical Architecture 10
   Access Independent Connectivity 28
   Application Framework 24
   Applications and the Technical Architecture 17
   Architecture Concept Models 14
   Architecture Frameworks 30
   Architecture Implementation Models 14
   Architecture Modeling Principles 13
   Architecture Specifications 15, 23
   Authentication 32
   Authorization 32
   Availability 32
   Browsing 24
   Compliance with MITA Principles 16
   Confidentiality 32
   Content Adaptation 29
   Content Formats 29
   Data Synchronization 29
   Device Management 29
   Directories 31
   End-to-End Solutions 20
   Integrity 32
   Internet Protocols 25
   Introduction 23
   Key System Specifications 27
   Location 28
   Messaging 25
   Methodology 13
   Mobile Internet Interfaces 32
   Naming, Numbering and Addressing 27
   Objectives 11
   Operating Systems 26
   Operation Support Systems 30
   Platform Support 26
   Presence 27
   Principles 16
   Privacy 30
   Quality of Service 33
   Reachability 27
   Reference Implementations 15
   Rich Call 24
   Security 31
   Service Discovery 28

Service Enablers 43
Specifications 23
Systems Software Architecture 35
UI 24
Web Domain 81
Work Phases 11
Mobile IPv6 77, 209, 210
   Address Autoconfiguration 212
   Address Configuration 216
   address space 212
   binding 211
   Binding Security Association Establishment 226
   Binding Update 211, 217, 219
   care-of address 211
   Context Features for Transfer 224
   Context Transfer Framework 225
   Design Points 212
   Detection of a mobile node 216
   Duplicate Address Detection 213
   fast handover 223
   global addressability 209
   home agent 211
   Home Agent Discovery 220
   ICMPv6 215
   IETF 371
   Ingress Filtering 219
   link-local address 212
   Localized Mobility Management 226
   motivation 209
   Protocol Overview 216
   Renumbering 221
   Route Optimization 218
   Router Advertisement 212
   scalability 218
   Seamless Mobility 223
   Security 213
   Status 228
   Tunneling 217
   use of the Binding Update 219
Mobile Media 241
Mobile Payment 48, 325
   Local Environment 327
   Payment Concepts 327
   Personal Environment 327
   Remote Environment 326

   Security 328
   Transaction Environments 326
   Wireless Identity Module 329
Mobile Service Brokers 40
Mobile Web Service Interfaces 39
Monolithic 59
multicast 77, 89
multicast group 90
Multimedia Messaging Service 44, 275, 276
   Applications 282
   Inter-Network Routing 280
   Interoperability between Mobile Devices 283
   Interoperability with Internet Applications 282
   Message Adaptation 282
   message delivery 279
   Message Encapsulation 281
   migration path 276
   Mobile Device Capabilities Negotiation 284
   Multimedia Messaging Service Center 277
   Network Architecture 278
   Standard 278
Multimedia Messaging Service Center 44, 277
Multimedia Sessions 81
   3GPP IP Multimedia Subsystem 231
   efficiency 233
   Real-Time Transport Control Protocol 82
   Real-Time Transport Protocol 82
   Resource Reservation Setup Protocol 82
   Session Announcement Protocol 83
   Session Initiation Protocol 81
   SIP network 84
   SIP session establishment 233
   Stream Control Transport Protocol 82
   Transmission Control Protocol 82
   User Datagram Protocol 82
   Voice over IP 81
Multiprotocol Label Switching 81, 94, 96
Multipurpose Internet Mail Extensions 73

N

Naming, Numbering and Addressing 27
Network Address Translator 75
Network News Transfer Protocol 73
Nokia Media Terminal 57
Non-repudiation 300

O

Object Management Group 437
   Modeling Specifications 437
Open Mobile Alliance 43, 351
   Benefits 352
   charter 351
   principles 352
   Scope of the Alliance 351
Open Source Development Lab 413
   Carrier Grade Linux Work Group 415
   project requirements 414
Open source software 64
Operating Systems 26
Operation Support Systems 30

P

Packet Data Control Protocol 174
Palm OS 141
Pearl DFRD 146
Personal Digital Assistant 141
Personal Profile 237
Platform Support 26
Pocket PC 141
Pocket PC 2002 141
PowerPC 59
Presence 27, 49, 287
Privacy 30
Public Key Cryptography 258
Public Key Infrastructure
   Applications Using PKI 310
   Certificate Details 308
   Certificate Revocation 308
   Digital Certificates 306
   History 306
   Issuing a Certificate 307
   PKI Technology 306
   Secure E-Mail 310
   Secure Web Applications 310
   Transitive Trust 309
   Using a Certificate 307
   Virtual Private Networks 310
   Wireless Public Key Infrastructure 311

Q

QoS 93
Quality of Service 33, 81, 93
   Best Effort 94
   Differentiated Services 95
   Integrated Services 94
   Multiprotocol Label Switching 94, 96
   Technologies 93
Quartz DFRD 146

R

Radio Resource Control 175
Raymond, Eric S. 64
Reachability 27
Real-Time Transport Control Protocol 82
Real-Time Transport Protocol 81, 82
Remote Authentication Dial-In User Service 259
Remote Method Invocation 112
Resource Reservation Setup Protocol 82
Rich Call 9, 24
Rivest 302
Rivest, Shamir, Adleman 302
RosettaNet 435
Router Advertisement 212

S

SAML 434
scoping of addresses 79
SDP next generation 83
Seamless Mobility 223
Secure E-Mail 310
Secure Web Applications 310
Security 31
Security Services 299
   Access control 300
   Asymmetric Encryption 301
   Authentication 300
   authorization 300
   Confidentiality 300
   Cryptography 300
   Digital Signatures 304
   Encryption 301
   Hash Algorithms 302
   Message Authentication Codes 303

Non-repudiation 300
Symmetric Encryption 301
Semantic Web 127, 431
  DARPA Agent Markup Language 130
  Extensible Markup Language 128
  Languages of the Semantic Web 128
  Resource Description Framework 128
  Role of Ontologies 127
  Using DAML 130
  W3C Web Ontology 130
Service Announcement Protocol 92
Service Availability Forum 421
Service Brokers 117
Service consumer 111
Service Description Protocol 92
Service Discovery 28
Service Enablers 43
  Authentication 47
  Browsing 45
  Data Synchronization 50
  Delivery 47
  Device Management 45
  Device Profile 46
  Digital Rights Management 49
  Group Management 51
  ID 47
  Instant Messaging 49
  Java Technology 44
  Mobile Digital Rights Management 49
  Mobile Payment 48
  Presence 49
  Streaming 50
  User Profile Management 50
Service enablers
  fragmentation 43
  Multimedia Messaging Service 44
Service provider 110
Session Announcement Protocol 83
Session Initiation Protocol 81, 287
  3GPP Architecture 293
  ACK 86
  address format 85
  BYE 86
  CANCEL 86
  E.164 85
  HyperText Transfer Protocol 81

INVITE 86
methods 86
OPTIONS 86
presence service 293
Proxy server 85
REGISTER 86
Registrar server 85
requests 86
Service Creation Model 290
Services 288
session establishment 86
Simple Mail Transfer Protocol 81
SIP IM/P in 3GPP 293
SIP IM/P in IETF 291
SIP Message 287
SIP network 84
termination 86
User Agents 84
Voice over IP 81
Shamir 302
Short Message Service 275
Simple Mail Transfer Protocol 73, 81
Simple Object Access Protocol 109, 113
Single Sign-On 260
SIP network 84
soft handover 159
Sparc 59
Stallman, Richard 58
  GNU C Compiler 58
Standardization 343
  Application Development Support 349
  Bluetooth SIG 409
  Carrier Grade Linux Work Group 413
  Challenges 343, 344
  Enabling Different Technology Evolution Speeds 347
  Enabling More Added Value to Consumers 348
  End-to-End Aspects 344
  European Telecommunications Standards Institute 391
  How to Focus Standardization Work Items 346
  Institute of Electrical and Electronics Engineers 401
  International Telecommunications Union 375
  Internet Engineering Task Force 369
  Java Community Process 423

Liberty Alliance 439
Location Inter-Operability Forum 443
Managing Growing Complexity 346
Mobile Commerce Fora 445
Object Management Group 437
Open Mobile Alliance 351
Open Source Development Lab 413
Principles 349
Service Availability Forum 421
Third Generation Partnership Project 359
vertical interfaces 344
Web Services Fora 433
Wireless Application Protocol Forum 355
Wireless Ethernet Compatibility Alliance 407
Wireless Village 449
World Wide Web Consortium 429
Stateless Address Autoconfiguration 77
Stream Control Transmission Protocol (SCTP) 70
Stream Control Transport Protocol 82
Streaming 50
superdistribution 314
Symbian 141
Symbian Operating System 143
   Architecture of Symbian OS, version 7 144
   Crystal 146
   Device Family Reference Designs 145
   features 143
   Generic Technologies 143
   Pearl 146
   Quartz 146
Symbian Platform 141
   Application Development 147
   Application Suite 147
   Java execution environment 147
   Motivation 141
   Overview 142
   Software Development Kits 147
Symbian Operating System 141
Symmetric Encryption 301
SyncML 243
   Interoperability 245
   Security 245
   synchronization session 244
   transport bindings 244
SyncML Technology 244
Systems Software Architecture 35, 36

Application Programming Interfaces 38
Applications 38
Content 38
Design Principles 35
Mobile Service Brokers 40
Mobile Web Service Interfaces 39
Network Servers 41
Service Provisioning Infrastructure 38
Terminal 37

T

Tanenbaum, Andrew 58
TechCom 451
Terminal Emulation Protocol (Telnet) 74
Third Generation Partnership Project 44, 359, 396
   Guests 362
   Individual Members 362
   Market Representatives 361
   Methods 367
   Observers 362
   Organization 359
   Partners 360
   Project Co-Ordination Group 363
   Support Functions 362
   Technical Specification Groups 363
   Working Procedures 367
Torvalds, Linus 58
Transitive Trust 309
Transmission Control Protocol 69, 82
Transport Protocols 69
   Stream Control Transmission Protocol 70
   Transmission Control Protocol 69
   User Datagram Protocol 70
Tunneling 217

U

UDDI 112, 434
UDDI.org 434
UMTS Terrestrial Radio Access Network 155
unicast 77
Universal Description, Discovery, and Integration 109
Universal Mobile Telecommunications System 155
   licenses 157
   Radio Access Network Architecture 158

Index **495**

Spectrum 156
WCDMA Basics 159
Unix 58
User Agent 84
User Datagram Protocol 72, 82
User Interface 24
User Profile Management 50

V

Virtual Private Networks 310
Voice over IP 81, 287
    signaling protocol/mechanism 287

W

W3C Web Ontology 130
WAP 2.0 45, 265, 357
    Data Synchronization 358
    External Functionality Interface 357
    Multimedia Messaging Service 358
    Persistent Storage Interface 358
    Pictogram 358
    Protocol Stack 357
    Provisioning 358
    User Agent Profile 357
    WAP Application Environment 357
    WAP Push 357
    Wireless Telephony Application 357
WAP Forum 45
WAP Specification 356
WCDMA Basics 159
Web Domain 81
    Internet Multimedia Architecture 81
    Semantic Web 127
    Session Initiation Protocol 81
    Web Services 109
Web Service Description Language 109
Web Service Technologies 112
    SOAP 112
    UDDI 112
    WSDL 112
    XML 112
Web Services 39, 109, 119, 431
    Advanced Web Services 120
    Binding 116
    Communication Framework 112
    Conversations 121

Data Encapsulation 112
Documentation 116
Electronic Business XML 119
Electronic Document Interchange Based Web
    Services 119
Key Participants 110
Message 116
PortTypes 116
Publishing Web Services Locally 119
Security Issues 122
Service 117
Service Brokers 117
Service consumer 111
Service Description 115
Service provider 110
Simple Object Access Protocol 109, 113
Transactions 120
Types 116
UDDI Service APIs 118
Universal Description, Discovery, and Integration
    109, 117
Web Service Description Language 109
Web Services Conversation Language 122
Web Services Inspection language 119
Workflow 121
XML Protocol 114
Web Services Conversation Language 122
Web Services Fora 433
    DAML-S 435
    OASIS 433
    RosettaNet 435
    UDDI.org 434
    UN/CEFACT 433
    Web Services Interoperability Organization 434
Web Services Inspection language 119
Web Services Interactive Applications 434
Well-Formed Content 267
Wi-Fi 407
Wideband Code Division Multiple Access
    Technology 155
    Basics 159
    capability classes 163
    CDMA principle used 159
    delay requirements 162
    fast power control 159
    handover 160

High Speed Downlink Packet Access 164
HSDPA 164
HSDPA Air Interface Performance 165
Mobile Device Capabilities 161
multiplexing principle 162
Physical Layer 161
Radio Access Network Architecture 158
setup connection 163
sharing resources 161
Spectrum 156
Typical capacities 166
Windows CE 141
Wireless 802 404
Wireless Application Protocol Forum 355
  design principles 355
  Mission 355
  Organization 356
  WAP 2.0 357
  WAP Specification 356
  Working Groups 356
Wireless Ethernet Compatibility Alliance 407
  Organization 408
  Testing Process 407
Wireless Identity Module 329
Wireless Local Area Networks 185, 186
  2.4 GHz frequency band 186
  Frequencies 185
  HIPERLANs 186
  IEEE 802.11 187
  purpose 185
Wireless Public Key Infrastructure 311
Wireless Village 295, 449
  Application Service Elements 296
  Decision Making in TechCom 451
  IMPS Solution 449
  Instant Messaging and Presence Service 295
  Interoperability Framework 295
  Issues in TechCom 451
  Organization 450
  Protocol Suite 296
  Public Key Infrastructure 299
  System Architecture 295
  Technical Specification Groups 450
  Working Procedures and Methods 451
World Wide Web Consortium 45, 429
  Cascading Style Sheets 430
  Current Activities 430
  Device Independence 431
  eXtensible Hypertext Markup Language 430
  Graphics 431
  Hypertext Markup Language 430
  Multimedia 431
  Operation and Organization 429
  Resource Description Framework 431
  Semantic Web 431
  Web Services 431
  XHTML Basic 430
  XHTML Modularization 430
WSDL 112

X

XHTML Basic 266, 430
XHTML Mobile Profile 266
  Cascading Style Sheets 269
  Document Type Definition 268
  Key Features 267
  Transformations 271
  Well-Formed Content 267
XHTML Modularization 430
XML 112
XML Protocol 114

MORE TITLES FROM IT PRESS

...ile Internet Technical ...itecture – The Complete ...age

...bile Internet will be much more than simply ...f accessing the Internet with a mobile device. ...e about integrating communication services ...everyday life. We will be able to use the ...Internet to help control our lives and to give ...e time to do the things we enjoy. The Mobile ...t will eventually change our way of life as ...e seen with the introduction of the telephone ...ision.

...y and the Internet will be smoothly unified ...Mobile World. This means solid integration ...y existing technologies by using a thorough ...anding of mobility and the unique character-...f mobile business. It is also necessary to ...a range of completely new technologies to ...new services and to meet the new challenges ...Mobile World. This must be done without ...essive complexity for Mobile Internet service ...ers.

...kia solution for covering these demands is ...bile Internet Technical Architecture (MITA).

...ages, CD-ROMs, ISBN 951-826-671-9

Technologies and Standardization

Nokia

The Technologies section describes the existing technologies that will form a solid basis for future innovations, development work and the products of the Mobile Internet. The reader will learn about the technologies that already exist either in research, in the product development phase or in actual products.

The Standardization section introduces a set of standardization organizations. It also discusses the role and challenges of standardization in the new Mobile World.

520 pages, CD, ISBN 951-826-668-9

Solutions and Tools

Nokia

The Solutions section gives insight into current Nokia solutions. It provides information on how various technologies are combined to produce systems and services today.

The Tools section addresses one of the corner stones of MITA: tools for application and service development. The section describes some examples on how applications and services can be developed with Nokia tools for the Mobile Internet.

510 pages, CD, ISBN 951-826-669-7

Visions and Implementations

Nokia

The Visions section describes a set of key functions that will be essential for the Mobile Internet. The reader is given a more future oriented view on how technologies will evolve. This includes topics,such as browsing, messaging, rich call, QoS and security.

The Implementations section gives a view on MIIA Reference Implementations that are used to validate new concepts, to gain experience and to smoothen the way for products based on the latest technologies.

520 pages, CD, ISBN 951-826-670-0

Professional Mobile Java with J2ME

Kontio

J2ME provides embedded software application developers the platform and operating system independence of Java programming language and thereby makes programmers' work easier. This book teaches how to develop applications using the hottest future technology: J2ME. Professional Mobile Java with J2ME covers J2ME architecture and explains in detail the functions of CLDC and MIDP. The book contains numerous practical examples of every topic covered and complete sample applications.

300 pages, CD, ISBN 951-826-554-2